D0138241

COUNTEREXAMPLES IN PROBABILITY AND STATISTICS

THE WADSWORTH & BROOKS/COLE STATISTICS/PROBABILITY SERIES

Richard A. Becker and John M. Chambers
Extending the S System

Richard A. Becker and John M. Chambers
S: An Interactive Environment for Data Analysis and Graphics

Peter J. Bickel, Kjell Doksum, and J. L. Hodges, Jr.
Festschrift for Erich L. Lehmann

George E. P. Box
The Collected Works of George E. P. Box
Volumes I and II, George C. Tiao, editor-in-chief

Leo Breiman, Jerome H. Friedman, Richard A. Olshen, and Charles J. Stone
Classification and Regression Trees

John M. Chambers, William S. Cleveland, Beat Kleiner, and Paul A. Tukey
Graphical Methods for Data Analysis

Franklin A. Graybill
Matrices with Applications in Statistics, Second Edition

Lucien M. Le Cam and Richard A. Olshen
Proceedings of the Berkeley Conference in Honor of Jerzy Neyman and Jack Kiefer, Volumes I and II

Joseph P. Romano and Andrew F. Siegel
Counterexamples in Probability and Statistics

John W. Tukey
The Collected Works of John W. Tukey
William S. Cleveland, editor-in-chief
Volume I: *Time Series*, 1949–1964, edited by David R. Brillinger
Volume II: *Time Series*, 1965–1984, edited by David R. Brillinger

COUNTEREXAMPLES IN PROBABILITY AND STATISTICS

Joseph P. Romano
University of California, Berkeley

Andrew F. Siegel
University of Washington, Seattle

Wadsworth & Brooks/Cole Advanced Books & Software
Monterey, California

Wadsworth & Brooks/Cole Advanced Books & Software
A Division of Wadsworth, Inc.

Printed in the United States of America
10 9 8 7 6 5 4 3 2 1

Library of Congress Cataloging-in-Publication Data
Romano, Joseph P., [date]
Counterexamples in probability and statistics.

Includes index.
1. Probabilities. 2. Mathematical statistics.
I. Siegel, Andrew F. II. Title.
QA273.R58 1985 519.2 85-19024
ISBN 0-534-05568-0

Sponsoring Editor: *John Kimmel*
Editorial Assistant: *Maria Rosillo Alsadi*
Production Editor: *Candyce L. Cameron*
Production Assistant: *Carol Rost*
Manuscript Editor: *Carol Reitz*
Permissions Editor: *Carline Haga*
Interior and Cover Design: *Katherine Minerva*
Art Coordinators: *Michele Judge, Sue C. Shepherd*
Interior Illustration: *Tim Keenan*
Typesetting: *Asco Trade Typesetting Ltd., Hong Kong*
Cover Printing and Printing and Binding: *Maple-Vail Book Manufacturing Group*

To our families.

PREFACE

One approaches a mathematical statement intending either to prove it must always hold true or to give evidence that it is false. Proofs are used to establish a statement's truth, whereas counterexamples are used as direct evidence that a statement is false. Counterexamples can also provide insight into a problem—for example, by showing which hypothesis needs to be strengthened in order to achieve a true result, by helping to establish a result as "best possible," or by clarifying the need for a particular choice of definition. A less technical but perhaps more important theme is that in statistics we really have no set principles that work in all situations. In the words of Kiefer (1977):

> "Statistics is not so simple a subject that it can be codified in terms of a simple recipe that yields satisfactory methods in all problems."

Counterexamples in Probability and Statistics is best used as a supplementary text or reference work to complement standard books on probability and statistical inference. We have had the pleasure of studying analysis and topology with books of counterexamples at hand for constant reference. It is hoped that students, teachers, and researchers may benefit as well from this work, which follows in a tradition of collections of counterexamples (Gelbaum and Olmsted 1964; Steen and Seebach 1970; Maar 1973; Khaleelulla 1982).

How does one set out to find a counterexample to a statement? There are several approaches to resolving a conjecture. Often one goes back and forth between attempting to prove it is true and attempting to disprove it by a counterexample. This iterative process usually converges to a resolution because (1) the current difficulties encountered while trying to prove the statement may suggest how a counterexample may be constructed, and (2) the current difficulties encountered while trying to construct a counterexample may show more clearly how the hypotheses succeed in forcing their conclusion.

When looking for a counterexample, we strongly recommend looking at the simplest cases first because if a trivial counterexample exists, it often shows clearly what is wrong with the conjecture under consideration (not to mention the time saved!). If a simple counterexample cannot be found, only then should you move toward more complicated situations. In this work we provide both simple and direct counterexamples as well as more realistic situations to indicate that the problem cannot be cleared up simply by ruling out the trivial cases.

Because we wanted this work to be as self-contained as possible, each chapter begins with an introduction including pertinent definitions and results for that subject area. The chapters are ordered according to difficulty, and also proceeding from basic material to refinements and more specialized areas. Finally, the appendix includes a list of further examples for which details may be found in the journal literature.

This project started at Princeton University as the senior thesis of the first author under the supervision of the second author. We gratefully acknowledge the helpful assistance we have had in assembling this work. In particular, we would like to thank C. Sugahara; those who reviewed the manuscript, Persi Diaconis, Richard M. Dudley, David Freedman, Douglas G. Kelly, Erich L. Lehmann, Colin L. Mallows, Carl N. Morris, and C. R. Rao; the U.S. Army Research Office; the National Science Foundation; and the staff at Brooks/Cole, especially John Kimmel and Candy Cameron.

Joseph P. Romano
Andrew F. Siegel

CONTENTS IN BRIEF

CONTENTS

CHAPTER ONE
PROBABILITY SPACES 1

CHAPTER THREE
MOMENTS OF RANDOM VARIABLES 38

CHAPTER FOUR
PROPERTIES OF REAL RANDOM VARIABLES 53

CHAPTER SIX
CONDITIONAL EXPECTATION, MARTINGALES, AND ALMOST SURE CONVERGENCE OF SUMS OF INDEPENDENT RANDOM VARIABLES 126

CHAPTER SEVEN
PROPERTIES OF STATISTICAL EXPERIMENTS 154

CHAPTER NINE
OPTIMALITY OF ESTIMATORS 190

CHAPTER TEN
HYPOTHESIS TESTING 232

CHAPTER ONE
PROBABILITY SPACES

Introduction

The fundamental notion underlying the theory of probability and statistics is that of a "probability space." The triple $(\Omega, \mathbf{F}, P)$ is called a "probability space" if the following conditions hold: The "sample space" is any non-empty set Ω whose elements ω are called "outcomes." The "σ-field" is a collection $\mathbf{F}$ of subsets of Ω (a σ-field will be defined below) called "events" or "measurable sets"; the "simple events" or "elementary events" are the singleton sets $\{\omega\}$ that are in $\mathbf{F}$. The "probability measure" is a set function P defined on $\mathbf{F}$ that has the following properties:

(i) For any event E in $\mathbf{F}$, $0 \leq P(E) \leq 1$.

(ii) $P(\Omega) = 1$ and $P(\phi) = 0$, where ϕ denotes the empty set.

(iii) If $E_1, E_2, \ldots$ is a pairwise disjoint collection of events in $\mathbf{F}$, then

$$P\left(\bigcup_{j=1}^{\infty} E_j\right) = \sum_{j=1}^{\infty} P(E_j).$$

In order for a collection $\mathbf{F}$ of subsets of Ω to be a σ-field, it must satisfy the following properties:

(i) Both ϕ and Ω are in $\mathbf{F}$.

(ii) If E is in $\mathbf{F}$, then so is its complement $E^c = \Omega - E$.

(iii) If $E_1, E_2, \ldots$ is any countable collection of sets in $\mathbf{F}$, then their union and intersection also belong in $\mathbf{F}$.

We will often use the σ-field $\mathbf{B}$, the class of Borel subsets of the real numbers. This is defined to be the smallest σ-field on the reals that contains all intervals; therefore, $(\mathbf{R}, \mathbf{B})$ is the natural measurable space to use when we want all intervals (and sets easily derived from intervals) to be measurable.

The "support" of a probability measure P defined on $(\mathbf{R}, \mathbf{B})$ consists of all points x for which

$$P[(x - \varepsilon, x + \varepsilon)] > 0 \qquad \text{for all } \varepsilon > 0.$$

Note that the support of a probability measure is a closed set. More generally, the support of a probability measure on a topological space can be defined as the set of all points for which every open neighborhood has positive measure.

A probability measure P is said to be "absolutely continuous" with respect to a probability measure Q [both defined on $(\Omega, \mathbf{F})$] if whenever $E \in \mathbf{F}$, then $Q(E) = 0$ implies that $P(E) = 0$. Probability measures P and Q are said to be "equivalent" if each is absolutely continuous with respect to the other, so that they have the same class of sets of measure zero. If, on the other hand, there exists a measurable set E for which $P(E) = 0$ and $Q(E^c) = 1$, then P and Q are said to be "mutually singular."

Two events A and B are said to be "stochastically independent" or just "independent" if

$$P(A \cap B) = P(A)P(B).$$

In general, a finite collection $E_1, E_2, \ldots, E_n$ of events is said to be independent if each subcollection satisfies the intersection/product property—that is, if

$$P(E_{k_1} \cap \cdots \cap E_{k_j}) = P(E_{k_1}) \cdots P(E_{k_j})$$

for each set $\{k_1, \ldots, k_j\}$ of distinct indices between 1 and n. An infinite collection (either countably or uncountably infinite) of events is said to be independent if each of its finite subcollections is independent.

The "conditional probability of A given B" is written as $P(A \mid B)$. It is defined by

$$P(A \mid B) = \frac{P(A \cap B)}{P(B)} \qquad \text{provided } P(B) > 0$$

and is left undefined if $P(B) = 0$.

If $\mathbf{C}$ is a class of subsets of Ω, then $\sigma(\mathbf{C})$ denotes the smallest σ-field of Ω that contains $\mathbf{C}$, called the "σ-field generated by $\mathbf{C}$." We say that classes of (measurable) events $\mathbf{C}_1$ and $\mathbf{C}_2$ are "independent" if $C_1 \in \mathbf{C}_1$ and $C_2 \in \mathbf{C}_2$ imply that $P(C_1 \cap C_2) = P(C_1) \cdot P(C_2)$.

A class of subsets (not necessarily a σ-field) is called a "π-system" if it is closed under the operation of taking finite intersections.

Two results frequently used in relation to a sequence of events are the Borel-Cantelli Lemmas. The first lemma says that if $E_1, E_2, \ldots$ is a sequence of events with $\Sigma_{i=1}^{\infty} P(E_i) < \infty$, then with probability one only finitely many of the events occur. A partial converse is given by the second Borel-Cantelli Lemma, which says that if the events $E_1, E_2, \ldots$ are independent and $\Sigma_{i=1}^{\infty} P(E_i) = \infty$, then with probability one infinitely many of the E_k occur. Set notation may be used to represent derived events like these as follows:

$$\{E_n \text{ i.o.}\} = \{E_n \text{ infinitely often}\} = \limsup_{n\to\infty} E_n = \bigcap_{n=1}^{\infty} \bigcup_{k=n}^{\infty} E_k$$

$$\{E_n \text{ f.o.}\} = \{E_n \text{ finitely often}\} = \{E_n \text{ infinitely often}\}^c$$

$$\{E_n \text{ ev.}\} = \{E_n \text{ eventually}\} = \liminf_{n\to\infty} E_n = \bigcup_{n=1}^{\infty} \bigcap_{k=n}^{\infty} E_k.$$

Given a sequence $E_1, E_2, \ldots$ of events defined on a probability space $(\Omega, \mathbf{F}, P)$, consider the σ-field

$$\mathbf{T} = \bigcap_{n=1}^{\infty} \sigma(E_n, E_{n+1}, \ldots).$$

This $\mathbf{T}$ is called the "tail σ-field" associated with the sequence $E_1, E_2, \ldots$ and its elements are called "tail events." Kolmogorov's Zero-One Law states that if an event E is in the tail σ-field associated with an independent sequence of events, then $P(E)$ must be either 0 or 1.

1.1. A probability space in which each simple event has probability zero.

Consider the probability space $((0, 1), \mathbf{L}, \mu)$ where the sample space is the unit interval, $\mathbf{L}$ denotes the class of Lebesgue-measurable subsets of $(0, 1)$, and μ is Lebesgue measure. In this case, we say that the probability measure μ denotes a uniform probability on $(0, 1)$. In particular, the probability of any interval in $(0, 1)$ is simply the length of the interval so that $\mu[(a, b)] = b - a$ whenever $0 < a \leq b < 1$. This probability space is well defined and is often considered to be the "standard" probability structure for the unit interval. However, for any real number x in $(0, 1)$, the simple event $\{x\}$ has Lebesgue measure zero; that is, $\mu(\{x\}) = 0$ because for every $\varepsilon > 0$, we have

$$\mu(\{x\}) \le \mu\left[\left(x - \frac{\varepsilon}{2}, x + \frac{\varepsilon}{2}\right)\right] = \varepsilon.$$

You might suppose that if the sample space were countable, it would follow that not all simple events could have probability zero. If all singleton sets were measurable, you would be correct. However, there do exist countable (even finite) sample spaces in which every simple event has probability zero. Consider the following case:

Sample space	$\Omega = \{a, b, c\}$
σ-field	$\mathbf{F} = \{\phi, \{a\}, \{b, c\}, \Omega\}$
Probability measure	$P(\phi) = 0, \quad P(\{a\}) = 0,$ $P(\{b, c\}) = 1, \quad \text{and} \quad P(\Omega) = 1.$

In this case there is only one simple event, namely $\{a\}$, because $\{b\}$ and $\{c\}$ are not measurable (they are not in the σ-field $\mathbf{F}$). This simple event $\{a\}$ has probability zero, and hence all simple events in this case have probability zero.

1.2. A probability space in which each simple event has probability one.

Naturally in such an example there can be only one simple event in the probability space and the rest of the space must have probability zero. The simplest example is provided by

$$\Omega = \{\omega\}, \qquad \mathbf{F} = \{\phi, \Omega\}, \qquad P(\phi) = 0, \qquad P(\Omega) = 1.$$

Another example is given by

$$\Omega = [0, 1], \qquad \mathbf{F} = \{\phi, \{0\}, (0, 1], \Omega\},$$
$$P(\phi) = P\{(0, 1]\} = 0, \qquad P(\{0\}) = P(\Omega) = 1.$$

1.3. Two events that are not disjoint but such that the probability of the union is equal to the sum of the probabilities.

The intersection need not be empty in order for the probability of the union to equal the sum of the probabilities of the events; the intersection need only have probability zero. Consider the interval (0, 1) with Lebesgue probability measure. Let $A = (0, \frac{1}{2}]$ and $B = [\frac{1}{2}, 1)$. Then

$$P(A \cup B) = P[(0, 1)] = 1 = \tfrac{1}{2} + \tfrac{1}{2} = P(A) + P(B),$$

although A and B are not disjoint because

$$A \cap B = \{\tfrac{1}{2}\} \neq \phi.$$

1.4. An event that is independent of itself.

Most pairs of independent events that we deal with represent distinct events; however, this need not be the case. In fact, any event with probability zero or one is independent of itself. This may be seen by observing that an event A is independent of itself (by definition) if and only if

$$P(A \cap A) = P(A)P(A)$$

or

$$P(A) = [P(A)]^2,$$

which we recognize as a quadratic equation with the two roots $P(A) = 0$ and $P(A) = 1$.

Examples of sets that are independent of themselves are therefore ϕ, Ω, N, and $\Omega - N$, where N is any event for which $P(N) = 0$.

1.5. Two independent but disjoint events.

We would ordinarily expect disjoint events A and B to be dependent because A and B cannot both occur, and so knowing that A has occurred rules out the possibility that B might have happened. However, the independence condition

$$P(A \cap B) = P(A)P(B)$$

will always be satisfied when $P(B) = 0$. We might choose $B = \phi$ and $A = \Omega$ to be our pair of independent but disjoint sets. Alternatively, we might choose B as the set of rational numbers in $(0, \frac{1}{2})$ and set $A = (\frac{1}{2}, 1)$ within the context of Lebesgue measure on the unit interval $(0, 1)$.

1.6. Three events that are pairwise independent but not independent.

Consider the tossing of a coin twice, and define the three events, A, B, and C, as follows:

A is "heads on the first toss."

B is "heads on the second toss."

C is "exactly one head and one tail (in either order) in the two tosses."

Then we have

$$P(A) = P(B) = P(C) = \tfrac{1}{2}$$

and

$$P(A \cap B) = P(B \cap C) = P(C \cap A) = \tfrac{1}{4} = \tfrac{1}{2} \cdot \tfrac{1}{2}.$$

It follows that any two of the events are pairwise independent. However, since the three events cannot all occur simultaneously, we have

$$P(A \cap B \cap C) = 0 \neq \tfrac{1}{8} = P(A)P(B)P(C).$$

Therefore the three events, A, B, and C, are not independent.

1.7. *N* events that are pairwise independent but not independent.

To generalize Example 1.6 to an arbitrary number N of events, consider the tossing of a coin $N - 1$ times, and define the events $E_1, \ldots, E_N$ as follows:

$$E_j = \begin{cases} \text{"}j\text{th toss of the coin was heads"} & \text{if } j \leq N - 1 \\ \text{"exactly one head and one tail (in either order) in the first two tosses"} & \text{if } j = N. \end{cases}$$

Then we have

$$P(E_j) = \tfrac{1}{2} \qquad \text{for } j = 1, \ldots, N,$$

and

$$P(E_i \cap E_j) = \tfrac{1}{4} \qquad \text{for } i \neq j.$$

As in Example 1.6, any two of the events $E_1, E_2, \ldots, E_N$ are pairwise independent. But, since all N events cannot occur simultaneously, we have

$$P(E_1 \cap E_2 \cap \cdots \cap E_N) = 0 \neq \frac{1}{2^N} = P(E_1)P(E_2)\cdots P(E_N).$$

Thus, these N events are not independent.

1.8. An infinite sequence of events such that every pair is independent, but no triple is independent.

Let I and J be independently distributed over $\{1, 2, \ldots\}$, representing the number of tosses of a fair coin needed in order to have heads occur just once (this is a Negative Binomial Distribution). Define events $A_1, A_2, \ldots$ as follows:

- If $I < J$, then none of the A_i happen.
- If $I \geq J$, then A_I and A_J and no others happen.

We see that A_i happens whenever $I \geq J$ and either $I = i$ or $J = i$, so that these events are well defined. Note that

$$P(A_i \cap A_j) = \frac{1}{2^{i+j}} \qquad \text{if } i \neq j$$

and

$$P(A_i) = P(A_i \text{ only}) + \sum_{j \neq i} P(A_i \cap A_j)$$

$$= \frac{1}{4^i} + \sum_{j \neq i} \frac{1}{2^{i+j}}$$

$$= \frac{1}{2^i}.$$

When $i \neq j$, we therefore have

$$P(A_i \cap A_j) = \frac{1}{2^{i+j}} = P(A_i) \cdot P(A_j)$$

and therefore A_i and A_j are independent. However, no triple (A_i, A_j, and

A_k, with distinct i, j, and k) is independent because three of these events can never occur simultaneously as constructed, and so

$$P(A_i \cap A_j \cap A_k) = 0 \neq \frac{1}{2^{i+j+k}} = P(A_i) \cdot P(A_j) \cdot P(A_k).$$

1.9. A sequence of any finite number of events, each with probability $\frac{1}{2}$, such that every proper subset is independent but the sequence itself is not.*

Let $k \geq 2$ be an integer. Define the events $A_1, \ldots, A_k$ so that the first $k - 1$ represent the event "heads" in a succession of independent tosses of a fair coin. However, the last one, A_k, is defined to happen if and only if an even number of heads have occurred in $A_1, \ldots, A_{k-1}$. Each event has probability $\frac{1}{2}$. (This may be seen for A_k by considering the first $k - 2$ tosses as known and then letting the last toss determine A_k itself.)

Any subset that does not contain A_k is clearly independent. If a proper subset contains A_k, then it must omit some other event. Because we may think of tossing the coin for this omitted event last, and because A_k depends directly on this toss, we see that any proper subset is independent. The entire sequence is quite dependent, however, because any event is determined by the other $k - 1$ events, since the total number of events that happen must always be odd.

1.10. For every $n \geq k \geq 2$ there are events $B_1, \ldots, B_n$ such that all the product rules hold except the one over $\{1, \ldots, k\}$.

Given events $B_1, \ldots, B_n$ and subsets $S \subset \{1, \ldots, n\}$, consider the possible "product rules" $P(\bigcap_{i \in S} B_i) = \prod_{i \in S} P(B_i)$. There are $2^n - n - 1$ nontrivial cases in which S has at least two elements. This example, due to Dudley, shows that all $2^n - n - 1$ cases must be checked in order to claim independence of $B_1, \ldots, B_n$.

Let $B_j^1 = B_j$ and let B_j^0 be the complement of B_j. Consider the 2^n "atoms" $B^S = \bigcap_{j=1}^n B_j^{S(j)}$ where $S(j) = 0$ or 1 for each j. Let each atom that intersects any B_j, where $j > k$, have probability $1/2^n$. (Here B_j may be any sets for which all 2^n atoms are nonempty.) Let the events $A_1, \ldots, A_k$ be defined as in Example 1.9. For each atom B^S such that $S(j) = 0$ for all $j > k$, set

*This counterexample was communicated by Richard M. Dudley, who credits S. Bernstein.

$$P(B^S) = P\left(\bigcap_{j=1}^{k} A_j^{S(j)}\right) \cdot 2^{k-n}.$$

For any proper subset S of $\{1, \ldots, k\}$ with r elements, where $0 \le r < k$, $P(\bigcap_{j \in S} B_j) = 2^{-r}(1 - 2^{k-n}) + 2^{k-n}P(\bigcap_{j \in S} A_j) = 2^{-r}$. If a set $M \subset \{1, \ldots, n\}$ intersects $\{k + 1, \ldots, n\}$ and has m elements, then $P(\bigcap_{j \in M} B_j) = 2^{-m}$. Thus, all product rules hold except that $P(\bigcap_{j=1}^k B_j) = 2^{-k}(1 - 2^{k-n}) + 2^{k-n}P(\bigcap_{j=1}^k A_j) \neq 2^{-k}$, as stated.

1.11. Three events such that the probability of the intersection is equal to the product of the probabilities even though the events are not independent.

Consider a single toss of a six-sided die and let events A, B, and C be defined as follows:

A is "the toss was even."

B is "the toss was odd."

C is "the toss was a seven."

Note that C is an event that never happens with a six-sided die. Therefore,

$$P(A \cap B \cap C) = 0 = \tfrac{1}{2} \cdot \tfrac{1}{2} \cdot 0 = P(A)P(B)P(C).$$

However, A and B are clearly not pairwise independent because

$$P(A)P(B) = \tfrac{1}{4} \neq 0 = P(A \cap B).$$

Therefore the events A, B, and C are not independent.

If it seems unfair to use an event with probability zero, we provide a different example, in which the product of the probabilities is not zero. Toss a fair coin three times and denote the result by (for example) HTH, which represents the sequence "heads, tails, heads." Define three events by

$A = \{\text{TTT, HTT, THH, HHH}\}$

$B = \{\text{TTH, THT, HHT, HHH}\}$

$C = \{\text{TTH, THT, HTH, HHH}\}$.

Note that

$$P(A) = P(B) = P(C) = \tfrac{1}{2}$$

and

$$P(A \cap B \cap C) = P(\{\text{HHH}\}) = \tfrac{1}{8}$$

so that the probability of the intersection is equal to the product of the probabilities. However, these are not independent events because, for example,

$$P(B \cap C) = P(\text{TTH, THT, HHH}) = \tfrac{3}{8} \neq \tfrac{1}{4} = P(A)P(B).$$

1.12. *N* events such that the probability of the intersection is equal to the product of the probabilities even though the events are not independent.

To construct such an example, we need only to choose events $E_1, E_2, \ldots, E_N$ that are not independent and choose E_N to have probability zero. The reasoning of the first part of Example 1.11 still applies. For example, let E_i mean "the result of the toss of a six-sided die was an even number" if $i = 1, 2, \ldots, N - 1$, but let E_N be the empty event "the toss was a seven." Then

$$P(E_1 \cap E_2 \cap \cdots \cap E_N) = 0 = P(E_1)P(E_2) \cdots P(E_N).$$

However,

$$P(E_1 \cap E_2) = \tfrac{1}{2} \neq \tfrac{1}{4} = P(E_1)P(E_2).$$

Therefore the events $E_1, E_2, \ldots, E_N$ are not independent.

1.13. Two independent events for which the conditional probability of *A* given *B* is not equal to the probability of *A*.

This surprising but trivial example is possible due to the fact that the conditional probability $P(A \mid B)$ is undefined when $P(B) = 0$ and so cannot be equal to $P(A)$. A concrete example is provided by the choice $A = \Omega$, the entire sample space, and $B = \phi$. They are independent because

$$P(A \cap B) = P(\phi) = 0 = P(A)P(B).$$

However, $P(A) \neq P(A \mid B)$ because this conditional probability is not defined.

1.14. Simpson's Paradox: What's good for the population can be bad for every subgroup.

The following version of Simpson's Paradox is similar to that presented in Lindley and Novick (1981).

Out of a population of 100 patients, 50 were given a treatment T and the remaining 50 were given a control. Let R denote recovery and $\bar{R}$ denote no recovery. The data are summarized here.

	R	$\bar{R}$		*Recovery Rate*
T	20	30	50	40%
$\bar{T}$	16	34	50	32%
	36	64	100	

The recovery rate for those treated, 40%, exceeds the recovery rate for the control group, indicating that the treatment is preferred.

However, if the population is further divided into Y (age 50 or less, let's say) and $\bar{Y}$ (older than 50), then the following frequencies are reported:

Y (*Younger*):

	R	$\bar{R}$		*Recovery Rate*
T	16	24	40	40%
$\bar{T}$	5	5	10	50%
	21	29	50	

$\bar{Y}$ (*Older*):

	R	$\bar{R}$		*Recovery Rate*
T	1	9	10	10%
$\bar{T}$	14	26	40	35%
	15	35	50	

We see that, for each subpopulation, the recovery rate for the controls exceeds that for the treated. In short, lack of treatment is good for the old and good for the young, but bad for the whole population!

The paradox is partially resolved by observing that mostly patients older than 50 were assigned to the treatment group; that is, age and treatment have been confounded. For a further discussion of the statistical implications of the paradox, the reader is referred to Lindley and Novick (1981) and the references given there.

1.15. Events for which the sum of the probabilities is infinite, but only finitely many occur with probability one.

This example says that the converse to the first Borel-Cantelli Lemma need not hold if the events E_k are not mutually independent. (But see Billingsley 1979, theorem 6.4, p. 73, for a weaker condition than mutual independence under which the converse does hold.) In fact, a sequence $E_1, E_2, \ldots$ may be constructed for which $\Sigma_{k=1}^{\infty} P(E_k)$ diverges but $P\{$only finitely many E_k occur$\} = 1$. Let $A_n = (0, 1/n)$ on the interval $(0, 1)$ with Lebesgue measure. Then $\Sigma P(A_n) = \Sigma 1/n = \infty$, but each ω in $(0, 1)$ occurs only in $A_1, \ldots, A_{[1/\omega]}$, where $[1/\omega]$ denotes the largest integer less than or equal to $1/\omega$. It follows that

$$P\{A_i \text{ finitely often}\} = 1.$$

1.16. Independent events with probabilities less than one, yet whose infinite intersection occurs with nonzero probability.

This example is obvious to any reader who is familiar with infinite products, but it may be surprising otherwise. Let the events $E_1, E_2, \ldots$ be independent, and let E_k for $k = 2, 3, \ldots$ be the event that card k is not chosen at random from a deck of k^2 cards labeled $1, 2, \ldots, k^2$. Thus, E_k has probability $1 - 1/k^2$, and

$$P(E_2 \cap E_3 \cap \cdots) = \prod_{k=2}^{\infty} P(E_k) = \prod_{k=2}^{\infty} \left(1 - \frac{1}{k^2}\right).$$

This product may be evaluated by cancellation:

$$\prod_{k=2}^{n} \left(1 - \frac{1}{k^2}\right) = \prod_{k=2}^{n} \frac{(k-1)(k+1)}{k \cdot k}$$

$$= \frac{1 \cdot 3}{2 \cdot 2} \frac{2 \cdot 4}{3 \cdot 3} \frac{3 \cdot 5}{4 \cdot 4} \cdots \frac{(n-1)(n+1)}{n \cdot n} = \frac{1}{2} \cdot \frac{n+1}{n}.$$

Taking the limit as $n \to \infty$, we find that the infinite intersection does indeed occur with positive probability:

$$P(E_2 \cap E_3 \cap \cdots) = \lim_{n\to\infty} \prod_{k=2}^{n} \left[\frac{1}{2}\frac{n+1}{n}\right] = \frac{1}{2}.$$

1.17. A countable collection of sequences of events such that the probability tends to one for each sequence of events, yet the probability of the intersection over the collection does not tend to one.

If the collection were finite, then no counterexample would be possible. This is because, given a finite collection of sequences $E_{1n}, E_{2n}, \ldots, E_{In}$ such that

$$\lim_{n\to\infty} P\{E_{in}\} = 1 \qquad \text{for } i = 1, \ldots, I,$$

we would then have

$$\lim_{n\to\infty} P\left\{\bigcap_{i=1}^{I} E_{in}\right\} = 1.$$

However, if we let the collection be infinite (so that $I = \infty$), then the probability of the intersection can be zero. Define sets that are either empty or the entire sample space as follows:

$$E_{in} = \begin{cases} \Omega & \text{if } n > i \\ \phi & \text{if } n \le i. \end{cases}$$

Then it is clear that for each fixed i, $P\{E_{in}\} \to 1$ as $n \to \infty$. However, the limit of the probability of the intersection is not one because

$$\lim_{n\to\infty} P\left\{\bigcap_{i=1}^{\infty} E_{in}\right\} = \lim_{n\to\infty} 0 = 0.$$

1.18. A discrete probability measure on the real numbers whose support is the entire real line.

By definition, the support of a discrete probability measure is not necessarily just the set of points at which it places mass. If the atomic mass points are too close together, they are not distinguished individually by the support set. For example, let $r_1, r_2, \ldots$ be an enumeration of the rational numbers, and let P be the probability measure that places mass $1/2^n$ at r_n. Then P is discrete but its support is the entire real number line because, for any real number x and for any $\varepsilon > 0$, we have

$$P\{(x - \varepsilon, x + \varepsilon]\} > 0$$

because the interval $(x - \varepsilon, x + \varepsilon]$ must contain at least one rational number, say r_n, and therefore must have a probability of at least $1/2^n$.

1.19. Two probability measures with the same support but that are not equivalent to each other.

If two probability measures are equivalent to each other (in the sense that each is absolutely continuous with respect to the other), then it follows that they must have the same support. The converse to this is not true, however, as shown by this counterexample.

As in Example 1.1, consider the probability space $([0, 1], \mathbf{L}, \mu)$ where μ denotes the Lebesgue measure on the unit interval. Also, let λ be the probability measure on the same space, placing mass $1/2^n$ at the nth rational number in $[0, 1]$. By an argument similar to Example 1.18, λ has support equal to $[0, 1]$, and therefore the supports of μ and λ are identical. However, μ and λ are not equivalent because neither is absolutely continuous with respect to the other. That they do not have the same sets of probability zero may be shown by the fact that

$$\mu(\{r_1\}) = 0 \quad \text{whereas} \quad \lambda(\{r_1\}) = \tfrac{1}{2}$$

and

$$\mu(\{\text{irrationals}\}) = 1 \quad \text{whereas} \quad \lambda(\{\text{irrationals}\}) = 0.$$

In fact, μ and λ are mutually singular.

1.20. A probability space in which no singleton sets $\{\omega\}$ are measurable.

Let the sample space Ω be any set with two or more elements; then the set $\{1, 2\}$ or the integers or the real numbers would be satisfactory. Let the σ-field be $\{\phi, \Omega\}$, the smallest possible σ-field for any space. Probabilities must now be defined as $P(\phi) = 0$ and $P(\Omega) = 1$. This is a valid probability space, and yet there are no measurable sets other than ϕ and Ω, by definition. In particular, singleton sets such as $\{\omega\}$ are not measurable, and there are no "simple events."

1.21. The union of a collection of σ-fields that is not itself a σ-field.

Although we know that the intersection of a collection of σ-fields (even an uncountable collection) must be a σ-field, this fails if we replace "intersection" with "union." Let $\Omega = \{a, b, c\}$ and consider the two σ-fields

$$G_1 = \{\{a\}, \{b, c\}, \phi, \Omega\} \quad \text{and} \quad G_2 = \{\{b\}, \{c, a\}, \phi, \Omega\}.$$

Their union is

$$G = G_1 \cup G_2 = \{\{a\}, \{b\}, \{b, c\}, \{c, a\}, \phi, \Omega\},$$

which is not a σ-field because the union

$$\{a\} \cup \{b\} = \{a, b\}$$

is not in G.

1.22. A separable σ-field containing a nonseparable sub-σ-field.

Consider $\Omega = \mathbf{R}$, the real numbers. The Borel sets $\mathbf{B}$ form a separable σ-field because they are generated by a countable class of sets—for example, by

$$B_i = (-\infty, r_i), \qquad i = 1, 2, \ldots$$

where $r_1, r_2, \ldots$ is an enumeration of the rational numbers.

Now consider the σ-field $\mathbf{F}$ that consists of the countable and co-countable sets; that is, a subset of $\mathbf{R}$ is a member of $\mathbf{F}$ if and only if either it or its complement is countable. It may be verified directly that $\mathbf{F}$ is a σ-field. To see that $\mathbf{F}$ is not separable, we will show that an arbitrary countable class of sets $A_1, A_2, \ldots$ in $\mathbf{F}$ cannot generate $\mathbf{F}$. Without loss of generality, by taking complements when necessary, we may suppose that each A_i is a countable set. The union of the A_i is then countable and therefore omits an uncountable number of real numbers. In particular, let s and t be two numbers that are omitted. It follows that each set in the σ-field generated by the A_i must contain either both s and t or neither s nor t. But $\{s\}$ is in $\mathbf{F}$, and therefore the collections $A_1, A_2, \ldots$ does not generate $\mathbf{F}$.

1.23. A tail event with probability strictly between zero and one.

Let $(\Omega, \mathbf{F}, P)$ be a probability space with $A \in \mathbf{F}$ that satisfies $0 < P(A) < 1$. Define the sequence of events $E_1, E_2, \ldots$ so that $E_i = A$ for all i. Define the tail σ-field $\mathbf{T}$ in the usual way; that is,

$$\mathbf{T} = \bigcap_{n=1}^{\infty} \sigma(E_n, E_{n+1}, \ldots).$$

Then A is clearly a tail event because $A \in \mathbf{T}$. However, $P(A)$ is neither zero nor one.

This does not violate the zero-one law because the sequence of events that form the σ-field is not independent in this example.

1.24. A tail σ-field formed from a nonindependent sequence of events, but for which every tail event has probability either zero or one.

This example shows that the converse to Kolmogorov's Zero-One Law does not hold. Let A be an event with $0 < P(A) < 1$, and consider the sequence of events

$$E_1 = A, \qquad E_2 = A, \qquad E_3 = E_4 = \cdots = \phi$$

where ϕ denotes the empty event. This is not an independent sequence of events because the first two are dependent. However, the tail σ-field consists of only ϕ and the sample space, and so each tail event has probability either zero or one.

One might still wonder whether an example can be given in which the sequence is not "eventually" independent (as this example is because ϕ is independent of itself). To construct one, first consider a probability space with a sequence of independent events $B_1, B_2, \ldots$ with $P(B_i) = \frac{1}{2}$ for all i. Define a new sequence of events E_i by interleaving new events between these:

$$B_1,\ B_2,\ B_1 \cup B_2,\ B_3,\ B_2 \cup B_3,\ B_4,\ B_3 \cup B_4, \ldots.$$

This new sequence is never independent, no matter how far out you go, because $B_n \cup B_{n+1}$ is dependent on $B_{n+1} \cup B_{n+2}$. However, the interleaved sequence $E_1, E_2, \ldots$ generates the same tail σ-field as the independent sequence $B_1, B_2, \ldots$, and so every tail event has probability either zero or one nonetheless.

1.25. Two probability measures defined on the same measurable space that agree on a class of sets but do not agree on the generated σ-field.

In general, if P_1 and P_2 are probability measures on $\mathbf{F}$ that agree on a class of subsets $\mathbf{C}$ in $\mathbf{F}$, then they agree on $\sigma(\mathbf{C})$ provided $\mathbf{C}$ is a π-system of sets. A proof is given in Billingsley (1979, theorem 3.3, p. 34). However, the conclusion need not hold without this added assumption.

Consider $\Omega = \{1, 2, 3, 4\}$ with the σ-field of all subsets of Ω. Let $C_1 = \{1, 2\}$, $C_2 = \{2, 3\}$, and $C_3 = \{3, 4\}$, and define the class of sets $\mathbf{C} = \{C_1, C_2, C_3\}$. Define probability measures P_1 and P_2 as follows:

ω	P_1	P_2
1	$\frac{1}{4}$	$\frac{1}{8}$
2	$\frac{1}{4}$	$\frac{3}{8}$
3	$\frac{1}{4}$	$\frac{1}{8}$
4	$\frac{1}{4}$	$\frac{3}{8}$

Then P_1 and P_2 agree on $\mathbf{C}$ because, for any $C \in \mathbf{C}$, we have $P_1(C) = P_2(C) = \frac{1}{2}$. However, the event $\{2\}$ is in $\sigma(\mathbf{C})$ because it is the intersection of C_1 with C_2. Since $P_1(\{2\}) = \frac{1}{4} \neq \frac{3}{8} = P_2(\{2\})$, it follows that P_1 and P_2 do not agree on $\sigma(\mathbf{C})$.

1.26. Two classes of events that are independent of each other, but whose generated σ-fields are dependent.

In general, if $\mathbf{C}_1$ and $\mathbf{C}_2$ are independent of each other and if each $\mathbf{C}_i$ is a π-system, then $\sigma(\mathbf{C}_1)$ and $\sigma(\mathbf{C}_2)$ are independent (see, for example, Billingsley 1979, theorem 4.2, pp. 43–44). This counterexample shows that the result is false in general.

Let $\Omega = \{1, 2, 3, 4\}$, let $\mathbf{B}$ be the σ-field of all subsets of Ω, and let P assign probability $\frac{1}{4}$ to each simple event $\{\omega\}$, where $\omega = 1, 2, 3, 4$. Let $\mathbf{C}_1 = \{\{1, 2\}\}$ and $\mathbf{C}_2 = \{\{2, 3\}, \{2, 4\}\}$. These are independent because if $C_1 \in \mathbf{C}_1$ and $C_2 \in \mathbf{C}_2$, then

$$P(C_1 \cap C_2) = \tfrac{1}{4} = P(C_1) \cdot P(C_2).$$

However, $\{2\} \in \sigma(\mathbf{C}_2)$ and $\{1, 2\} \in \sigma(\mathbf{C}_1)$, and these events are not independent because

$$P(\{1, 2\} \cap \{2\}) = \tfrac{1}{4} \neq \tfrac{1}{8} = P(\{1, 2\}) \cdot P(\{2\}).$$

1.27. A σ-field $\mathbf{G}$ generated by a class of sets $\mathbf{A}$ [so that $\mathbf{G} = \sigma(\mathbf{A})$], but $\sigma(\{A_1 \times A_2 : A_1, A_2 \in \mathbf{A}\})$ is not equal to $\sigma(\{G_1 \times G_2 : G_1, G_2 \in \mathbf{G}\})$.*

Care must be taken in extending measure spaces to Cartesian cross-products, as this example shows. Let $\mathbf{G}$ be the σ-field on $\mathbf{R}$ generated by

* This example was communicated by Clyde Sugahara.

the class $\mathbf{A} = \{[0, 1]\}$ of sets. Then $\mathbf{G} = \sigma(\mathbf{A}) = \{\phi, \mathbf{R}, [0, 1], \mathbf{R} - [0, 1]\}$. In the Cartesian cross-product space, $\mathbf{R} \times \mathbf{R} = \mathbf{R}^2$, we have

$$\sigma(\{A_1 \times A_2 : A_1, A_2 \in \mathbf{A}\}) = \{\phi, [0, 1] \times [0, 1], \mathbf{R}^2, \mathbf{R}^2 - ([0, 1] \times [0, 1])\}$$

which does not comprise all of $\sigma(\{G_1 \times G_2 : G_1, G_2 \in \mathbf{G}\})$ because, for example, it also contains the set $[0, 1] \times \mathbf{R}$.

1.28. A finitely additive probability that is not countably additive; hence, a finitely additive probability on a field need not extend to a countably additive probability on the generated σ-field.

Let $\Omega = \{1, 2, \dots\}$ and let $\mathbf{B}$ be all the subsets of Ω either that are finite or whose complement is finite. Then $(\Omega, \mathbf{B})$ is a measurable space. If $B \in \mathbf{B}$ is finite, let $P(B) = 0$, but otherwise let $P(B) = 1$. Then P is finitely additive on $\mathbf{B}$; however, it is not countably additive. To see why, let $B_i = \{i\}$. Then $B_i \in \mathbf{B}$, all B_i are disjoint, and their union is $\mathbf{B}$. If P were countably additive, we would have $1 = P(\Omega) = \Sigma\, P(B_i) = 0$, a contradiction.

Does this example contradict the Caratheodory Extension Theorem (which says that a σ-finite measure on a field has a unique extension to a measure on the generated σ-field)? The answer is no. The reason is that, although you start with a measure on a field, it must be countably additive to begin with. This is sometimes overlooked, perhaps because set operations in a field include only finite operations.

1.29. A nonmeasurable subset of the real numbers with respect to Lebesgue measure.

The point of this example is that there do exist subsets of the interval $[0, 1)$ outside of $\mathbf{B}$, the Borel sets, and that sets are omitted even when we pass by completion to the Lebesgue-measurable sets. In particular, Lebesgue measure can never be extended to the σ-field of all subsets of $[0, 1)$. Although for discrete countable probability spaces it is always possible to extend the measure to the σ-field of all subsets, in general this is not possible. The following construction gives a classic example.

Consider the probability space $([0, 1), \mathbf{L}, \mu)$, where $\mathbf{L}$ denotes the Lebesgue-measurable sets and μ denotes the Lebesgue measure. We will show that there exists a subset of $[0, 1)$ that is not Lebesgue-measurable. This will show that we cannot extend a probability measure on $[0, 1)$ to all subsets and still retain the property that the measure of an interval in $[0, 1)$ is its length. But see Example 1.31 for an interesting rebuttal.

The general way to construct a nonmeasurable set is to invoke the Axiom of Choice, which says that if **C** is any collection of nonempty sets, then there exists a function f defined on **C** that assigns to each set A in **C** an element $f(A)$ in A (see, for example, Royden 1968, p. 18).

It is also useful to define "addition modulo 1" of x and y, where x and y are both in $[0, 1)$:

$$(x + y)_1 = \begin{cases} x + y & \text{if } x + y < 1 \\ x + y - 1 & \text{if } x + y \geq 1. \end{cases}$$

Thus, if each x in $[0, 1)$ is assigned the angle $2\pi x$, then addition modulo 1 corresponds to the addition of angles on the unit circle. In the construction of Lebesgue measure it is easily shown that it is invariant under translation modulo 1 (Royden 1968, p. 63). Stated precisely, if E is a Lebesgue-measurable set in $[0, 1)$, then for any x in $[0, 1]$, the set $(E + x)_1$ is also measurable and has the same measure as E.

Using these facts, we can construct a nonmeasurable set. Define an equivalence relation so that x and y are equivalent if and only if there exists a rational number r such that $(x + r)_1 = y$. Equivalence of x and y will be denoted by $x \sim y$. By the standard theory of equivalence relations, $[0, 1)$ is partitioned into disjoint equivalence classes. Thus, any two elements of a given equivalence class differ by a rational number, and any two elements from distinct equivalence classes differ by an irrational number. Invoking the Axiom of Choice, we find there exists a set E that contains exactly one element from each equivalence class. (Note that there is an uncountable number of equivalence classes because each equivalence class is countable, and hence the "choice set" E is uncountable.) If $r_0, r_1, r_2, \ldots$, with $r_0 = 0$, is an enumeration of the rational numbers in $[0, 1)$, define $E_i = (E + r_i)_1$. Note that $E = E_0$.

We now claim that the sets $E_0, E_1, E_2, \ldots$ represent a disjoint collection whose union is all of $[0, 1)$, which may be seen as follows. Suppose x is in both E_i and E_j. Then we may write $x = x_i + r_i = x_j + r_j$, with x_i and x_j in E. But then $x_i - x_j = r_j - r_i$ is rational, so $x_i \sim x_j$. And E is assumed to have only one element from each equivalence class, implying that $i = j$. Therefore, the sequence $E_0, E_1, \ldots$ is indeed disjoint. Furthermore, each real number x in $[0, 1)$ is in some equivalence class and so differs from some element in E by some rational number r_i. But then x is in E_i, showing that the union of the sequence $E_0, E_1, \ldots$ is all of $[0, 1)$.

It is now straightforward to see that E is a nonmeasurable set. Suppose, anticipating a contradiction, that it is measurable. Then, since Lebesgue measure is translation invariant modulo 1, each E_i must be measurable and have the same measure; that is, $\mu(E_i) = \mu(E)$. But then

$$\mu([0, 1)) = \sum_{i=0}^{\infty} \mu(E_i) = \sum_{i=0}^{\infty} \mu(E).$$

Note that the right-hand side is either zero or infinity, depending on whether or not $\mu(E)$ is zero. Either of these possibilities contradicts the fact that $\mu([0, 1)) = 1$. We conclude that E is not measurable.

It is important to observe that we have used the Axiom of Choice in a crucial way to show the existence of a set that is not otherwise directly constructible. Note also that we have made no use of Lebesgue measure other than translation invariance and countable additivity.

1.30. A probability space on the real numbers for which every subset is measurable.

It may seem a surprise to you, if you have spent time worrying about nonmeasurable subsets of the real numbers, that there do exist perfectly valid probability measures on the reals for which every subset is measurable. For example, consider the real numbers together with the σ-field of all subsets and the probability measure that places all of its mass at 0 so that

$$P(E) = \begin{cases} 1 & \text{if 0 is in } E \\ 0 & \text{if 0 is not in } E. \end{cases}$$

In fact, any probability measure concentrated on a countable subset of the real numbers may be extended so that every subset of the reals is measurable. But, as shown in Example 1.29, it is not possible to extend Lebesgue measure so that all subsets are measurable; at least it is not possible if we assume the Axiom of Choice. But see the next example for a surprise when the Axiom of Choice is omitted from the model for set theory.

1.31. A set of axioms under which every subset of the real numbers is Lebesgue-measurable.

If you ever thought that the classical nonmeasurable set of Example 1.29 was not completely satisfying, you are not alone. One crucial step involves using the Axiom of Choice to select one element from each of an uncountably infinite collection of sets without providing a rule for this selection process. This axiom is used because selection would not otherwise be possible; it is often used in mathematics to justify operations that would not otherwise be humanly possible. An excellent introductory treatment of the Axiom of Choice and its role in set theory is provided by Halmos (1960).

Mathematicians who want their logical foundations to include only operations that could be done in a finite amount of time may omit the Axiom of Choice from the underpinnings of their mathematical universe. If this is done, mathematics becomes much more constructive and certain paradoxes (concerning objects that exist but are not directly constructible) disappear. In particular, it is possible to assert that

EVERY SUBSET OF THE REAL NUMBERS
IS LEBESGUE-MEASURABLE

without getting into any trouble, provided the Axiom of Choice is omitted. This surprising result is from Solovay (1970). This means essentially that if the axioms of set theory (omitting the Axiom of Choice) are consistent (that is, cannot lead to a contradiction), then adding the statement "every subset of the real numbers is Lebesgue-measurable" will not introduce a contradiction.

In the words of Solovay (1970, p. 1):

> We show that the existence of a non-Lebesgue measurable set cannot be proved in Zermelo-Frankel set theory (ZF) if use of the axiom of choice is disallowed. In fact, even adjoining an axiom DC to ZF, which allows countably many consecutive choices, does not create a theory strong enough to construct a non-measurable set.

Unfortunately, omitting the Axiom of Choice also rules out the proofs of many classical results of advanced mathematics, and as Solovay points out (p. 3), "Of course, the axiom of choice is true, and so there are non-measurable sets." Nonetheless, his result points out how important this axiom is in the construction of a non-Lebesgue-measurable set (see Example 1.29).

1.32. An uncountable set with Lebesgue measure zero.

All countable subsets of [0, 1] have Lebesgue measure zero. Because uncountable sets are larger in some sense, we might expect them to all have positive measure (provided, of course, they are measurable). The Cantor Set is an example of an uncountable measurable set with measure zero, showing that uncountable sets may be quite small in a measure-theoretic sense.

From the interval [0, 1], we will remove a countable number of intervals, guaranteeing that the remaining set is measurable. First, remove the middle third, $(\frac{1}{3}, \frac{2}{3})$. Second, remove the middle third of each of the two remaining intervals, $(\frac{1}{9}, \frac{2}{9})$ and $(\frac{7}{9}, \frac{8}{9})$. Continue so that at the kth stage we remove 2^{k-1} intervals, each of length $1/3^k$. In the limit, the Cantor Set remains.

The measure of the Cantor Set is

$$1 - \sum_{k=1}^{\infty} \frac{2^{k-1}}{3^k} = 1 - \frac{1}{3} \sum_{k=0}^{\infty} \left(\frac{2}{3}\right)^k = 1 - \frac{\frac{1}{3}}{1 - \frac{2}{3}} = 0.$$

To see that the Cantor Set is uncountable, express each ω in [0, 1] as a ternary (base 3) expansion:

$$\omega = .a_1 a_2 a_3 \cdots = \sum_{k=1}^{\infty} \frac{a_k}{3^k}$$

where each a_k is 0, 1, or 2. The Cantor Set can be described as the set of ω that can be written using only the digits 0 and 2 (all others are eliminated by removing the intervals). Because an uncountable number of sequences may be formed using two symbols (consider the binary expansion representations of all numbers between zero and one), it follows that the Cantor Set is uncountably infinite.

CHAPTER TWO
RANDOM VARIABLES, DENSITIES, AND DISTRIBUTION FUNCTIONS

Introduction

A real-valued point function $X(\cdot)$ defined on the probability space $(\Omega, \mathbf{F}, P)$ is called a "real random variable" (or sometimes just a "random variable") if the set $\{\omega \in \Omega : X(\omega) \leq x\} \in \mathbf{F}$ for every real number x. We would like to be able to assign a probability to the event "X is less than or equal to x" and so must require that this set be in the σ-field of measurable events. In fact, the point function $F(x) = P\{\omega \in \Omega : X(\omega) \leq x\} = P(X \leq x)$ is called the "cumulative distribution function" of the random variable X. Knowledge of the cumulative distribution function of a random variable allows us to compute probabilities for any Borel-measurable set E, other than just those of the form $\{\omega \in \Omega : X(\omega) \leq x\}$ for some real x, using the relationship:

$$P(X \in E) = \int_E dF$$

where the integral is interpreted as a Lebesgue-Stieltjes integral with respect to the cumulative distribution function F. Thus, we will often specify the cumulative distribution function of X without any reference to the probability measure P.

In practice, two types of cumulative distribution functions are of particular interest. The first is when F corresponds to a random variable that assumes at most a countable number of values $a_1, a_2, \ldots$ with probabilities $p_1, p_2, \ldots$. In this case, the cumulative distribution function has jumps at points $a_1, a_2, \ldots$ and stays constant in any interval that does not contain any of these points. The jump at a_i is of height $P(X = a_i) = p_i$. Such a random variable X is called a "discrete random variable," and the function $p(a) = P(X = a)$ is called the "discrete probability function" of X. The second type of cumulative distribution function is called "absolutely continuous" (and the random variable itself is called "continuous") if the cumulative distribution function can be written as follows:

$$F(x) = \int_{-\infty}^{x} f(t)\,dt$$

where the integration is with respect to Lebesgue measure. In this case, the function $f(x)$ is called a "probability density function" of X. If $f(x)$ is continuous at x, then we have $F'(x) = f(x)$.

The "Gaussian distribution" (also sometimes referred to as the "normal distribution") with parameters μ (a real number, the mean) and σ^2 (a positive real number, the variance) is defined as the distribution with density function

$$\frac{1}{\sigma(2\pi)^{1/2}} \exp\left[-\frac{[(x-\mu)/\sigma]^2}{2}\right].$$

In the special case $\sigma = 0$, there is no density function, but such a Gaussian distribution is degenerate and places probability one at μ. The "standard Gaussian distribution" is the Gaussian distribution with $\mu = 0$ and $\sigma^2 = 1$.

We can extend the notion of a random variable to vector-valued random variables in a natural way. The collection of k single-valued real functions $X_1(\cdot), \ldots, X_k(\cdot)$ that map Ω into $\mathbf{R}^k$ is called a "k-dimensional random variable" if the set $\{\omega : X_1(\omega) \leq x_1, \ldots, X_k(\omega) \leq x_k\}$ is in $\mathbf{F}$ for all vectors $(x_1, \ldots, x_k)$. The cumulative distribution function in this case is defined to be

$$F(x_1, \ldots, x_k) = P\{X_1 \leq x_1 \text{ and} \ldots \text{and } X_k \leq x_k\}.$$

The cumulative distribution function F_i of X_i alone, referred to as the "marginal distribution of X_i," is given by:

$$F_i(x) = F(\infty, \ldots, \infty, x, \infty, \ldots, \infty)$$

where x is in the ith position. This function is the probability that X_i (by itself, without regard to the values of the other components) is less than or equal to x. If a marginal distribution has a density function, this density is called the "marginal density." The special case $k = 2$ is referred to as the "bivariate" case.

We say that $(X_1, \ldots, X_n)$ has a "multivariate Gaussian distribution" if every linear combination has a (univariate) Gaussian distribution. (See, for example, Chapter 2 of Anderson 1958 for further details.)

Even more generally, consider a map X from the measurable space $(\Omega, \mathbf{F})$ to the measurable space $(\Lambda, \mathbf{G})$ where in each case we have specified the sample space and the σ-field of measurable sets. We say that X is a "random object," "measurable function," or a "Λ-valued random variable" if

$$\{\omega \in \Omega : X(\omega) \in A\} \in \mathbf{F} \qquad \text{for every } A \in \mathbf{G}.$$

For example, when Λ is the real line and $\mathbf{G}$ is the Borel σ-field, a measurable function will be a real random variable, as was previously defined. Thus a real random variable is a special case of a measurable function, when the measurable function maps to the real numbers with the Borel sets. A measurable function X from a probability space $(\Omega, \mathbf{F}, P)$ to a measurable space $(\Lambda, \mathbf{G})$ induces a probability measure Q on $(\Lambda, \mathbf{G})$ in the following natural way:

$$Q(A) = P\{\omega : X(\omega) \in A\} \qquad \text{for each } A \in \mathbf{G}.$$

We refer to Q as the "P law" of the "P distribution" of X. The sets $\{\omega : X(\omega) \in A\}$, where $A \in \mathbf{G}$, form a sub-σ-field of $\mathbf{F}$ called the "σ-field generated by X" or the "σ-field induced by X."

Now consider random variables (X, Y) and a measurable set S with $P(X \in S) \neq 0$. Then the ratio

$$\frac{P\{\omega : X(\omega) \in S \text{ and } Y(\omega) \leq y\}}{P\{\omega : X(\omega) \in S\}} = F(y \mid X \in S)$$

is interpreted as the "conditional probability" of $Y \leq y$ given the condition that X is in S. The function $F(y \mid X \in S)$ is called the "conditional distribution function" of Y given that $X \in S$.

Suppose P is a probability measure defined on $\Omega = \mathbf{R}^2$ equipped with the usual Borel σ-field $\mathbf{F}$. Define the coordinate maps

$$X(x, y) = x \quad \text{and} \quad Y(x, y) = y \qquad \text{for } (x, y) \in \mathbf{R}^2.$$

If (X, Y) has a joint density function $f(x, y)$ with respect to Lebesgue measure in the plane, then

$$P(A) = \iint_A f(x, y)\, dx\, dy \qquad \text{for } A \in \mathbf{F}.$$

If $f_X(x)$ is a marginal density of X, then the ratio

$$h(y \mid x) = \frac{f(x, y)}{f_X(x)}$$

is called a "conditional probability density function" of Y given that $X = x$. We must be careful about the interpretation of $h(y \mid x)$, particularly if $P\{X = x\} = 0$. The important property of $h(y \mid x)$ is the following: If A is a Borel set in $\mathbf{R}$ and $B \in \mathbf{F}$, let $B(x) = \{y : (x, y) \in B\}$. Then

$$P[\{X \in A\} \cap B] = \int_A \int_{B(x)} h(y \mid x) f_X(x)\, dy\, dx.$$

Precisely, if $g(x) = \int_{B(x)} h(y \mid x)\, dy$, then $g(x)$ is a version of the conditional probability of B given that $X = x$. A more general approach to conditional expectations and conditional probability is deferred to Chapter 6.

2.1. A real-valued function, on a probability space, that is not a random variable.

Consider the probability space $(\Omega, \mathbf{F}, P)$ where

$$\Omega = \{1, 2, 3\} \quad \text{and} \quad \mathbf{F} = (\Omega, \phi, \{1, 2\}, \{3\}\}.$$

Then $\{1\}$ is not a measurable set, and therefore the identity function, with $f(1) = 1, f(2) = 2$, and $f(3) = 3$, which maps from $(\Omega, \mathbf{F})$ to $(\mathbf{R}, \mathbf{B})$, is not a measurable function. Therefore f is not a random variable.

This example is less artificial than it might appear. The restriction that $\{1\}$ not be measurable would be important if, for example, this were a sub-σ-field that represented only partial information. In betting schemes (and economic analyses) it is important to "hide" the future so that information about it is only partially available. In this sense, $\{1, 2\}$ might be measurable (corresponding to the potential partial information that either 1 or 2 will occur), but more precise information would be unavailable.

For another example, consider the probability space $([0, 1), \mathbf{L}, \mu)$ consisting of the Lebesgue-measurable sets with Lebesgue measure in $[0, 1)$. Let E denote the nonmeasurable set as constructed in Example 1.29. Define the function

$$X(\omega) = \begin{cases} 0 & \text{if } \omega \in E \\ 1 & \text{otherwise.} \end{cases}$$

As defined, X is not a random variable because, in particular,

$$\{\omega \in [0, 1) : X(\omega) \leq \tfrac{1}{2}\} = E,$$

which is not a measurable set with respect to $\mathbf{L}$.

2.2. A real random variable X and a real function Y both defined on a common probability space such that $P(X = Y) = 1$, but Y is not a real random variable.

The point of this example is that the function need not be measurable even if it is equal almost everywhere to a measurable function. However, the function would have to be measurable if it mapped from a "complete" measure space to the real numbers. ("Completeness" here is the property that every subset of a set of measure zero is measurable.) Without this assumption, measurability need not follow.

Consider $(\Omega, \mathbf{F}, P)$ with $\Omega = \{-1, 0, 1\}$, $\mathbf{F} = \{\phi, \Omega, \{1\}, \{-1, 0\}\}$, and $P(\{1\}) = 1$. Define X and Y as follows:

ω	$X(\omega)$	$Y(\omega)$
-1	1	3
0	1	2
1	1	1

Because the event $\{\omega : X(\omega) = Y(\omega)\} = \{1\} \in \mathbf{F}$ has probability one, we have $X = Y$ with probability one. Because X is constant, it is measurable. However, Y is not measurable because $\{\omega : Y(\omega) \leq 2\} = \{0, 1\}$, which is not in $\mathbf{F}$. Note, however, that it is always possible to enlarge the σ-field and extend the measure to a complete space.

2.3. A cumulative distribution function that is continuous but does not have a density.

For further details, the reader is referred to Gelbaum and Olmsted (1964, p. 96). The classic example of a function that is continuous but not absolutely continuous is called the "Lebesgue Function" or the "Cantor Function"; it may be constructed as follows: For every x in $[0, 1]$, write x in a ternary (i.e., base 3) expansion:

$$x = \sum_{i=1}^{\infty} \frac{a_i}{3^i} = .a_1 a_2 a_3 \ldots$$

where each $a_i = a_i(x)$ is 0, 1, or 2. The Cantor Set is defined as those x in $[0, 1]$ that have an expansion with $a_i \neq 1$ for all $i = 1, 2, \ldots$. Let $n = n(x)$ be the first index for which $a_n = 1$, or ∞ if no 1 occurs in any expansion of x. The Cantor Function defined on $[0, 1]$ is then given by

$$F(x) = \sum_{i=1}^{n(x)} \frac{a_i(x)}{2^{i+1}} + \frac{1}{2^n}.$$

This function has the unusual property that it is a continuous distribution function [provided we define $F(x)$ to be zero if $x \leq 0$ and one if $x \geq 1$]. However, it has a vanishing derivative almost everywhere because it is flat on the intervals that are removed to form the Cantor Set. Since this derivative (equal almost everywhere to the zero function) is not a density function, it follows that the Cantor Function defines a distribution that is not absolutely continuous, even though the cumulative distribution function is continuous.

2.4. A distribution that is neither discrete nor continuous.

It is very easy to construct a distribution that is neither discrete nor continuous by mixing two distributions together. For example, toss a fair coin. If it comes up heads, the random variable takes the value zero. If the coin comes up tails, the random variable is defined to be an independent observation of a standard Gaussian random variable. The resulting mixture of standard Gaussian with degenerate mass at zero is not discrete because it is not concentrated on a countable set. It is not continuous because of its mass at zero.

Although most of the standard distributions used in statistics are either discrete (binomial, Poisson, hypergeometric, and so on) or continuous (Gaussian, chi-squared, t, F, and so on), there are nonetheless some moderately common distributions that are neither discrete nor continuous. The class of compound Poisson distributions (which are infinitely divisible), where the underlying distribution is not discrete, falls into this category. The compound Poisson distribution with parameter λ and underlying distribution F is defined as the result of the following two-stage procedure:

1. Observe K from a Poisson distribution with mean λ.
2. Sum K independent observations from F, and report this value as the random variable. If $K = 0$, then report zero.

With the particular choice of F as the chi-squared distribution with two degrees of freedom, the resulting compound Poisson distribution may be interpreted as the "noncentral chi-squared distribution with zero degrees of freedom" (Siegel, 1979). This may be interpreted as a distribution in the "extended Gaussian family" of distributions that is neither discrete nor continuous.

2.5. An unbounded density function.

Let the random variable X have the distribution function $F(x)$ given by:

$$F(x) = \begin{cases} 0 & \text{if } x < 0 \\ x^{1/2} & \text{if } 0 \leq x \leq 1 \\ 1 & \text{if } x > 1. \end{cases}$$

Then X has a density function $f(x)$ given by

$$f(x) = \begin{cases} \dfrac{1}{2x^{1/2}} & \text{if } 0 \leq x \leq 1 \\ 0 & \text{otherwise.} \end{cases}$$

Note that $f(x)$ is unbounded for x near zero. In fact, as x approaches zero by positive values, $f(x)$ tends toward infinity slowly enough so that the density function still integrates to one.

An alternative example is the density of the chi-squared distribution with one degree of freedom.

2.6. A discontinuous density function.

There is no requirement that a density function be continuous. For example, the density function of the exponential distribution $f(x) = \exp(-x)I(x \geq 0)$ is discontinuous at zero. The uniform distribution $U(0, 1)$ is discontinuous at its endpoints. Perhaps a more interesting example is provided by mixing two uniform distributions, $U(0, 1)$ and $U(0, 2)$, with equal probabilities. The resulting mixture has a density with a discontinuity at one, which is well inside the support of the distribution.

2.7. A continuous density that does not tend asymptotically to zero.

Most of the continuous densities $f(x)$ on the real line that we deal with tend to zero in the limit as x tends to ∞ or to $-\infty$. However, this need not always be the case.

Let Y be geometrically distributed, so that $P(Y = n) = 2^{-n}$, where $n = 1, 2, \ldots$, and let X (conditional on the value of Y) have a Gaussian distribution with mean Y and variance $n^{-2}2^{-2n}$. Then the (unconditional) distribution of X will be a mixture of these Gaussian distributions and will have a density that sums them. In particular, $f(n) \geq (2\pi)^{-1/2}2^{-n}n2^{n} \geq 0.3989n$, and so the density $f(x)$ takes on larger and larger values as x tends to infinity. Of course, between these peaks f does tend to be very close to zero, but these very narrow and very high peaks are sufficient to guarantee that $\lim_{x\to\infty} f(x) \neq 0$ because the limit does not exist.

2.8. Two different densities that generate the same distribution.

Let X be a random variable with the uniform distribution $U(0, 1)$, so that its distribution function is given by

$$F(x) = \begin{cases} 0 & \text{if } x \leq 0 \\ x & \text{if } 0 \leq x < 1 \\ 1 & \text{if } x \geq 1. \end{cases}$$

A probability density function of X is any function f that satisfies

$$F(x) = \int_{-\infty}^{x} f(t)\, dt.$$

Clearly, the function $f(t) = I_{(0,1)}(t)$, the indicator function of the unit interval, is a probability density function for X. But, trivially, if we let $g(t) = I_{[0,1]}(t)$, then g is also a probability density function, even though it differs from f at the points zero and one. In fact, any function that is equal to f almost everywhere with respect to Lebesgue measure will be a probability density function of X.

The point is that changing the density value on a set of measure zero (with respect to the carrier measure, Lebesgue measure in this case) does not change the value of the integral and is therefore also a density function for the same distribution.

2.9. An infinite family of bivariate distributions with the same marginal distributions.

Whereas a joint distribution determines the marginal distributions, the converse is false. For example, consider the sample space $\Omega = \{0, 1\}$. For any ε with $0 \leq \varepsilon \leq \frac{1}{4}$, suppose the jointly distributed random variable (X, Y) has the discrete probability function $p(x, y)$ shown in the following table:

		Y	
		0	1
X	0	$\frac{1}{4} + \varepsilon$	$\frac{1}{4} - \varepsilon$
	1	$\frac{1}{4} - \varepsilon$	$\frac{1}{4} + \varepsilon$

Then the marginal probabilities of X and Y are independent of ε and are given by

$$P(X = 0) = P(X = 1) = \tfrac{1}{2}$$
$$P(Y = 0) = P(Y = 1) = \tfrac{1}{2}.$$

Thus, there are infinitely many joint probability distributions with these marginals. For an example of a family of bivariate distributions in the plane that have bivariate densities with the same marginal densities, see Mood, Graybill, and Boes (1974, p. 142).

2.10. A bivariate distribution that is not bivariate Gaussian but that has Gaussian marginal distributions.

If (X, Y) are jointly distributed random variables with a bivariate Gaussian distribution, then it is well known that the marginal distributions are also Gaussian (including, possibly, the degenerate distribution concentrated at one point). The converse is false, however; that is, there do exist bivariate distributions that are not Gaussian but whose marginals are Gaussian. (Kowalski 1973 gives some examples.) We give two examples here.

a. Consider the bivariate Gaussian distribution with mean at the origin and covariance matrix equal to the identity matrix. Its density function is

$$g(x, y) = \frac{1}{2\pi} \exp\left\{-\frac{x^2 + y^2}{2}\right\}$$

for any real x and y. Now let (X, Y) have a joint density given by

$$f(x, y) = \begin{cases} 2g(x, y) & \text{if } xy \geq 0 \\ 0 & \text{otherwise.} \end{cases}$$

Then the marginal distribution of X is given by

$$\int_{-\infty}^{\infty} f(x, y)\, dy = \begin{cases} \displaystyle\int_{-\infty}^{0} 2g(x, y)\, dy & \text{if } x \le 0 \\ \displaystyle\int_{0}^{\infty} 2g(x, y)\, dy & \text{if } x > 0 \end{cases}$$

$$= \begin{cases} \dfrac{1}{\pi} \exp\left(-\dfrac{x^2}{2}\right) \displaystyle\int_{-\infty}^{0} \exp\left(-\dfrac{y^2}{2}\right) dy & \text{if } x \le 0 \\ \dfrac{1}{\pi} \exp\left(-\dfrac{x^2}{2}\right) \displaystyle\int_{0}^{\infty} \exp\left(-\dfrac{y^2}{2}\right) dy & \text{if } x > 0 \end{cases}$$

$$= \frac{1}{(2\pi)^{1/2}} \exp\left(-\frac{x^2}{2}\right),$$

which we recognize as the density of the standard Gaussian distribution. Similarly, the marginal distribution of Y is the standard Gaussian distribution. The bivariate distribution is not Gaussian, however, since it is not supported on the entire real plane and is also not supported on a one-dimensional subspace.

b. This example is from Roussas (1973, p. 80). Let (X, Y) have joint density given by

$$f(x, y) = \frac{1}{2\pi} \exp\left[-\frac{(x^2 - y^2}{2}\right] + \frac{1}{4\pi e} x^3 y^3 I_{[-1,1]}(x) I_{[-1,1]}(y).$$

It is straightforward to verify that f is a proper density function because it is nonnegative and integrates to one. Since the second term satisfies

$$\int_{-1}^{1} \frac{1}{4\pi e} x^3 y^3\, dy = \int_{-1}^{1} \frac{1}{4\pi e} x^3 y^3\, dx = 0,$$

it follows that the marginal distributions of X and Y are Gaussian. However, it is easily checked that this joint density is not a bivariate Gaussian density.

2.11. Two Gaussian random variables whose sum does not have a Gaussian distribution.

In general, if $(X_1, \ldots, X_n)$ has a multivariate Gaussian distribution, then any linear combination $a_1 X_1 + \cdots + a_n X_n$ will also have a Gaussian distribution, even if the X_i are not independent. However, a linear combination of Gaussian random variables need not be Gaussian in general if individual (marginal) rather than multivariate Gaussianity holds.

Let X have the standard Gaussian distribution with mean zero and variance one. Observe X; then toss a fair coin and define Y as follows:

$$Y = \begin{cases} X & \text{if the toss is "heads"} \\ -X & \text{if the toss is "tails."} \end{cases}$$

By symmetry of X about the origin, it follows that Y also has a standard Gaussian distribution. However, the sum $X + Y$ has a positive probability mass of $\frac{1}{2}$ at zero, since this sum is zero whenever the toss is tails. On the other hand, $X + Y$ is not degenerate because it is continuously distributed whenever the toss is heads. Such a mixture of a discrete and a continuous distribution cannot have a Gaussian distribution. Note that (X, Y) does not have a bivariate Gaussian distribution!

2.12. Three pairwise independent, Gaussian random variables that are not trivariate Gaussian.

Three mutually independent Gaussian random variables must be jointly trivariate Gaussian; however, pairwise independence is not sufficient to guarantee this. Let X, Y, and Z_0 be independent Gaussian random variables with mean zero and variance one. Define

$$Z = |Z_0| \operatorname{sgn}(XY)$$

where

$$\operatorname{sgn}(t) = \begin{cases} 1 & \text{if } t > 0 \\ 0 & \text{if } t = 0 \\ -1 & \text{if } t < 0. \end{cases}$$

Then Z has a Gaussian distribution because $\operatorname{sgn}(XY)$ is 1 or -1 with probability $\frac{1}{2}$, independent of Z_0. To verify the pairwise independence of X and Z, for example, observe that $\operatorname{sgn}(XY)$ is statistically independent of X. The other two pairs are also seen to be independent. However, (X, Y, Z) is not trivariate Gaussian because if it were, then (due to pairwise independence) X, Y, and Z would have to be mutually independent. This is impossible because

$$P(X > 0,\ Y > 0, \text{ and } Z < 0) = 0 \neq \tfrac{1}{8} = P(X > 0) \cdot P(Y > 0) \cdot P(Z < 0).$$

An example of $n > 3$ pairwise independent, Gaussian random variables that are not multivariate Gaussian may be constructed by adjoining $n - 3$ independent Gaussian random variables independently to the three of this example.

2.13. A bivariate distribution without a density but whose marginal distributions possess densities.

If (X, Y) has a joint density with respect to Lebesgue measure on $\mathbf{R}^2$, then the marginals have densities with respect to Lebesgue measure on $\mathbf{R}$. However, this counterexample shows that the converse to this statement is false.

Let $X = Y$ where both have a standard Gaussian distribution. Then the marginal distributions of (X, Y) each have the standard Gaussian density. The bivariate distribution is concentrated on the line $y = x$, however, and therefore does not have a density with respect to two-dimensional Lebesgue measure in this plane. It does have a density with respect to Lebesgue measure on the line $y = x$.

2.14. A discontinuous bivariate density with continuous marginal densities (from Anderson 1958, p. 37).

Consider the bivariate Gaussian density function given by

$$g(x, y) = \frac{1}{2\pi(1 - \rho^2)^{1/2}} \exp\left\{-\frac{x^2 - 2\rho xy + y^2}{2(1 - \rho^2)^{1/2}}\right\}.$$

Define the bivariate density function f as follows:

$$f(x, y) = \begin{cases} g(-x, y) & \text{if } 1 \leq |x| \leq 2 \text{ and } 1 \leq |y| \leq 2 \\ g(x, y) & \text{otherwise.} \end{cases}$$

If we assume that $\rho \neq 0$, then $f(x, y)$ is not continuous when $|x| = 1$, $|x| = 2$, $|y| = 1$, or $|y| = 2$. In effect, values of $g(x, y)$ have been swapped in the four squares. (For example, in the first quadrant the square is bounded by the lines $x = 1$, $x = 2$, $y = 1$, and $y = 2$.) Integrating against y or x is then the same whether or not the values have been swapped. Thus the marginal distributions of X and Y are Gaussian; in particular, they are continuous.

Another example is provided by any continuous density g with $g(0) > 0$ and $g(-x) = g(x)$ for all x. Define

$$f(x, y) = \begin{cases} 2g(x)g(y) & \text{if } xy > 0 \\ 0 & \text{if } xy \le 0. \end{cases}$$

Then f is discontinuous at (0, 0), but the marginals (both are g) are continuous.

2.15. A continuous bivariate density with a discontinuous marginal density.

Let X have an exponential distribution with mean one. Conditional on X, let Y have a Gaussian distribution with mean $1/X$ and variance 1. Then (X, Y) has a bivariate density function $f(x, y) = g(x)h(y \mid x)$ given as follows:

$$f(x, y) = \begin{cases} \dfrac{1}{(2\pi)^{1/2}} \exp\left\{-x - \dfrac{[y - (\frac{1}{x})^2]}{2}\right\} & \text{if } x > 0 \\ 0 & \text{otherwise.} \end{cases}$$

Note that $f(x, y)$ is continuous everywhere in the plane. To see this, suppose $(x_n, y_n) \to (x, y)$. If x is nonzero, then it is clear that $f(x_n, y_n) \to f(x, y)$. If $x = 0$, then $f(x_n, y_n) \to f(x, y) = 0$ regardless of how many x_n values are positive. However, the marginal density $f(x) = e^{-x}I(x > 0)$ has a jump at zero and is therefore not a continuous function.

2.16. A continuous function from $\mathbf{R}^2$ to [0, 1], increasing in each coordinate separately and also satisfying $\lim_{x\to-\infty} F(x, y) = 0$ for all y, $\lim_{y\to-\infty} F(x, y) = 0$ for all x, and $\lim_{x,y\to\infty} F(x, y) = 1$, and yet F is not a distribution function on R^2.

If F is a distribution function—say, for the random variable (X, Y)—then for every $x_1 < x_2$ and $y_1 < y_2$ we must have

$$\begin{aligned} &P(x_1 < X \le x_2 \quad \text{and} \quad y_1 < Y \le y_2) \\ &\quad = F(x_2, y_2) - F(x_2, y_1) - F(x_1, y_2) + F(x_1, y_1) \ge 0. \end{aligned}$$

Now define

$$F(x, y) = \begin{cases} 0 & \text{if } x \le 0 \text{ or } y \le 0 \\ \min[1, \max(x, y)] & \text{otherwise.} \end{cases}$$

It is straightforward to verify that F satisfies the stated monotonicity and limiting properties. However, if $0 \leq x_1 \leq y_1 < x_2 \leq y_2 \leq 1$, then

$$F(x_2, y_2) - F(x_2, y_1) - F(x_1, y_2) + F(x_1, y_1) = y_2 - x_2 - y_2 + y_1 < 0,$$

showing that F is not a distribution function.

2.17. Two random variables that are real-valued functions of a real variable, but whose composition is not a random variable.

The composition of the functions f and g is defined as the function $h(x) = f[g(x)]$. Although many operations on random variables result in (measurable) random variables, it may be surprising to find out that function composition need not.

Let the function f map from $(\mathbf{R}, \mathbf{P}(\mathbf{R}))$ (that is, from the real numbers with the σ-field of all subsets) to the reals. Define $f = I_N$, the indicator function of the non-Lebesgue-measurable set constructed in Example 1.29. Note that f is a random variable because $f^{-1}(A)$ is in $\mathbf{P}(\mathbf{R})$ for all Borel sets A. Define $g(x) = x$, a function from $(\mathbf{R}, \mathbf{B})$ (the reals with the Borel sets) to the reals. Note that f and g represent random variables that are real-valued functions of a real variable. However, their composition is not a random variable because

$$[f(g)]^{-1}(\{1\}) = g^{-1}[f^{-1}(\{1\})] = g^{-1}(N) = N,$$

which is not measurable with respect to $(\mathbf{R}, \mathbf{B})$, the space that g maps from.

As you see, the "trick" is that the σ-field of the domain of the function is not specified when we require that the function be a random variable, but the σ-field of the codomain (the range) is required to be the Borel sets. This leads to a counterexample even in the case where one σ-field is the Borel sets $\mathbf{B}$ and the other is the Lebesgue-measurable sets $\mathbf{L}$.

Let E denote a subset of $\mathbf{R}$ that is Lebesgue-measurable but not Borel-measurable. Such a set exists because of a cardinality argument (Halmos 1950, p. 67), which says that the cardinality of the Borel sets is that of the continuum (i.e., of the real numbers) but that the cardinality of the Lebesgue-measurable sets is larger because it includes all subsets of the Cantor Set, which itself has the cardinality of the continuum. Define

$$f(x) = I_E(x), \quad \text{mapping from } (\mathbf{R}, \mathbf{L})$$

$$g(x) = x, \quad \text{mapping from } (\mathbf{R}, \mathbf{B}).$$

These are both random variables from their respective spaces to the reals. However, the composition, $f(g)$, is not a random variable from the domain of g because

$$[f(g)]^{-1}(\{1\}) = g^{-1}[f^{-1}(\{1\})] = g^{-1}(E) = E$$

is not measurable with respect to the Borel sets, which is the σ-field of the domain of g.

CHAPTER THREE
MOMENTS OF RANDOM VARIABLES

Introduction

The mathematical expectation of a random variable X with cumulative distribution function $F(x)$ is defined as follows:

$$E(X) = \int_{-\infty}^{\infty} x \, dF(x)$$

where the integral is interpreted as the Lebesgue-Stieltjes integral of x with respect to $F(x)$. When X has a probability density function $f(x)$, then this reduces to the Lebesgue integral:

$$E(X) = \int_{-\infty}^{\infty} xf(x) \, dx.$$

When X has a discrete probability distribution that takes on values x_1, $x_2, \ldots$, the Lebesgue-Stieltjes integral reduces to a summation as follows:

$$E(X) = \sum_i x_i P(X = x_i).$$

In general, if $g(X)$ is a Borel-measurable function of X, we may define the expectation of $g(X)$ as follows:

$$E[g(X)] = \int_{-\infty}^{\infty} g(x) \, dF(x).$$

By definition, the "rth moment" of X is $E(X^r)$, provided that it exists. When $r = 1$, the first moment is called the mean of X. The variance of X, denoted

$\mathrm{Var}(X)$, is defined as follows:

$$\mathrm{Var}(X) = E\{[X - E(X)]^2\} = E(X^2) - [E(X)]^2.$$

If we are considering an n-dimensional random variable $(X_1, \ldots, X_n)$ and if $h(X_1, \ldots, X_n)$ is a Borel-measurable real-valued function of $(X_1, \ldots, X_n)$, then the expectation of $h(X_1, \ldots, X_n)$ is the following multiple integral:

$$E[h(X_1, \ldots, X_n)] = \int_{\mathbf{R}^n} h(x_1, \ldots, x_n)\, dF(x_1, \ldots, x_n).$$

For example, $h(X_1, \ldots, X_n) = \min(X_1, \ldots, X_n)$ is the minimum or first order statistic from a sample of size n. To define the "order statistics," let $X_1, \ldots, X_n$ be n random variables. If they are rearranged in ascending order of magnitude, such that $X_{(1)} \le X_{(2)} \le \cdots \le X_{(n)}$, then $X_{(j)}$ $(j = 1, 2, \ldots, n)$ is the jth order statistic from a sample of size n.

To find the moments of a random variable X, it is useful to consider its moment-generating function and its characteristic function. The "moment-generating function" is defined as follows:

$$m_X(t) = E(e^{tX}).$$

If $m_X(t)$ exists, then the rth moment of X is given by

$$E[X^r] = \left.\frac{d^r m_X(t)}{dt^r}\right|_{t=0}.$$

Similarly, the "characteristic function" of X, which always exists, is defined as follows:

$$\phi_X(t) = E(e^{itX}).$$

The rth moment of X (if it exists) may then be obtained by

$$E[X^r] = i^{-r} \left.\frac{d^r \phi_X(t)}{dt^r}\right|_{t=0}.$$

3.1. A random variable with infinite mean.

We give two examples; the first makes use of a discrete random variable (the "ill-luck paradox"; see Feller 1971, p. 15). Let $X_0, X_1, X_2, \ldots$ be a sample from a continuous distribution; that is, all the X_i's are independent and identically distributed with a continuous distribution. Suppose X_0 can be interpreted as the luck experienced by person P_0 as measured in dollars—say, from some random event. Define the random variable $N = \min\{n : X_n < X_0\}$; that is, N is the number of people $P_1, \ldots, P_N$ observed until finally a person P_N experiences worse luck. We may consider N to be a measure of P_0's bad luck, since it measures how long it takes until someone else experiences even worse luck. To find the mean of N, note that the event $\{N = n\}$ is equivalent to the event $\{N > n - 1\} - \{N > n\}$. Also, the event $\{N > j\}$ occurs if and only if the minimal term of the sequence $X_0, X_1, \ldots, X_j$ occurs at X_0. By symmetry, $P\{N > j\} = 1/(j + 1)$. Thus,

$$P\{N = n\} = \frac{1}{n} - \frac{1}{n + 1} = \frac{1}{n(n + 1)}.$$

We may now compute the expectation:

$$E(N) = \sum_{n=1}^{\infty} nP\{N = n\} = \sum_{n=1}^{\infty} \frac{1}{n + 1} = \infty.$$

Indeed, P_0 has very bad luck because the expected number of people to be observed until someone has luck as bad is infinite! The paradoxical nature here becomes even more apparent when we see that the random variable $\min\{n : X_n > X_0\}$ also has an infinite expectation. This then says that the expected time until someone experiences better luck is also infinite.

Here is another example to show that a continuous random variable may also have an infinite mean. Let X have the density function

$$f(x) = \frac{1}{x^2} I(x \geq 1).$$

Then the expected value of X is

$$E(X) = \int_1^{\infty} \left(\frac{1}{x}\right) dx = \infty.$$

3.2. A random variable whose mean does not exist.

Let the continuous random variable X have the Cauchy distribution centered at the origin with density given by

$$f(x) = \frac{1}{\pi(1 + x^2)}, \qquad -\infty < x < \infty.$$

The mean of X is then

$$E(X) = \int_{-\infty}^{\infty} \frac{x\,dx}{\pi(1 + x^2)} = \frac{1}{2\pi} \log(1 + x^2)\Big|_{-\infty}^{\infty}$$

and this integral does not exist because

$$\int_{0}^{\infty} xf(x)\,dx = \int_{-\infty}^{0} xf(x)\,dx = \infty.$$

Furthermore, no higher moments exist. Although a graph of the function $f(x)$ resembles the Gaussian density with mean zero (but is more peaked), its tails approach the axis much more slowly. This indicates higher probabilities for extremely large or small values, and as a consequence the mean does not exist.

This kind of behavior is not restricted to only continuous random variables. Here is an example of a discrete random variable whose mean does not exist:

$$P(X = n) = \frac{3}{\pi^2 n^2}, \qquad n = \cdots, -2, -1, 1, 2, \ldots.$$

The mean of the positive values of X is ∞, and the mean of the negative values is $-\infty$; therefore, the mean of X does not exist.

3.3. Two random variables such that the expectation of the sum is not equal to the sum of the expectations.

Given two random variables X and Y,

$$E(X + Y) = E(X) + E(Y)$$

provided all the expectations exist and are finite. This equation does not hold in general, however, as is shown by the example.

Let $X = -Y$, where X has a Cauchy distribution (as defined in Example 3.2). Then the left-hand side exists and is finite:

$$E(X + Y) = E(0) = 0,$$

but the right-hand side is undefined (and therefore cannot be said to equal zero) because the expectations $E(X)$ and $E(Y)$ do not exist.

3.4. A finite expectation of a sum of three random variables that is not determined by the marginal distributions.

We will exhibit two triples of random variables such that the marginal distributions are equal from one set to another, but the expectation of the sum is different from one triple to another even though these expectations are both finite. This counterexample is from "An Unexpected Expectation" by Simmons (1977), and further details may be found there.

Let U be uniformly distributed on $(0, 1)$. Note that $1 - U$ and $|2U - 1|$ are also uniformly distributed on $(0, 1)$. First, let $X = Y = \tan(\pi U/2)$ and let $Z = -2\tan(\pi U/2)$. Then

$$E(X + Y + Z) = E(0) = 0.$$

Next, redefine these three random variables. Let $X = \tan(\pi U/2)$, $Y = \tan[\pi(1 - U)/2]$, and $Z = -2\tan(\pi|2U - 1|/2)$. The marginal distributions are unchanged, but the joint distribution is different. Note that if $0 < U < \frac{1}{2}$, then $Z = \tan(\pi U/2) - \operatorname{ctn}(\pi U/2)$, but if $\frac{1}{2} < U < 1$, then $Z = \operatorname{ctn}(\pi U/2) - \tan(\pi U/2)$. Using these and other trigonometric simplifications, we find that

$$X + Y + Z = \begin{cases} 2\tan\dfrac{\pi U}{2} & \text{if } 0 < U < \frac{1}{2} \\ 2\operatorname{ctn}\dfrac{\pi U}{2} & \text{if } \frac{1}{2} < U < 1. \end{cases}$$

It follows that $X + Y + Z$ is positive with probability one and that the expectation of this sum is nonzero! The exact value may be calculated, and we find that

$$E(X + Y + Z) = \frac{4}{\pi}\log(2) = 0.8825424\ldots.$$

It is worth noting that this kind of pathological behavior cannot occur when the marginal distributions have finite expectations. Also, as Simmons pointed out, it cannot occur with just two summands.

3.5. Two random variables such that the expectation of their product is not the product of the expectations.

Although it is interesting to note that $E(XY) = E(X)E(Y)$ holds whenever X and Y are independent provided these expectations are all finite (or, more generally, if X and Y are uncorrelated; see Chapter 4), the result is false in general. For example, consider a toss of a six-sided die. Let X take on the value one if the die comes up an even number and zero if the die comes up an odd number. Let Y take on the actual value that the die comes up. Then

$$E(X) = P(\text{“die comes up even”}) = \tfrac{1}{2}$$

$$E(Y) = \sum_{j=1}^{6} \frac{j}{6} = \frac{7}{2}.$$

Therefore $E(X)E(Y) = \frac{7}{4}$. However,

$$E(XY) = \sum_{j=1}^{6} E(XY \mid Y = j)P(Y = j) = \frac{2 + 4 + 6}{6} = 2$$

$$\neq \frac{7}{4} = E(X)E(Y).$$

3.6. A random variable such that the expectation of its reciprocal is not the reciprocal of its expectation.

Let X be uniformly distributed on (0, 1) so that its density function is $f(x) = I(0 < x < 1)$. Then

$$E(X) = \int_0^1 x\,dx = \tfrac{1}{2}$$

so that the reciprocal of the expectation is $1/[E(X)] = 2$. However, the expectation of the reciprocal is very different:

$$E\left(\frac{1}{X}\right) = \int_0^1 \left(\frac{1}{x}\right) dx = \infty.$$

In general, therefore, we cannot expect $E[g(X)]$ to be equal to $g[E(X)]$. With $g(x) = 1/x$, this will hold if and only if X and $1/X$ are uncorrelated because it is equivalent to $1 = E[X \cdot (1/X)] = E(X)E(1/X)$. If X is positive and nondegenerate, then $E(1/X)$ can never be equal to $1/E(X)$ because of Jensen's inequality and the fact that $1/x$ is a strictly convex function when $x > 0$. In case you have decided that $E(1/X)$ is never equal to $1/E(X)$ when X is nondegenerate, however, here is a counterexample involving a negative value for which they are equal:

$$X = \begin{cases} -1 & \text{with probability } \frac{1}{9} \\ \frac{1}{2} & \text{with probability } \frac{4}{9} \\ 2 & \text{with probability } \frac{4}{9}. \end{cases}$$

For this particular distribution, it is easily verified that $E(X) = E(1/X) = 1$.

3.7. Two random variables such that the expectation of the minimum is not the minimum of the expectations.

Reconsider the ill-luck paradox presented in Example 3.1, where X_0, X_1, X_2, ... is a sample from a continuous distribution and N is the first index for which $X_N < X_0$, representing the time until someone has worse luck than person 0. Also define M to be the first index for which $X_M > X_0$, the time until someone has better luck than person 0. As we saw in Example 3.1,

$$E(N) = E(M) = \infty.$$

However, X_1 will be either larger or smaller than X_0 with probability one, so the minimum of M and N is one. Therefore,

$$E[\min(M, N)] = E(1) = 1 \neq \infty = \min[E(M), E(N)].$$

However, it is always true that

$$E[\min(M, N)] \leq \min[E(M), E(N)]$$

provided these expectations exist.

3.8. A random variable whose first moment exists but no higher moments exist.

Although it follows that whenever the kth moment of a random variable exists, so do all moments of positive lower order, in fact the converse is false. For example, let the random variable X have a density given by

$$f(x) = \frac{3}{2} x^{-5/2} I_{(1,\infty)}(x).$$

Then the expected value of X is

$$E(X) = \frac{3}{2} \int_1^\infty x^{-3/2}\, dx = -3x^{-1/2} \Big|_1^\infty = 3.$$

However, for integer values of $k > 1$, we find that

$$E(X^k) = \frac{3}{2}\int_1^{\infty} x^{k-(5/2)}\,dx = \frac{\frac{3}{2}}{k-\frac{3}{2}}x^{k-(3/2)}\bigg|_1^{\infty} = \infty.$$

In fact, for this example the moment of order k does exist, although it is infinite. We may modify the example slightly to achieve a case where the higher moments would be of the form $\infty - \infty$ and therefore would not exist. To do this, let the density have the same basic form but be symmetric about zero:

$$g(x) = \frac{3}{4}x^{-5/2}I_{(1,\infty)}(|x|).$$

More generally, the Student's t-distribution with $r + 1$ degrees of freedom has moments of order $0, 1, \ldots, r$, but no higher moments exist. (See, for example, Mood, Graybill, and Boes, 1974, p. 542.)

3.9. A random variable X and a positive number p such that $\lim_{x\to\infty} x^p P(|X| > x) = 0$, but $E(|X|^p) = \infty$.

It is true that if $p > 0$ and

$$\lim_{x\to\infty} x^p P(|X| > x) = 0,$$

then the rth moment exists; that is, $E(|X|^r) < \infty$ for all r such that $0 \le r < p$. However, this inequality must be weak; it does not extend to $r = p$.

Let $p = 1$, and suppose X has cumulative distribution function

$$F(x) = \begin{cases} 1 - \dfrac{2\log(2)}{x\log(x)} & \text{if } x \ge 2 \\ 0 & \text{if } x < 2. \end{cases}$$

Then we have

$$x^p P(|X| \ge x) = x\frac{2\log(2)}{x\log(x)} \to 0 \qquad \text{as } x \to \infty.$$

However, the pth moment is not finite here because

$$E(|X|) = \int_2^{\infty} x\,d\left\{1 - \frac{2\log(2)}{x\log(x)}\right\}$$

$$= 2\log(2)\int_2^\infty x\frac{1+\log(x)}{[x\log(x)]^2}\,dx$$

$$\geq 2\log(2)\int_2^\infty \frac{1}{x\log(x)}\,dx = \infty.$$

3.10. A random variable whose moment-generating function does not exist.

Although the characteristic function of a random variable always exists, the moment-generating function may not. For example, suppose the random variable X has the Cauchy distribution with density given by

$$f(x) = \frac{1}{\pi(1+x^2)}.$$

The integral

$$\int_{-\infty}^{\infty} e^{tx}\frac{1}{\pi(1+x^2)}\,dx$$

is infinite for any $t \neq 0$ because the integrand itself tends to ∞ either as $x \to \infty$ or as $x \to -\infty$. Thus the moment-generating function does not exist in this example.

3.11. A random variable, all of whose moments exist, but whose moment-generating function does not exist.

Existence (finiteness) of the moment-generating function for some $t > 0$ implies that all moments exist and are finite; however, the converse is false. Consider the lognormal distribution, which is the distribution of $Y = e^X$ where X has a Gaussian distribution. Suppose X has mean zero and variance one, so that Y has the standard lognormal distribution. The moments of Y exist for all orders $n = 1, 2, 3, \ldots$ because

$$E(Y^n) = E(e^{nX}) = \frac{1}{(2\pi)^{1/2}}\int_{-\infty}^{\infty} e^{nx}e^{-x^2/2}\,dx$$

$$= \frac{1}{(2\pi)^{1/2}}\int_{-\infty}^{\infty} \exp\left[-\frac{(x-n)^2}{2} + \frac{n^2}{2}\right]dx = e^{n^2/2}.$$

However, the moment-generating function does not exist because if $t > 0$, then

$$E(e^{tY}) = E[e^{t(e^X)}] = \frac{1}{(2\pi)^{1/2}} \int_{-\infty}^{\infty} e^{te^x - x^2/2}\, dx$$

$$\geq \frac{1}{(2\pi)^{1/2}} \int_0^{\infty} \exp\left[t\left(1 + x + \frac{x^2}{2} + \frac{x^3}{6}\right) - \frac{x^2}{2}\right] dx$$

$$= \infty$$

because the exponential term is a third-degree polynomial in x for which the x^3 term has a positive coefficient; this exponential must then tend to ∞ as $x \to \infty$. Therefore the moment-generating function of Y does not exist.

3.12. A random variable whose first five moments are equal to those of a standard Gaussian, but which does not have a Gaussian distribution.

Matching a few moments is not sufficient, by itself, to guarantee that two distributions are equal. If X has a standard Gaussian distribution, then its first five moments are $E(X) = 0$, $E(X^2) = 1$, $E(X^3) = 0$, $E(X^4) = 3$, and $E(X^5) = 0$. Consider the discrete distribution Y on three points:

$$Y = \begin{cases} 3^{1/2} & \text{with probability } \frac{1}{6} \\ -3^{1/2} & \text{with probability } \frac{1}{6} \\ 0 & \text{with probability } \frac{2}{3}. \end{cases}$$

Then, by symmetry, it follows that all odd moments of Y are zero. We calculate directly to see that $E(Y^2) = 1$ and $E(Y^4) = 3$. Thus the first five moments of Y match those of X. The distributions are very different, however; not only is Y not Gaussian, but it is not even a continuous distribution.

In fact, for any finite n there exists a discrete (and hence non-Gaussian) distribution whose first n moments are equal to those of a standard Gaussian. This cannot be extended to all moments because the Gaussian is completely determined by its moments.

3.13. A random variable whose first and third moments are both zero, but which is not symmetrically distributed.

Although the first and third moments (if they exist) are zero for any distribution that is symmetric about zero, these conditions are not sufficient to ensure symmetry. Let X have the following distribution:

$$X = \begin{cases} -2 & \text{with probability } \frac{2}{5} \\ 1 & \text{with probability } \frac{1}{2} \\ 3 & \text{with probability } \frac{1}{10}. \end{cases}$$

Then $E(X) = 0$ and $E(X^3) = 0$ are directly verified, despite the fact that the distribution of X is not symmetric about any center.

3.14. A distribution not determined uniquely by its moments.

A distribution determines a unique set of moments, assuming they exist. It is interesting to consider the reverse problem. Given a set of moments, can we determine a unique distribution with these moments? It is true that a distribution is uniquely determined by its moments if the distribution is supported on some bounded interval (Feller 1971, p. 225). However, the result is far from true in general, although various conditions have been proposed to ensure that it will be true in particular cases.

Suppose X has the lognormal distribution with density given by

$$f_1(x) = \{\sigma(x-a)(2\pi)^{1/2}\}^{-1} \exp\left\{-\frac{[\log(x-a)-m]^2}{2\sigma^2}\right\} I_{(a,\infty)}(x)$$

for some positive σ, with fixed values of a and m. Let Y have the density given by

$$f_2(x) = f_1(x)\left[1 + \sin\left\{2\pi \frac{\log(x-a)-m}{\sigma^2}\right\}\right].$$

Then both $f_1(x)$ and $f_2(x)$ determine the same set of moments given by

$$E(X^r) = \sum_{i=0}^{r} \binom{r}{i} a^{r-i} \exp\left(mi + \frac{i^2\sigma^2}{2}\right), \qquad r = 0, 1, \ldots.$$

The reader is referred to Patel, Kapadia, and Owen (1976, p. 10) and to Heyde (1963) for further details. However, in the special case when $\sigma = 1$, $a = 0$, and $m = 0$, it would suffice to show that

$$\int_0^\infty x^r f_1(x) \sin[2\pi \log(x)]\, dx = 0 \qquad \text{for } r = 0, 1, \ldots$$

because then f_2 would be a nonnegative probability density function with the same moments as $f_1(x)$. Changing variables so that $s = \log(x) - r$, we see that this integral becomes

$$(2\pi)^{-1/2} \exp\left(\frac{r^2}{2}\right) \int_{-\infty}^{\infty} \exp\left(\frac{-s^2}{2}\right) \sin(2\pi s)\, ds.$$

This integral clearly vanishes, since the integrand is odd and bounded absolutely by the integrable function $\exp(-s^2/2)$. Thus the result is established. A similar change of variables shows that the result holds for arbitrary σ, m, and a. In general, then, a distribution is not completely determined by its moments even when those moments exist as finite real numbers. Fu (1984) also gives an example.

3.15. A class of distributions with identical moment sequences (Feller 1971, p. 227).

Example 3.15 is an extreme example showing that a distribution is not determined uniquely by its moments, for it gives a whole class of distributions with the same moments. Consider the class of probability densities indexed by α in (0, 1) defined by

$$f_\alpha(x) = \frac{1}{24} \exp(-x^{1/4})[1 - \alpha \sin(x^{1/4})] I_{(0,\infty)}(x).$$

Tedious integration by parts yields

$$\int_0^{\infty} x^k \exp(-x^{1/4}) \sin(x^{1/4})\, dx = 0 \qquad \text{for } k = 0, 1, \ldots.$$

So, independent of α, the kth moment of a random variable with density $f_\alpha(x)$ is given by

$$\int_0^{\infty} \left(\frac{x^k}{24}\right) \exp(-x^{1/4})\, dx.$$

Thus all the densities $f_\alpha(x)$ have identical moment sequences.

3.16. A probability density that is not an even function even though all odd moments vanish.

It is easy to see that the odd moments of an even density function vanish, if they exist. However, it is surprising that all the odd moments of a density function may vanish even if the density is not an even function. Again, consider the class of probability density functions from Example 3.15 given by

$$f_\alpha(x) = \frac{1}{24}\exp(-x^{1/4})[1 - \alpha\sin(x^{1/4})]I_{(0,\infty)}(x)$$

for any α in (0, 1). Define the mixture distribution

$$f(x) = \begin{cases} \dfrac{f_\alpha(x)}{2} & \text{if } x \geq 0 \\ \dfrac{f_\beta(-x)}{2} & \text{if } x < 0 \end{cases}$$

for any α and β in (0, 1) with $\alpha \neq \beta$. Because the moments of $f_\alpha(x)$ and $f_\beta(x)$ are identical, it follows that all odd moments of $f(x)$ are zero. On the other hand, since $\alpha \neq \beta, f(x)$ is not an even function.

3.17. A distribution whose characteristic function is differentiable at zero but whose first moment does not exist.

Recall that the existence of the kth moment of a random variable X guarantees the existence of the kth derivative of the characteristic function of X evaluated at the origin. The converse is true for even-order derivatives (see Chung 1974, Theorem 6.4.1). This example shows that the converse does not extend to odd-order derivatives.

Suppose the random variable X has the following characteristic function:

$$\phi(t) = C\sum_{n=2}^{\infty}\frac{\cos(nt)}{n^2\log(n)}$$

where $C > 0$. This series, formally differentiated term by term, converges uniformly. Therefore $\phi'(t)$ exists and is continuous for all t, in particular when $t = 0$ (Zygmund 1947, p. 272). Note that $\phi(t)$ is the characteristic function of the distribution corresponding to masses of $C/[2n^2\log(n)]$ placed at n and $-n$, where $n = 2, 3, 4, \ldots$. The constant C is, of course, the normalizing constant so that the sum of the probability masses is one. (It may be verified that this series of probability masses is in fact convergent.) Thus, since the series

$$\sum_{n=2}^{\infty}\frac{1}{n\log(n)} = \infty,$$

as can be seen by applying the integral test, we see that

$$E(|X|) = \sum_{n=2}^{\infty} \frac{2|n|C}{2n^2 \log(n)} = \infty$$

and therefore the first moment does not exist because this distribution is symmetric about zero.

3.18. A random variable whose expectation does not exist, yet the expectations of certain order statistics from a sample do exist.

Although the expectation of the order statistics of a sample from a cumulative distribution function $F(x)$ will exist if the expectation of the distribution function $F(x)$ exists, the converse is false. We will give two examples; in the first the expectation is infinite, whereas in the second the expectation does not exist at all.

First, consider the random variable X with probability density function

$$f(x) = x^{-2} I_{(1,\infty)}(x).$$

The corresponding cumulative distribution function is given by $F(x) = 1 - 1/x$ for $x > 1$ and is zero otherwise. The expected value of X is infinite because

$$E(X) = \int_1^\infty x f(x)\, dx = \int_1^\infty \frac{1}{x}\, dx = \infty.$$

However, we will show that the expectations of all order statistics in a sample of size n exist except for the largest, $X_{(n)}$, the maximum of the sample. In general, if the sample is from a distribution $F(x)$ and density $f(x)$, then the density of the jth-order statistic is easily found (Mood, Graybill, and Boes 1974, p. 254) to be

$$f_{X_{(j)}}(x) = \frac{n!}{(j-1)!\,(n-j)!} [F(x)]^{j-1} [1 - F(x)]^{n-j} f(x).$$

For example, if $j = 1$, then the density of $\min(X_1, \ldots, X_n)$ is given by

$$f_{X_{(1)}}(x) = n x^{-n-1} \qquad \text{if } x > 1$$

and is zero otherwise. Thus, assuming $n > 1$, we have

$$E[X_{(1)}] = \int_1^\infty n x^{-n}\, dx = \frac{n}{1-n} x^{1-n} \Big|_1^\infty = \frac{n}{n-1}.$$

In general, since

$$f_{X_{(j)}}(x) = \frac{n!}{(j-1)!(n-j)!}\left[1 - \frac{1}{x}\right]^{j-1}\left(\frac{1}{x}\right)^{n-j+2}$$
$$< \frac{n!}{(j-1)!(n-j)!}\left(\frac{1}{x}\right)^{n-j+2},$$

we have

$$0 < E[X_{(j)}] < \frac{n!}{(j-1)!(n-j)!}\int_1^{\infty}\left(\frac{1}{x}\right)^{n-j+1} dx.$$

Because the integral exists for any $j < n$, it follows that the expectation of the jth-order statistic for any $j < n$ exists and is finite.

For a second example, in which the mean does not exist at all, consider the Cauchy distribution with density function

$$f(x) = \frac{1}{\pi(1 + x^2)}.$$

As we know from Example 3.2, the mean of the Cauchy distribution does not exist. However, the expectation of the jth-order statistic of a sample of size n may be expressed, after using the trigonometric substitution $x = \tan(u)$, by

$$E[X_{(j)}] = \frac{n!}{\pi^n(j-1)!(n-j)!}\int_{-\pi/2}^{\pi/2}[\tan(u)]\left[\frac{\pi}{2} + u\right]^{j-1}\left[\frac{\pi}{2} - u\right]^{n-j} du.$$

The reader is referred to David (1970, p. 26) for the details, but it is easy to show that all expectations exist except when $j = 1$ or $j = n$. This is quite believable in view of the fact that the Cauchy density behaves in its left and right tails like the right tail of the density in the first example just above, where only the maximum-order statistic was infinite.

CHAPTER FOUR
PROPERTIES OF REAL RANDOM VARIABLES

Introduction

Two real random variables X and Y are said to be "equal in distribution," which is written as $X \stackrel{d}{=} Y$ or $\mathbf{L}(X) = \mathbf{L}(Y)$ and expresses the fact that the distribution laws are equal. Random variables $X_1, \ldots, X_k$ are "independent" if their joint distribution function can be factored as the product of the marginal distributions—that is, if

$$F_{X_1,\ldots,X_k}(x_1, \ldots, x_k) = F_{X_1}(x_1) \cdot \cdots \cdot F_{X_k}(x_k).$$

Random variables $X_1, \ldots, X_k$ are called a "sample" from $F(x)$ if they are independent and identically distributed according to F. A weaker notion than "independent and identically distributed" is "exchangeable." A collection of random variables $X_1, \ldots, X_k$ is said to be exchangeable if, for every permutation $(n_1, \ldots, n_k)$ of the integers $(1, \ldots, k)$, we have

$$(X_1, \ldots, X_k) \stackrel{d}{=} (X_{n_1}, \ldots, X_{n_k}).$$

We define the "covariance" of real random variables X and Y to be

$$\operatorname{Cov}(X, Y) = E\{[X - E(X)][Y - E(Y)]\}.$$

When X and Y have finite positive variances, the "correlation" of X and Y is defined as

$$\operatorname{Cor}(X, Y) = \frac{\operatorname{Cov}(X, Y)}{\sigma_X \sigma_Y},$$

where σ_X and σ_Y are the standard deviations of X and Y, respectively. The Cauchy-Schwarz inequality, sometimes called the "correlation inequality," states that $|\text{Cor}(X, Y)| \le 1$ with equality if and only if $Y = a + bX$ with probability one for some real constants a and b. (Note that $b \ne 0$ because of our assumption that the variances are positive.) Random variables are said to be "uncorrelated" if $\text{Cor}(X, Y) = 0$.

We have considered $E(X)$, the mean of X, which is used as a measure of location of the distribution of X. Alternative measures of location are the mode and median of X. When $f(x)$ is a continuous density or a discrete probability function of X, a "mode" is a point at which $f(x)$ attains a maximum. Note that a cumulative distribution function F is called "unimodal" at a point x if the graph of F is convex in $[-\infty, x]$ and concave in $[x, \infty]$; intervals of constancy are not excluded. A "median" of X is defined as any point between

$$a = \inf\{x : F(x) \ge \tfrac{1}{2}\} \quad \text{and} \quad b = \sup\{x : F(x) \le \tfrac{1}{2}\},$$

where $F(x)$ is the cumulative distribution function of X. One common choice is $(a + b)/2$; note that for a continuous distribution, a and b are equal and there is a unique median. The dispersion of X is often measured by $\text{Var}(X)$. An alternative measure is the "mean absolute deviation from the mean" defined by $E[|X - E(X)|]$.

Given real random variables X and Y, we are often interested in the distribution of their sum $X + Y$. If X and Y are independent with characteristic functions $\phi_X(t)$ and $\phi_Y(t)$, respectively, then the characteristic function of $X + Y$ is given by a product: $\phi_{X+Y}(t) = \phi_X(t)\phi_Y(t)$. Because there is a one-to-one relationship between cumulative distribution functions and characteristic functions, we can use the product of the characteristic functions to determine the distribution of $X + Y$. Equivalently, if X and Y are independent with densities $f_X(x)$ and $f_Y(y)$, then the density of $X + Y$ is the convolution of the densities f_X and f_Y, denoted by $f_{X+Y} = f_X * f_Y$.

Suppose two random variables X and Y are jointly distributed with some known distribution. Given $X = x$, one might reasonably want to find some function $g(x)$, a predictor, that is close to Y in some sense. A measure of closeness is the mean squared error criterion $E\{[g(X) - Y]^2\}$. The function $g(X)$ that minimizes the mean squared error is $g(x) = E(Y \mid X = x)$, the conditional expectation of Y given $X = x$. This g is called the "best predictor" of Y based on X. (Also see Chapter 6 for more general details and examples of conditional expectations.)

The distribution of X is said to be "infinitely divisible" if, for each $n = 1, 2, \ldots$, there exists a distribution function F_n such that if $X_1, \ldots, X_n$ are independent and identically distributed from F_n, then $X \overset{d}{=} X_1 + \cdots + X_n$. In terms of characteristic functions, ϕ is the characteristic function of an infinitely divisible distribution if, for each n, there exists a characteristic function ϕ_n such that $\phi(t) = [\phi_n(t)]^n$. Chung (1974, p. 239) notes that an infinitely divisible characteristic function $\phi(t)$ is never zero for real values of t.

A distribution function F is said to represent a "stable law" if convolutions from F still have distribution F after a suitable change of location and scale. That is, whenever $X_1, X_2, \ldots$ are independent and identically distributed with distribution F, for each n there exist real $a_n \neq 0$ and b_n such that

$$\frac{X_1 + \cdots + X_n}{a_n} + b_n \sim F.$$

Note that every stable law is infinitely divisible because F_n may be chosen to be the distribution of $X_1/a_n + b_n/n$.

4.1. Two random variables that are equal in distribution, but whose difference is not equal in distribution to the zero distribution.

Suppose X and Y are independent and each has a standard Gaussian distribution with mean zero and variance one. Then they are equal in distribution. However, the distribution of $X - Y$ is Gaussian with mean zero and variance two, which is clearly not the zero distribution. In fact, $X - Y$ is equal in distribution to zero if and only if $X = Y$ with probability one. This example shows that in general one cannot simply apply the same operation to both sides of an "equal in distribution" symbol and expect to preserve the equality because $X \stackrel{d}{=} Y$ even though $X - Y \stackrel{d}{\neq} 0$.

4.2. Random variables such that the sum of the modes is not the mode of the sum.

In general, the expectation operator is a linear operator. In particular, the mean of a sum is equal to the sum of the means provided they all exist and are finite. This is not the case for modes, which are also used as a measure of location. We provide two examples: one for discrete and one for continuous random variables.

For the first example, let X and Y be independent, each with a discrete probability function defined by

$$P(X = 0) = P(Y = 0) = \tfrac{3}{5}$$
$$P(X = 1) = P(Y = 1) = \tfrac{2}{5}.$$

Then $\text{mode}(X) = \text{mode}(Y) = 0 = \text{mode}(X) + \text{mode}(Y)$. However,

$$P(X + Y = 0) = \tfrac{9}{25}$$
$$P(X + Y = 1) = \tfrac{12}{25}$$
$$P(X + Y = 2) = \tfrac{4}{25}.$$

Therefore $\text{mode}(X + Y) = 1 \neq 0 = \text{mode}(X) + \text{mode}(Y)$.

For the second example, let X and Y be independent random variables with the exponential density function given by

$$f(t) = e^{-t} I_{[0,\infty)}(t).$$

The mode of this distribution, its highest point, is at $t = 0$ because the exponential function is positive and decreasing on $[0, \infty)$ and is zero in $(-\infty, 0)$. Therefore,

$$\text{mode}(X) + \text{mode}(Y) = 0 + 0 = 0.$$

The distribution of the sum of two independent exponentials may be found directly, by using properties of the Gamma distribution family, or by noting that the standard exponential distribution is half of a chi-squared deviate with two degrees of freedom and so their sum is half of a chi-squared deviate with four degrees of freedom. In any case, the density of the sum $X + Y$ is found to be

$$f_{X+Y}(t) = te^{-t} I_{[0,\infty)}(t).$$

To find the mode of $X + Y$, we will maximize $f_{X+Y}(t)$ for $t \geq 0$. Setting the derivative equal to zero, we find that

$$f'_{X+Y}(t) = (1 - t)e^{-t} = 0 \qquad \text{when } t = 1$$

and therefore the mode occurs at $t = 1$ because $f''(1) < 0$. Thus,

$$\text{mode}(X + Y) = 1 \neq 0 = \text{mode}(X) + \text{mode}(Y).$$

4.3. A random variable such that the reciprocal of the median is not a median of the reciprocal.

If a real random variable X is always positive (i.e., $X > 0$), then $1/\text{median}(X)$ is a median of $1/X$ because the reciprocal operation perfectly reverses the ordering. If X takes on both positive and negative values, however, then the reciprocal operation reverses the positive and the negative values separately and does not perfectly reverse the ordering, and the median operator need not commute with the reciprocal operator.

For example, let X be uniformly distributed on the interval $(-1, 5)$. Then we have

$$\frac{1}{\text{median}(X)} = \frac{1}{2} \neq \frac{1}{3} = \text{median}\left(\frac{1}{X}\right)$$

which follows because $P(1/X > \frac{1}{3}) = P(0 < X < 3) = \frac{1}{2}$.

4.4. Random variables such that the median of the sum is not the sum of the medians.

As in Example 4.3, suppose X and Y are independent and that each has an exponential distribution with cumulative distribution function

$$F(t) = \begin{cases} 1 - e^{-t} & \text{if } t \geq 0 \\ 0 & \text{otherwise.} \end{cases}$$

Then the medians of X and of Y can be found by solving

$$F(t) = 1 - e^{-t} = \tfrac{1}{2}$$

and we find that

$$\text{median}(X) + \text{median}(Y) = 2\log(2).$$

The sum $X + Y$ has density te^{-t} for $t \geq 0$, as we found in Example 4.2. Integrating this, we find the cumulative distribution function of the sum to be

$$F_{X+Y}(t) = 1 - e^{-t} - te^{-t} \qquad \text{for } t \geq 0.$$

To see whether $2\log(2)$ is the median of this distribution, we compute

$$F_{X+Y}[2\log(2)] = 1 - e^{-2\log(2)} - 2\log(2)e^{-2\log(2)}$$

$$= 1 - \frac{1}{4} - \frac{\log(2)}{2} = 0.403\ldots.$$

Because this value, $0.403\ldots$, is not equal to $\frac{1}{2}$ and because F_{X+Y} is a continuous strictly increasing function for positive real arguments, it follows that

$$\text{median}(X + Y) \neq \text{median}(X) + \text{median}(Y).$$

4.5. Independent and identically distributed random variables such that each term has the same distribution as the sample mean.

A trivial example may be given using any degenerate distribution concentrated at one point. However, there are also more general and surprising examples.

In general, if $X_1, \ldots, X_n$ is a sample from a distribution with finite positive variance σ^2, then

$$\text{Var}(\bar{X}) = \text{Var}\left[\frac{X_1 + \cdots + X_n}{n}\right] = \frac{\sigma^2}{n}.$$

This says that the variance of the sample mean, a measure of its dispersion, will be less than that of any of the X_i terms. Intuitively, we expect the distribution of the sample mean to be more concentrated about the expectation of the distribution, reflecting the idea that it combines information from the entire sample and should therefore be a more precise estimate than any individual term. The point of this example is that this need not be the case for distributions without a variance.

Let X have a Cauchy distribution centered at zero with scale parameter $s > 0$ and a density function given by

$$f_s(x) = \frac{s}{\pi(s^2 + x^2)}.$$

The characteristic function of this distribution is

$$\phi_s(t) = E(e^{itX}) = \int_{-\infty}^{\infty} \frac{se^{itx}}{\pi(s^2 + x^2)}\,dx = e^{-s|t|}$$

where the integral can be evaluated using the Residue Theorem from complex analysis (see, for example, Conway 1978, p. 112).

Now suppose $X_1, \ldots, X_n$ are independent random variables with the standard Cauchy distribution $f_1(x)$ with $s = 1$. The distribution of the sum $Y = X_1 + \cdots + X_n$ may be found by noting that the characteristic function of Y is the product of the characteristic functions of the X_i, so that

$$E(e^{itY}) = [\phi_1(t)]^n = e^{-(n)|t|} = \phi_n(t).$$

Using the unique correspondence between characteristic functions and distributions, we see that Y has a Cauchy distribution with scale parameter n. The distribution of $\bar{X} = Y/n$ is therefore Cauchy with scale parameter 1, which is the same distribution as X_1 (and is also the same distribution as the other terms).

From a practical viewpoint, this says that the sample mean does no better in estimating the central location of a Cauchy distribution than does a single observation, even when (say) a million values are averaged. Note, however, that the median (rather than the mean) does do better than a single observation.

4.6. Independent and identically distributed random variables such that the sample mean is more disperse than each term (Feller 1971, p. 52).

Suppose Y has a standard Gaussian distribution with mean zero and variance one and with cumulative distribution function F_Y. Let $X = 1/Y^2$. The cumulative distribution function of X may be computed as follows, for $t > 0$:

$$F_X(t) = P(X \leq t) = P\left(|Y| \geq \frac{1}{t^{1/2}}\right) = 2\left[1 - F_Y\left(\frac{1}{t^{1/2}}\right)\right].$$

The density of X is then given, for $t > 0$, by

$$f_X(t) = \frac{e^{-1/(2t)}}{(2\pi)^{1/2} t^{3/2}} I_{(0,\infty)}(t).$$

Now consider the more general family of densities given by

$$f_\alpha(t) = \frac{\alpha e^{-\alpha^2/(2t)}}{(2\pi)^{1/2} t^{3/2}} I_{(0,\infty)}(t)$$

with the corresponding cumulative distribution functions

$$F_\alpha(t) = 2\left[1 - F_Y\left(\frac{\alpha}{t^{1/2}}\right)\right] I_{(0,\infty)}(t).$$

Detailed calculations show that this family is closed under convolution and that

$$f_\alpha * f_\beta = f_{\alpha+\beta}.$$

Then, for example, if X, X_1, and X_2 are independent random variables with density function $f_1(t)$, we have

$$P\left[\frac{X_1 + X_2}{4} \leq t\right] = F_2(4t)$$

$$= 2\left\{1 - F_Y\left[\frac{2}{(4t)^{1/2}}\right]\right\} = F_1(t)$$

for positive t. Therefore, $(X_1 + X_2)/4$ has the same distribution as X does. In this sense, the average $(X_1 + X_2)/2$ is actually more disperse than X.

In general, the convolution formula implies that

$$\frac{X_1 + \cdots + X_n}{n^2} \stackrel{d}{=} X_1$$

so that the average, $\bar{X}$, satisfies

$$\bar{X} \stackrel{d}{=} nX_1$$

and we see that the averages are increasing by order n instead of converging to a limit. For this family of distributions, increasing the sample size makes $\bar{X}$ worse and worse as an estimate of location.

More generally, suppose $X_1, X_2, \ldots, X_n$ is a sample from a stable law with index α, with $0 < \alpha \leq 2$. Then

$$\frac{X_1 + \cdots + X_n}{n} \stackrel{d}{=} n^{(1-\alpha)/\alpha} X_1 .$$

If $0 < \alpha < 1$, then the distribution of the sample mean is more disperse than the distribution of X_1, since the factor $n^{(1-\alpha)/\alpha}$ is greater than one and it even tends to infinity as n increases. Note that F_1 given here is a stable law of index $\frac{1}{2}$. The Cauchy distribution corresponds to $\alpha = 1$, whereas the Gaussian distribution corresponds to $\alpha = 2$.

4.7. A convolution of two unimodal continuous distributions that is not unimodal (Feller 1971, Problem V.25, p. 168).

Suppose X_1 has the uniform density $f_1(x)$ on (a_1, b_1), and X_2 has the uniform density $f_2(x)$ on (a_2, b_2). It is straightforward to verify that the convolution of f_1 and f_2 is triangular on $(a_1 + a_2, b_1 + b_2)$. (See, for example, Mood, Graybill, and Boes 1974, p. 186.)

Using this fact, we will find independent random variables X and Y with densities f_X and f_Y that are unimodal, yet the density of $X + Y$ is not unimodal. Let X and Y be independent, each with density given by the following mixture of two uniforms:

$$f(x) = \varepsilon \frac{u(x/a)}{a} + (1 - \varepsilon) \frac{u(x/b)}{b}$$

where $u(x) = I_{(0,1)}(x)$ is the uniform density, $0 < a < b$, and $0 < \varepsilon < 1$. This is simply a mixture of the uniform distributions $U(0, a)$ with probability ε and $U(0, b)$ with probability $1 - \varepsilon$. Since intervals of constancy are not excluded in the definition of a unimodal distribution, $f(x)$ is clearly unimodal. We will now show, however, that $g = f * f$, the density of $X + Y$, is not unimodal.

Observe that $f(x) > \varepsilon u(x/a)/a$ and the convolution of $u(x/a)/a$ with itself is triangular on $(0, 2a)$ with a peak of $1/a$ at a. It thus follows that $g(a) > \varepsilon^2/a$. Similarly, $g(b) > \varepsilon^2/b$. Furthermore, since the convolution $f * f$ is greater than the convolution of $(1 - \varepsilon)u(x/b)/b$ with itself, it follows that the integral of g from b to $2b$ is greater than $(1 - \varepsilon)^2/2$. It will follow that there exists a minimum between a and b. For example, set $a = 1$ and $b = 9$. If no such minimum exists, then the integral from a to b is at least $(b - a)g(b) > 8\varepsilon^2/2$. But the integral from b to $2b$ is at least $(1 - \varepsilon)^2/2$, making the sum greater than one if $\varepsilon = \frac{1}{2}$, a contradiction because g must be a density function. Hence, $g(x)$ is not unimodal, since there exists a minimum of $g(x)$ between a and b.

4.8. A convolution of two singular distributions that is absolutely continuous (from Feller 1971, Section V.4).

The convolution of an absolutely continuous distribution with any other distribution must result in an absolutely continuous distribution. This example shows that an absolutely continuous distribution can result even if neither distribution that enters into the convolution is absolutely continuous.

Let $X_1, X_2, \ldots$ be independent and identically distributed taking the value zero or one with probability $\frac{1}{2}$. Then

$$\sum 2^{-i} X_i$$

is uniformly distributed on $[0, 1]$ and may be interpreted as the binary representation of a number. Breaking this sum up into its even and odd components:

$$U = 2^{-2}X_2 + 2^{-4}X_4 + 2^{-6}X_6 + \cdots$$

$$V = 2^{-1}X_1 + 2^{-3}X_3 + 2^{-5}X_5 + \cdots$$

we obtain two independent random variables (because they depend on different X_i terms). U and V each has a distribution of the Cantor type, which is singular with respect to Lebesgue measure because their supports have measure $1 - \frac{1}{2} - \frac{1}{4} - \frac{1}{8} - \cdots = 0$ due to the missing terms in the summations. Nevertheless, the convolution of these two singular distributions gives rise to uniform (Lebesgue) measure on [0, 1], which is indeed absolutely continuous.

4.9. Random variables that are pairwise independent but not independent.

Recall the three events A, B, and C from Example 1.6 that were pairwise independent but not mutually independent. Define random variables to be the indicator functions of these events, so that $X = I_A$, $Y = I_B$, and $Z = I_C$. Then the pairwise independence of X, Y, and Z follows from the pairwise independence of the underlying events. Similarly, nonindependence of the sets forces nonindependence of the random variables.

4.10. An infinite sequence of random variables such that every pair is independent, but no triple is independent.

Reconsider Example 1.8, in which a sequence of events $E_1, E_2, \ldots$ was found in which every pair is independent but no triple is independent because

$$P(E_i) = \frac{1}{2^i}$$

$$P(E_i \cap E_j) = \frac{1}{2^{i+j}} \qquad \text{if } i \neq j$$

$$P(E_i \cap E_j \cap E_k) = 0 \qquad \text{if } i, j, \text{ and } k \text{ are distinct.}$$

Define the sequence of random variables to be the indicator functions of these events, so that

$$X_i = I_{E_i}.$$

It is now straightforward to verify that every pair of random variables from this sequence is independent but that no triple is independent.

A related example is provided by Feller (1959, p. 1252). Based on permutations, a sequence of random variables of any finite length is found such that pairwise independence holds but mutual independence does not.

4.11. Two random variables that are uncorrelated but not independent.

Although two independent random variables are uncorrelated (provided finite second moments exist), the converse is false, as this counterexample shows. Let X and Y be the sum and difference, respectively, of two coin tosses where $1 =$ heads and $0 =$ tails. We have the following table of joint probabilities:

		X		
		0	1	2
Y	-1	0	$\frac{1}{4}$	0
	0	$\frac{1}{4}$	0	$\frac{1}{4}$
	1	0	$\frac{1}{4}$	0

In particular, we may compute

$$E(X) = 1, \quad E(Y) = 0, \quad \text{and} \quad E(XY) = 0.$$

This implies that $\text{Cov}(X, Y) = E(XY) - E(X)E(Y) = 0$. Thus, X and Y are uncorrelated. However, X and Y are clearly not independent because, for example,

$$P(X = 0 \quad \text{and} \quad Y = 0) = 0 \neq (\tfrac{1}{4})(\tfrac{1}{2}) = P(X = 0)P(Y = 0).$$

4.12. Two random variables that are uncorrelated even though one may be predicted perfectly from the other.

The point of this example is to further expose the fallacy that uncorrelated random variables are independent. In fact, as shown here, uncorrelated random variables may be directly related by a functional relationship and, hence, may be dependent. The lesson is that correlation is strictly a measure of the linear aspects of dependence and may not be sensitive to nonlinearities.

Let X have a uniform distribution on $(-a, a)$ for some positive a. Consider the random variables X and X^2. Clearly, these two random variables are not independent because, conditional on $X = x$, we have $X^2 = x^2$ with probability one. Due to symmetry considerations, however, X and X^2 are uncorrelated. The covariance of X and X^2 is given by

$$\begin{aligned}\operatorname{Cov}(X, X^2) &= E(X \cdot X^2) - E(X)E(X^2)\\ &= E(X^3) - E(X)E(X^2).\end{aligned}$$

Since the distribution of X is symmetric about the origin and has support on a bounded set, all odd moments of X vanish. So $E(X) = E(X^3) = 0$, which forces $\operatorname{Cov}(X, X^2) = 0$, and so X and X^2 are uncorrelated. In fact, whenever the first and third moments of a random variable X vanish, it will be true that X and X^2 are uncorrelated. This seems especially surprising because there is a direct functional dependence of X^2 on X. Nevertheless, the best linear predictor of X^2 based on X is a constant.

4.13. Two independent random variables such that the expectation of the quotient is not the quotient of the expectations.

Although the expectation $E(XY)$ of the product of two random variables is equal to the product $E(X)E(Y)$ of their expectations whenever X and Y are independent (or, more generally, uncorrelated), this is false for quotients even if we assume independence of X and Y. For example, let X and Y be independent and exponentially distributed random variables with density given by $f(t) = \lambda \exp(-\lambda t) I_{(0,\infty)}(t)$. Then,

$$E(X) = E(Y) = \int_0^\infty t\lambda \exp(-\lambda t)\, dt = \frac{1}{\lambda}.$$

On the other hand, the density of X/Y is easily computed (see Mood, Graybill, and Boes 1974, p. 208) and is found to be $g(t) = (1 + t)^{-2} I_{(0,\infty)}(t)$. Therefore

$$\begin{aligned}E\left(\frac{X}{Y}\right) &= \int_0^\infty \frac{t}{(1+t)^2}\, dt = \int_0^\infty \frac{dt}{1+t} - \int_0^\infty \frac{dt}{(1+t)^2}\\ &= \left[\log(1+t) + \frac{1}{1+t}\right]_0^\infty = \infty \neq 1 = \frac{E(X)}{E(Y)}.\end{aligned}$$

It remains true, of course, that

$$E\left[\frac{X}{Y}\right] = E(X)E\left[\frac{1}{Y}\right]$$

whenever X and Y are independent and the expectations exist.

4.14. Two independent integrable random variables such that the quotient does not have an expectation even though it is always finite.

Let X and Y be independently distributed, each having a standard Gaussian distribution with mean zero and variance one. The ratio X/Y is almost always finite, and if we require that Y never be zero, then Y still has a standard Gaussian distribution independent of X, and X/Y is always finite. By symmetry considerations, the distribution of X/Y is the same as that of $-X/Y$, and so one would naturally expect that $E(X/Y)$ would be zero.

In fact, however, $E(X/Y)$ does not exist because X/Y has the Cauchy distribution (see Example 3.2). To see this, note first that $X/|Y|$ has the same distribution as X/Y; then observe that $|Y|$ has the same distribution as the square root of a chi-squared distribution with one degree of freedom because $|Y|$ is the square root of Y^2, which is the sum of the square of one standard Gaussian deviate. It follows that $X/|Y|$ has a t-distribution with one degree of freedom, which can be verified to have the same density as the Cauchy distribution.

4.15. Two random variables for which the expectation of the product is equal to the product of the expectations, but they are not independent.

Although for independent random variables we will have $E(XY) = E(X)E(Y)$ whenever these expectations exist and are finite, this relationship may still hold even if they are not independent. For example, consider Example 4.14 and let X be uniform on $(-a, a)$ for some $a > 0$, and let $Y = X^2$. Then

$$E(XY) = E(X^3) = 0 = 0 \cdot E(Y) = E(X)E(Y)$$

even though Y is not independent of X.

4.16. Uncorrelated Gaussian random variables that are not independent.

If (X, Y) has a bivariate Gaussian distribution and neither X nor Y is degenerate, then X and Y are independent if and only if they are uncorrelated. However, if we require only that X and Y each be (marginally) Gaussian, then this conclusion need not hold.

Let X have a standard Gaussian distribution with mean zero and variance one, and let H denote the independent toss of a fair coin. Define

$$Y = \begin{cases} X & \text{if } H \text{ comes up heads} \\ -X & \text{if } H \text{ comes up tails.} \end{cases}$$

By symmetry, Y also has a standard Gaussian distribution. They are uncorrelated because each has expectation zero and

$$E(XY) = E[E(XY \mid H)] = \frac{E(X^2)}{2} + \frac{E(-X^2)}{2} = 0.$$

However, X and Y are not independent because, for example,

$$P(|X| > 1 \quad \text{and} \quad |Y| < 1) = 0 \neq P(|X| > 1)P(|Y| < 1)$$

which follows because $|X|$ is always equal to $|Y|$, but both probabilities on the right-hand side are positive.

4.17. Random variables that are exchangeable but not independent.

In general, any collection of independent and identically distributed random variables is exchangeable, but this example shows that the converse is false.

Suppose (X, Y) assumes the four points $(0, 1)$, $(0, -1)$, $(1, 0)$, and $(-1, 0)$ with equal probabilities of $\frac{1}{4}$ at each. By symmetry, we clearly have (X, Y) and (Y, X) equal in distribution; however, X and Y are not independent. To see this, observe that

$$P(X = 1) = \tfrac{1}{4}, \qquad P(Y = 0) = \tfrac{1}{2}.$$

However,

$$P(X = 1)P(Y = 0) = \tfrac{1}{8} \neq \tfrac{1}{4} = P(X = 1 \quad \text{and} \quad Y = 0).$$

Note, however, that a trivial calculation shows that X and Y are uncorrelated.

4.18. Random variables that are independent but not exchangeable.

Exchangeability may be thought of as a weaker condition than "independent and identically distributed," but it is neither weaker nor stronger than independence alone (see also Example 4.17).

Let X and Y be independent, and suppose X has a Poisson distribution with mean two and Y has a standard Gaussian distribution. Then X and Y are independent but are not exchangeable because of the different natures of their marginal distributions.

4.19. Random variables that are identically distributed but not exchangeable.

Let X and Y be independently distributed, each with the standard Gaussian distribution with mean zero and variance one, and define $Z = Y$. Then X, Y, and Z are identically distributed. They are not exchangeable because (X, Y, Z) has a correlation of zero between the first two components, whereas (Y, Z, X) has a correlation of one between its first two components.

4.20. Random variables that are exchangeable but not uncorrelated.

Let (X, Y) be uniformly distributed over the set $A = \{(x, y) : x \geq 0, y \geq 0,$ and $x + y \leq 1\}$. By symmetry, X and Y are clearly exchangeable. To see that X and Y are not independent, note that

$$P(X \leq \tfrac{1}{2} \quad \text{and} \quad Y \leq \tfrac{1}{2}) = \int_0^1 \int_0^1 2I_A(x, y)\, dx\, dy = \tfrac{1}{2}.$$

However,

$$P(X \leq \tfrac{1}{2}) = \int_0^{1/2} \int_0^{1-x} 2\, dy\, dx = \tfrac{3}{4}.$$

Therefore,

$$P(X \leq \tfrac{1}{2})P(Y \leq \tfrac{1}{2}) = \tfrac{9}{16} \neq \tfrac{1}{2} = P(X \leq \tfrac{1}{2} \quad \text{and} \quad Y \leq \tfrac{1}{2}).$$

In fact, we can show that X and Y are correlated. Note that $\text{Cov}(X, Y) = E(XY) - E(X)E(Y)$. The marginal density of X is given by

$$\int_0^{1-x} 2\, dy = 2(1 - x)I_{[0,1]}(x).$$

Thus,

$$E(X) = \int_0^1 x \cdot 2(1 - x)\, dx = \tfrac{3}{4}.$$

Also,

$$E(XY) = \int_0^1 \int_0^{1-x} xy\,dy\,dx = \tfrac{1}{24}.$$

Thus, $\mathrm{Cov}(X, Y) = \frac{1}{24} - \frac{9}{16} = -\frac{25}{48} \neq 0$, so that X and Y are correlated.

A trivial example is provided by (X, Y) when $X = Y$ and X is not degenerate. Another example is given by $(X + Y, X + Z)$, which is exchangeable but not uncorrelated whenever X, Y, and Z are independent and identically distributed but not degenerate.

4.21. Two marginal distributions for which no joint distribution exists with a correlation of -1.

Suppose X and Y each have marginal exponential distributions. By the Cauchy-Schwarz inequality, $\mathrm{Cor}(X, Y) = -1$ implies that $X = -mY + b$ with probability one for some m and b with $m > 0$. This is clearly impossible, since X and Y take on nonnegative values with probability one.

4.22. A family of bivariate distributions such that the range of possible correlations is a small subset of $[-1, 1]$.

Suppose X and Y are bivariate Gaussian with means μ_X and μ_Y, respectively, and with covariance matrix

$$\begin{bmatrix} \sigma_X^2 & \rho\sigma_X\sigma_Y \\ \rho\sigma_X\sigma_Y & \sigma_Y^2 \end{bmatrix}.$$

Note that the correlation of X and Y may take on any value in $[-1, 1]$. Now consider the transformation defined by $W = \exp(X)$ and $Z = \exp(Y)$. This produces a set (W, Z) of bivariate lognormal random variables. In this case, as shown by deVeaux (1976), the following inequality holds:

$$\frac{e^{-\sigma_X\sigma_Y} - 1}{\{(e^{\sigma_X^2} - 1)(e^{\sigma_Y^2} - 1)\}^{1/2}} \leq \mathrm{Cor}(W, Z) \leq \frac{e^{\sigma_X\sigma_Y} - 1}{\{(e^{\sigma_X^2} - 1)(e^{\sigma_Y^2} - 1)\}^{1/2}}.$$

This has some striking implications. If, for example, we restrict ourselves to the family of distributions with $\sigma_X = 1$ and $\sigma_Y = 4$ but we allow any values for the means and the correlation between X and Y, then the correlation between W and Z is constrained to lie in the interval from -0.000251 to

0.01372! Such a result raises a serious question in practice about how to interpret the correlation between lognormal random variables. Clearly, small correlations may be very misleading because a correlation of 0.01372 indicates, in fact, that W and Z are perfectly functionally (but nonlinearly) related.

4.23. A symmetric matrix with positive determinant that is not a covariance matrix.

Consider the matrix given by

$$A = \begin{bmatrix} 1 & 2 & 3 \\ 2 & 1 & 2 \\ 3 & 2 & 1 \end{bmatrix}.$$

Although A is a symmetric matrix with positive determinant equal to 8, A could not possibly be a covariance matrix. This is easy to see because if it were, then the correlation between the first and second elements would have to be $2/(1 \cdot 1) = 2$, which is impossible because correlations must be between -1 and 1.

In general, in order for a symmetric matrix A to be a covariance matrix for some multivariate distribution, it is required that A be a nonnegative definite matrix.

4.24. A positive definite quadratic form of a multivariate Gaussian distribution that does not have a chi-squared distribution.

Although many quadratic forms that arise in multivariate statistics have chi-squared distributions, it is not sufficient that the form be positive definite. Let (X, Y) be bivariate Gaussian with mean zero and variance one for each component, and with correlation zero. Consider the quadratic form

$$(X, Y)\begin{bmatrix} 1 & 0 \\ 0 & 2 \end{bmatrix}\begin{bmatrix} X \\ Y \end{bmatrix} = X^2 + 2Y^2.$$

Even when a constant scale factor is allowed, this quantity does not have a chi-squared distribution with any (even fractional) number of degrees of freedom! This may be seen by comparing the characteristic function of this quantity with that of a chi-squared distribution. Although a sum of independent chi-squared variables is chi-squared, a weighted sum need not be.

4.25. Random variables X, Y, and Z such that Y is positively correlated with both X and Z, but X and Z are negatively correlated with each other.

Let (X, Y, Z) have the multivariate Gaussian distribution with covariance matrix

$$A = \begin{bmatrix} 1 & \frac{1}{2} & -\frac{1}{4} \\ \frac{1}{2} & 1 & \frac{1}{2} \\ -\frac{1}{4} & \frac{1}{2} & 1 \end{bmatrix}.$$

Note that A is positive definite so that the distribution exists and is well defined. A simple way to check that A is positive definite is to check that the three inner matrices

$$1, \quad \begin{bmatrix} 1 & \frac{1}{2} \\ \frac{1}{2} & 1 \end{bmatrix}, \quad \text{and} \quad A$$

all have positive determinants (Williamson, Crowell, and Trotter 1972, p. 379).

Now, $\mathrm{Cor}(X, Y) = \frac{1}{2} > 0$ and $\mathrm{Cor}(Y, Z) = \frac{1}{2} > 0$; however, $\mathrm{Cor}(X, Z) = -\frac{1}{4} < 0$. Thus the property of being "positively correlated with" is not transitive (and neither is the property of being "negatively correlated with").

4.26. Random variables X, Y, and Z such that X is uncorrelated with both Y and Z, but Y is correlated with Z.

Let (X, Y, Z) have the multivariate Gaussian distribution with covariance matrix

$$A = \begin{bmatrix} 1 & \frac{1}{2} & 0 \\ \frac{1}{2} & 1 & 0 \\ 0 & 0 & 1 \end{bmatrix}.$$

Then, $\mathrm{Cor}(X, Z) = \mathrm{Cor}(Y, Z) = 0$, yet $\mathrm{Cor}(X, Y) = \frac{1}{2}$. Thus the property of being "uncorrelated with" is not transitive.

4.27. Random variables that are perfectly correlated, yet one is not a linear function of the other.

Suppose X is uniformly distributed on $(0, 1)$. Define

$$Y = \begin{cases} X & \text{if } X \text{ is an irrational number} \\ 0 & \text{otherwise.} \end{cases}$$

Since $X = Y$ with probability one, it follows that $\mathrm{Cor}(X, Y) = 1$. However, Y is not a linear function of X.

The point is that the Cauchy-Schwarz inequality says that Y is a linear function of X *with probability one*. Even some of the best texts fail to state the Cauchy-Schwarz inequality accurately with regard to this subtlety.

4.28. A function of two independent random variables, X and Y, that is independent of X but is not almost surely a function of Y alone.

Suppose X and Y are independent random variables and that $g(x, y)$ is a given measurable function. Then $g(X, Y)$ defines a random variable. If X and $g(X, Y)$ are statistically independent, then one might be tempted to conclude that g is (almost surely) a function of Y alone because its independence from X suggests that it must depend on something else. Although it would be tempting, this conclusion need not follow because almost sure behavior cannot generally be deduced from distributional hypotheses.

Let X and Y be independent and identically distributed with the standard Gaussian distribution. Let

$$g(x, y) = \begin{cases} 1 & \text{if } xy > 0 \\ 0 & \text{otherwise.} \end{cases}$$

Note that g is the indicator function of the first and third quadrants. To see that $g(X, Y)$ is independent of X, note that a regular conditional distribution for $g(X, Y)$ given $X = x$ is the distribution that places mass $\frac{1}{2}$ at one and at zero, regardless of the value of x. However, g is clearly not a function of y alone, even almost surely.

4.29. Random variables such that the characteristic function of their sum is the product of their characteristic functions, yet they are not independent.

If two random variables are independent, then the characteristic function of their sum must be the product of their characteristic functions. The converse of this is false in general.

Suppose X and Y are independent random variables and each has the Cauchy distribution with density

$$f(x) = \frac{1}{\pi(1 + x^2)}.$$

Let $Z = X$ so that, trivially, X and Z are dependent. From Example 4.5 we know that

$$\frac{X + Y}{2} \stackrel{d}{=} X$$

and therefore,

$$X + Y \stackrel{d}{=} 2X \stackrel{d}{=} X + Z.$$

Thus the characteristic function of $X + Z$ is equal to the characteristic function of $X + Y$. Using the independence of X and Y, we find

$$\phi_{X+Z}(t) = \phi_{X+Y}(t) = \phi_X(t)\phi_Y(t) = \phi_X(t)\phi_Z(t).$$

Thus we see that the characteristic function of the sum $X + Z$ is equal to the product of their respective characteristic functions even though X and Z are dependent because $X = Z$.

4.30. Random variables such that the density of their sum is the convolution of their densities, yet they are not independent (Feller 1971, Problem III.1, p. 99).

In general, if X and Y are independent random variables with densities f and g, then the sum $X + Y$ has a density given by the convolution $f * g$. The converse, however, does not necessarily hold. Example 4.29 provides a counterexample, but we will give another example here that may be approached directly without any reference to characteristic functions.

Suppose the sample space S is the union of the region of the plane bounded by the quadrilateral with vertices $(0, 0)$, $(0, \frac{1}{2})$, $(1, 1)$, and $(\frac{1}{2}, 1)$ with the region bounded by the triangle with vertices $(\frac{1}{2}, 0)$, $(1, 0)$, and $(1, \frac{1}{2})$. See Figure 4.30.1. If (X, Y) is uniformly distributed over S, then the joint density of (X, Y) is given by

$$f_{X,Y}(x, y) = 2I_S(x, y).$$

It is easy to check that the marginal distributions of X and Y are uniform, for example, by verifying visually that a horizontal or a vertical line that intersects S will have a constant total length of $\frac{1}{2}$ within S.

If X' and Y' are *independent* random variables, each uniform on $(0, 1)$, then the density of their sum, found by the convolution of the densities of X' and Y', is the triangular density (see, for example, Mood, Graybill, and Boes

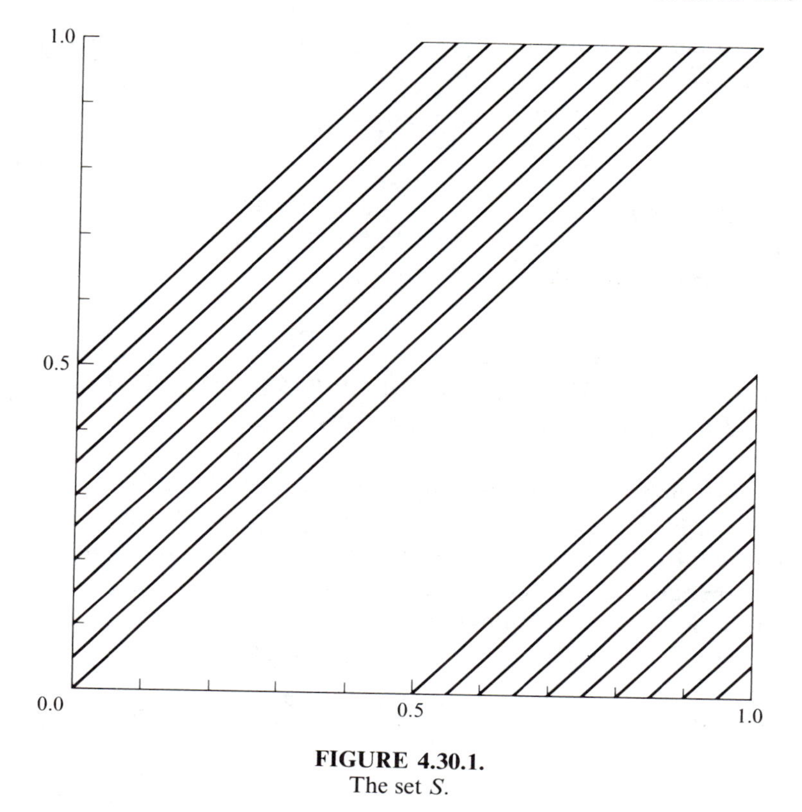

FIGURE 4.30.1.
The set S.

1974, p. 186) given by

$$g_{X'+Y'}(t) = tI_{(0,1)}(t) + (2 - t)I_{[1,2)}(t).$$

However, if X and Y have the joint density function $f_{X,Y}(x, y)$ as previously given, then X and Y are not independent. To see this, note that knowledge of the value of X restricts the possible values of Y.

On the other hand, it is still the case that $X + Y$ has the joint density function $g_{X'+Y'}$, and we will now show that $X' + Y' \stackrel{d}{=} X + Y$. To find the distribution of $X + Y$, note that if (X, Y) is uniform on S and (U, V) is uniform on the complement of S relative to $(0, 1) \times (0, 1)$, the unit square, then by symmetry we have $(X, Y) \stackrel{d}{=} (V, U)$. Thus,

$$X + Y \stackrel{d}{=} U + V.$$

So if X' and Y' are independent uniformly distributed random variables on $(0, 1)$, we can represent $X' + Y'$ as a mixture:

$$X' + Y' = \begin{cases} X + Y & \text{with probability } \frac{1}{2} \\ U + V & \text{with probability } \frac{1}{2}. \end{cases}$$

Then $X' + Y'$ is a mixture of two random variables with the same distribution; hence,

$$X' + Y' \stackrel{d}{=} X + Y.$$

So, in fact, $X + Y$ has the density $g_{X'+Y'}$, as claimed earlier. Therefore the density of the sum $X + Y$ is the convolution of their densities, even though X and Y are not independent.

4.31. Two distinct characteristic functions that are equal in an interval (adapted from Feller 1971, p. 506).

It is well known that two probability distributions are equal if and only if their characteristic functions are equal. This example shows that it would not be enough for the characteristic functions to be equal in some interval; they must be equal everywhere. This is somewhat surprising because two analytic functions that are equal in an interval must be equal everywhere. A counterexample is possible because not every characteristic function is analytic.

Consider the random variable with a discrete distribution given as follows:

$$P(X = x) = \begin{cases} \dfrac{1}{2} & \text{if } x = 0 \\ \dfrac{2}{\pi^2(2n-1)^2} & \text{if } x = (2n-1)\pi \text{ or } x = -(2n-1)\pi, \\ & \text{where } n = 1, 2, 3, \ldots \\ 0 & \text{otherwise.} \end{cases}$$

Note that the support of X is $\ldots, -5\pi, -3\pi, -\pi, 0, \pi, 3\pi, 5\pi, 7\pi, \ldots$ and that X is symmetric so that its characteristic function is real. The characteristic function ϕ of X is

$$\phi(t) = E(e^{itX}) = \frac{1}{2} + \sum_{n=1}^{\infty} \frac{4}{\pi^2(2n-1)^2} \cos[(2n-1)\pi t].$$

This sum contains a standard Fourier series (Gradshteyn and Ryzhik 1965, formula 1.444.6, p. 39):

$$\sum_{n=1}^{\infty} \frac{\cos[(2n-1)x]}{(2n-1)^2} = \frac{\pi}{4}\left(\frac{\pi}{2} - |x|\right) \qquad \text{for } -\pi \le x \le \pi$$

and this sum is periodic with period 2π. Therefore,

$$\phi(t) = \frac{1}{2} + \frac{4}{\pi^2}\left[\frac{\pi}{4}\left(\frac{\pi}{2} - |\pi t|\right)\right]$$

$$= 1 - |t| \qquad \text{for } -1 \leq t \leq 1$$

and $\phi(t)$ is periodic with period 2 outside $[-1, 1]$.

Consider also a discrete random variable Y given as follows:

$$P(Y = y) = \begin{cases} \dfrac{4}{\pi^2(2n-1)^2} & \text{if } y = (2n-1)\pi/2 \text{ or } y = -(2n-1)\pi/2, \\ & \text{for } n = 1, 2, \ldots \\ 0 & \text{otherwise.} \end{cases}$$

The distribution of Y is supported on $\ldots, -5\pi/2, -3\pi/2, -\pi/2, \pi/2, 3\pi/2, 5\pi/2, \ldots$ and is also symmetric about zero. In fact, Y is the distribution of $X/2$ conditional on X being nonzero. The characteristic function θ of Y is

$$\theta(t) = E(e^{itY}) = \sum_{n=1}^{\infty} \frac{8}{\pi^2(2n-1)^2} \cos\left[(2n-1)\pi\frac{t}{2}\right].$$

Using the same Fourier sum as before, we find that

$$\theta(t) = 1 - |t| \qquad \text{for } -2 \leq t \leq 2$$

and θ is periodic with period 4 outside $[-2, 2]$.

These two characteristic functions ϕ and θ are plotted in Figure 4.31.1. It is clear that they coincide in the interval $[-1, 1]$.

4.32. Two distinct characteristic functions whose absolute values are equal at all real values.

Consider the characteristic functions $\phi(t)$ and $\theta(t)$ from Example 4.31. It is clear that $|\phi(t)| = |\theta(t)|$ for all real values of t. However, $\phi(t) \neq \theta(t)$ in many places—for example, when $t \in (1, 3)$—and so ϕ and θ are distinct. In fact, their supports are different because ϕ is the characteristic function of a distribution with mass at zero, whereas θ is the characteristic function of a distribution with no mass at zero.

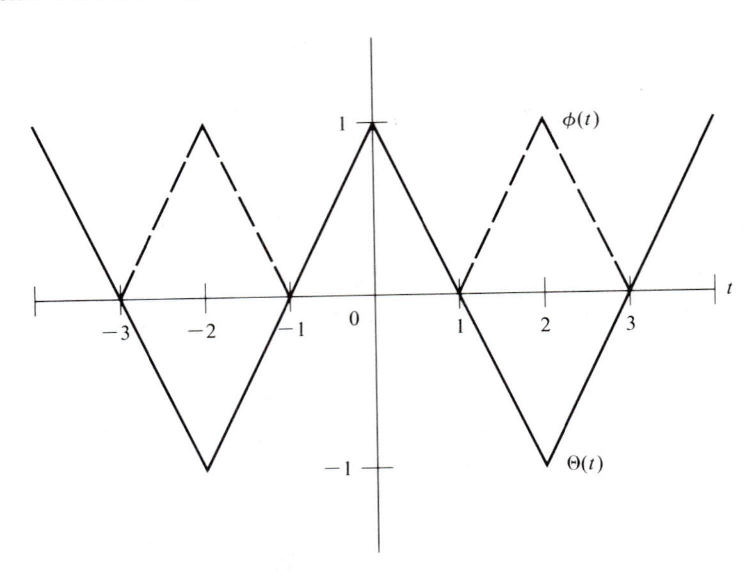

FIGURE 4.31.1.
The characteristic functions ϕ and θ.

4.33. Convolutions $F * G$ and $F * H$ that are equal even though G and H are not equal (Feller 1971, p. 506).

Consider the density function

$$f(x) = \frac{1 - \cos(x)}{\pi x^2} \qquad \text{for all real } x$$

and let F denote the corresponding cumulative distribution function. The characteristic function, η, of this distribution may be found by applying the Fourier inversion formula to the triangular density. The result is that

$$\eta(t) = \begin{cases} 1 - |t| & \text{if } -1 \leq t \leq 1 \\ 0 & \text{otherwise.} \end{cases}$$

Next, recall the characteristic functions ϕ and θ from Example 4.31, and let G and H denote their respective cumulative distribution functions. Because $\phi(t)$ and $\theta(t)$ are equal when $-1 \leq t \leq 1$ and $\eta(t)$ vanishes outside this interval, we have

$$\eta(t)\phi(t) = \eta(t)\theta(t)$$

for all real values of t. It follows that $F * G = F * H$ despite the fact that $G \neq H$.

4.34. Identically distributed random variables such that their difference does not have a symmetric distribution.

If X and Y are independent and identically distributed, then $X - Y$ has a symmetric distribution about zero. The independence assumption cannot be dropped in general. (However, if X and Y are exchangeable, then $X - Y$ does have a symmetric distribution.)

For example, let X and X have the joint probability function shown in the table:

		Y: -1	0	1	
X	-1	$\frac{1}{9}$	$\frac{1}{9}$	0	$\frac{2}{9}$
	0	0	$\frac{4}{9}$	$\frac{1}{9}$	$\frac{5}{9}$
	1	$\frac{1}{9}$	0	$\frac{1}{9}$	$\frac{2}{9}$
		$\frac{2}{9}$	$\frac{5}{9}$	$\frac{2}{9}$	

We see that the marginal distributions of X and Y are identical. The distribution of $X - Y$ is not symmetric, however, because

$$\begin{aligned} P(X - Y = 2) &= P(X = 1 \quad \text{and} \quad Y = -1) = \tfrac{1}{9} \neq 0 \\ &= P(X = -1 \quad \text{and} \quad Y = 1) = P(X - Y = -2). \end{aligned}$$

4.35. Two real random variables, each symmetric about zero, but whose sum is not symmetrically distributed.

In general, when X and Y are independent and each has a symmetric distribution about zero, then $X + Y$ has a distribution symmetric about zero. The result is false without the assumption of independence.

For example, let X and Y have the joint discrete probability function shown in the table:

		Y: -1	0	1
X	-1	$\frac{1}{12}$	$\frac{1}{6}$	$\frac{1}{12}$
	0	$\frac{1}{12}$	$\frac{1}{6}$	$\frac{1}{12}$
	1	$\frac{1}{6}$	0	$\frac{1}{6}$

Then X and Y are both symmetric with a distribution that places mass $\frac{1}{3}$ at each of $-1, 0$, and 1. However, $X + Y$ is not symmetric. To see this, observe that if $X + Y$ were symmetric, it would have to be symmetric about $E(X + Y) = 0$ (because the expectation exists). But $P(X + Y = 2) = \frac{1}{6} \neq \frac{1}{12} = P(X + Y = -2)$, which is a contradiction.

4.36. Two random variables, each symmetric about zero, but whose sum is symmetric about a nonzero center (from Chen and Shepp 1983).

Whenever X and Y are symmetric about zero with $E(X) = E(Y) = 0$ and if it is true that $X + Y$ is symmetric, then the sum must be symmetric about $E(X + Y) = 0$. A counterexample shows that $X + Y$ can be symmetric about $c \neq 0$ if we do not assume that X and Y each has a mean (or, more generally, satisfies the weak law of large numbers).

A complete example is presented by Chen and Shepp (1983); we will give the basic idea for its construction. Let

$$\phi(s, t) = E(e^{isX+itY})$$

denote the joint characteristic function of (X, Y). Now, a real random variable is symmetric about d if and only if its characteristic function $\phi(t)$ divided by $\exp(itd)$ is real (see Chung 1974, Theorem 6.2.6). Hence, if $\phi(s, 0)$ is real so that X is symmetric about zero, $\phi(0, t)$ is real so that Y is symmetric about zero, and $\phi(s, s)$ divided by $\exp(isc)$ is real so that $X + Y$ is symmetric about c, then an example could be constructed. Clearly, such a function ϕ exists. Chen and Shepp argue that there is indeed a characteristic function with these properties.

4.37. A sum of independent and identically distributed random variables that is symmetric about zero, even though the individual distributions are not.

Whenever independent and identically distributed random variables X and Y are symmetric about zero, it follows that $X + Y$ is also. The converse is false. The following construction of a counterexample is similar to Lemma 3 of Freedman and Diaconis (1982a). Here, however, an explicit density function is given. The construction of a similar density was personally communicated to one of the authors by LeCam.

First, note that a real random variable is symmetric about zero if and only if its characteristic function is real. If X and Y (independent) each has characteristic function ϕ, so that $X + Y$ has characteristic function ϕ^2, then the hypothesis is that ϕ^2 is real. We must find a characteristic function ϕ whose square is real, but ϕ itself is not everywhere real, so that ϕ is sometimes purely imaginary.

We will choose X and Y to have a density of the form $f = g + \delta h$, where g is a density function but h integrates to zero. The corresponding characteristic function is $\phi = \xi + \delta\eta$. We will choose g so that its characteristic function ξ is real and vanishes outside of $[-1, 1]$, whereas η will be purely imaginary and will vanish on $[-1, 1]$. Define $g(x) = [1 - \cos(x)]/(\pi x^2)$. From Example 4.33, we know that the characteristic function is

$$\phi(t) = \begin{cases} 1 - |t| & \text{if } -1 \le t \le 1 \\ 0 & \text{otherwise.} \end{cases}$$

For any $a > 4$, let

$$h(x) = -ig(x)[g(x)e^{iax} - g(x)e^{-iax}] = 2g^2(x)\sin(ax).$$

Note that

$$|h(x)| = \left|\frac{2[1 - \cos(x)]^2}{\pi^2 x^4}\right| \le \frac{1}{2\pi^2}\min\left(1, \frac{4}{x^4}\right)$$

so that h is integrable. But h is an odd function, so that $\int h = 0$ as required. To show that $f = g + \delta h$ is a density, it remains to show that f is nonnegative. Observe that

$$f(x) = g(x) + \delta h(x) = g(x)[1 + 2\delta g(x)\sin(ax)]$$

$$= g(x)\left[1 + \frac{2\delta[1 - \cos(x)]}{\pi x^2}\sin(ax)\right].$$

Because $|[1 - \cos(x)]/x| \le 1$ and $|[\sin(ax)]/x| \le a$, we may set $\delta = 1/a$ and find that

$$f(x) \ge g(x)\left[1 - 2\delta\frac{a}{\pi}\right] = g(x)\left[1 - \frac{2}{\pi}\right] \ge 0.$$

The argument will be complete if we show that the Fourier Transform of η (itself the Fourier Transform of h) is purely imaginary and vanishes off of $[-1, 1]$. But the Fourier Transform of $\eta/(2\pi)$ is h, which is the product of the Fourier Transforms of $\xi(t)/(2\pi)$ and $-[i/(2\pi)]\cdot[\xi(t - a) - \xi(t + a)]$. Thus, $\eta(t)/(2\pi)$ is the convolution of $\xi(t)/(2\pi)$ with $-[i/(2\pi)]\cdot[\xi(t - a) - \xi(t + a)]$, using the uniqueness and inversion theorems for characteristic functions. Hence $\eta(t)$ is purely imaginary. Since $a > 4$, this convolution $\eta(t)$ equals zero if $-1 \leq t \leq 1$. Note, however, that $\eta(t)$ is not always zero because $h(x)$ is not identically zero.

Thus, we have an example of a characteristic function ϕ such that as t increases from 0 to ∞ (or decreases from 0 to $-\infty$), $\phi(t)$ is real, then zero, then purely imaginary, and then zero. Because it is not always real, its corresponding distribution is not symmetric about zero. But its square is real, and so the convolution of the distribution with itself is symmetric about zero.

4.38. A characteristic function that is real in a neighborhood of zero but is not everywhere real.

Example 4.37 shows that this phenomenon is possible. A construction may also be found in Freedman and Diaconis (1982a, Lemma 4).

4.39. A distribution that is symmetric but is not the result of a symmetrization procedure (from Feller 1971, Section V.5).

If X and Y are independent and identically distributed, then we say that the distribution of $X - Y$ is obtained by "symmetrization" of the distribution of X. Whereas every symmetrization distribution is symmetric, not every symmetric distribution is the result of symmetrization of a distribution. For example, consider the distribution that places mass $\frac{1}{2}$ at -1 and at 1. To see that this distribution cannot be the symmetrization of any distribution F, note that the support of F must include one, two, or more points. But then the symmetrization distribution would have to have a support including (respectively) one, three, or more points. The required number, two, is not possible.

4.40. A characteristic function that never vanishes but is not infinitely divisible.

Although it is true that the characteristic function of an infinitely divisible distribution never vanishes, the converse is false. Let X be a Bernoulli trial with probability $p \neq 0$ or 1, so that $P(X = 1) = p$ and $P(X = 0) = 1 - p$. Then the characteristic function of X is

$$\phi(t) = (1 - p) + pe^{it}$$

which is never zero, even though the Bernoulli distribution is not infinitely divisible.

4.41. An infinitely divisible distribution that is not a stable law.

Although all stable laws are infinitely divisible, the converse is false. Consider X_1 and X_2, independent and identically distributed random variables with a Poisson distribution with mean one. Consider the normalized sum

$$Y = \frac{X_1 + X_2}{a} + b.$$

Can this have the same distribution? That is, can Y have a Poisson distribution with mean one for some choice of a and b? We would need $b = 0$ by a support argument and $a = 2$ in order to have the right expectation. But then $Y = \frac{1}{2}$ with positive probability, and therefore we cannot have a Poisson distribution. It follows that the Poisson distribution with mean one is not a stable law.

All Poisson distributions are infinitely divisible, however. To see that the Poisson distribution with mean one is infinitely divisible, let n be a positive integer. Because the Poisson distribution with mean one is the n-fold convolution of Poisson distributions with mean $1/n$, infinite divisibility follows.

4.42. Random variables X and Y for which the best predictor of Y given X is not the inverse function of the best predictor of X given Y.

Suppose (X, Y) has the bivariate Gaussian distribution with zero means, variances σ_X^2 and σ_Y^2, and correlation ρ. Then it can be shown that $g(x) = E(Y \mid X = x) = x\rho\sigma_Y/\sigma_X$ is the best predictor of Y based on X in the sense of minimum squared prediction error. Note that $E(Y \mid X = x)$ is actually linear and that the line $y = x\rho\sigma_Y/\sigma_X$ is the familiar "regression line" of y on x.

The best predictor $h(y)$ of X based on Y is found similarly: $h(y) = E(X \mid Y = y) = y\rho\sigma_X/\sigma_Y$. The lines $y = g(x)$ and $x = h(y)$ are not the same unless $\rho = 1$ or $\rho = -1$—that is, when X and Y are linearly dependent. This has the important implication in practical regression analysis that, given data $(x_1, y_1), \ldots, (x_n, y_n)$, the regression line of y on x is not the same as that of x on y unless all data points fall exactly on a line. Therefore the choice of the predictor (the independent variable) is crucial.

4.43. Random variables X and Y such that X predicts Y perfectly, but Y has no power in predicting X.

Suppose (X, Y) is uniform on $(0, 1) \times \{0\}$. In other words, X is uniform on $(0, 1)$ and Y is identically zero. Thus, for any value $X = x$, we can predict Y perfectly. On the other hand, knowing that $Y = 0$ gives no information for predicting X. In this case, regardless of the value of Y, the best predictor of X is $\frac{1}{2}$, the mean of X.

4.44. Nonconstant random variables X and Y such that Y is a function of X, but the best predictor of X given Y is zero.

Let X take on the values -1, 0, and 1, each with probability $\frac{1}{3}$, and set $Y = X^2$. Then Y is a function of X, but the function of Y with the smallest expected squared prediction error of X is always zero because if $Y = 0$ then we know $X = 0$, and if $Y \neq 0$ then (conditionally) X is equally likely to be 1 or -1. Note that Y does provide useful information about X (for example, if $Y \neq 0$, then we know for certain that $X \neq 0$), but the single best predictor does not make good use of this information.

CHAPTER FIVE
SEQUENCES OF RANDOM VARIABLES

Introduction

The examples in this chapter primarily concern the limiting behavior of a sequence of random variables. Several types of convergence will be studied, of which the four main types are listed below.

We will use the following terminology. A property is said to hold "almost surely" (a.s.) on a probability space $(\Omega, \mathbf{B}, P)$ with respect to the probability measure P if there is a set C in $\mathbf{B}$ with $P(C) = 0$ such that the property holds outside C. The "law" of a random object X, written $\mathbf{L}(X)$, means the probability distribution of X.

The four main types of convergence are as follows:

(i) Convergence almost surely—A sequence of random variables $X_1, X_2, \ldots$ defined on a common probability space $(\Omega, \mathbf{B}, P)$ is said to "converge almost surely" (or to "converge strongly") to a random variable X_∞, written $X_n \overset{\text{a.s.}}{\to} X_\infty$, if we have

$$P\left\{\omega \text{ in } \Omega : \lim_{n\to\infty} X_n(\omega) = X_\infty(\omega)\right\} = P\left\{\lim_{n\to\infty} X_n = X_\infty\right\}$$
$$= 1.$$

More generally, if $X_1, X_2, \ldots, X_\infty$ take values in a metric space (S, ρ) equipped with the Borel σ-field generated by the open sets of S and if ρ is the metric, then we say that X_n converges almost surely to X_∞ if

$$P\{\omega : \rho[X_n(\omega), X_\infty(\omega)] \to 0\} = 1.$$

(ii) Convergence in probability—A sequence of real random variables $X_1, X_2, \ldots$ defined on a common probability space $(\Omega, \mathbf{B}, P)$ is said to "converge in probability" to a random variable X_∞, written $X_n \xrightarrow{\text{P}} X_\infty$, if, for every positive ε, we have

$$\lim_{n\to\infty} P\{\omega : |X_n(\omega) - X_\infty(\omega)| > \varepsilon\} = \lim_{n\to\infty} P\{|X_n - X_\infty| > \varepsilon\}$$

$$= 0.$$

More generally, if $X_1, X_2, \ldots, X_\infty$ take values in a metric space (S, ρ) equipped with the Borel σ-field generated by the open sets of S and ρ is the metric, then we say that X_n converges in probability to X_∞ if, for every $\varepsilon > 0$, we have

$$P\{\omega : \rho[X_n(\omega), X_\infty(\omega)] > \varepsilon\} \to 0 \qquad \text{as } n \to \infty.$$

(iii) Convergence in distribution—A sequence of real random variables X_1, $X_2, \ldots$ is said to "converge in distribution" (or to "converge in law") to a random variable X_∞ that has distribution F_∞, written $X_n \xrightarrow{\text{d}} X_\infty$, if $F_n(x) \to F_\infty(x)$ as $n \to \infty$ pointwise for all x at which F_∞ is continuous, where F_n is the cumulative distribution function of X_n. An equivalent definition is $X_n \xrightarrow{\text{d}} X_\infty$ if $E[f(X_n)] \to E[f(X_\infty)]$ for all real-valued functions f that are bounded and continuous.

The notion of convergence in probability laws can be placed in a more general setting. Suppose S is a metric space and $\mathbf{S}$ is a σ-field in S. Often $\mathbf{S}$ is the Borel σ-field generated by the open sets of S. Suppose Q_1, $Q_2, \ldots$ are probability laws defined on $(S, \mathbf{S})$. We say that the sequence of probability laws Q_n converges weakly to Q_∞ if

$$\lim_{n\to\infty} \int_S f\,dQ_n = \int_S f\,dQ_\infty$$

for all bounded continuous measurable functions from $(S, \mathbf{S})$ to $(\mathbf{R}, \mathbf{B})$, where $\mathbf{B}$ is the Borel σ-field on $\mathbf{R}$. Note that no mention need be made of random objects. However, suppose $X_1, X_2, \ldots$ are random objects defined on (possibly different) probability spaces $(\Omega_n, \mathbf{B}_n, P_n)$, all with range space $(S, \mathbf{S})$. Suppose Q_n is the probability distribution induced by X_n and P_n. By a simple change of variable, Q_n converges weakly to Q_∞ if and only if $\lim E[f(X_n)] = E[f(X_\infty)]$ for all bounded continuous measurable functions from S to $\mathbf{R}$. We may write $X_n \xrightarrow{\text{d}} X_\infty$ in this case, although this notation is poor because the random objects X_n are auxiliary. To show that it is really the laws of the random objects that are converging, we may write instead $\mathbf{L}(X_n) \to \mathbf{L}(X_\infty)$. That we are dealing with a useful notion of convergence of probability laws stems from the following fact. When $(S, \mathbf{S}) = (\mathbf{R}^k, \mathbf{B}^k)$, k-dimensional Euclidean space equipped with the Borel σ-field, then $\mathbf{L}(X_n) \to \mathbf{L}(X_\infty)$ is equivalent to

$$\lim_{n\to\infty} Q_n(A) = Q_\infty(A)$$

for all sets A in $\mathbf{S}$ whose boundary has probability zero under Q_∞. In the case of dimension $k = 1$, this reduces further to the continuity restriction in our first definition of convergence in the distribution of real random variables.

If Q_n and Q are probabilities on a metric space S equipped with the Borel σ-field $\mathbf{S}$, we say a subclass $\mathbf{C}$ of $\mathbf{S}$ is a "convergence-determining class" if weak convergence of Q_n to Q is equivalent to $Q_n(A) \to Q(A)$ for all sets A in $\mathbf{C}$ whose boundary has probability zero under Q. $\mathbf{C}$ is called a "determining class" if two probabilities P and Q are identical whenever $P(A) = Q(A)$ for all $A \in \mathbf{C}$. A convergence-determining class is always a determining class, but the converse is false.

(iv) Convergence in pth mean—A sequence of real random variables X_1, X_2, ... defined on a common probability space $(\Omega, \mathbf{B}, P)$ is said to "converge in the pth mean" (or "converge in L^p") to a random variable X_∞, written $X_n \xrightarrow{\mathbf{L}^p} X_\infty$, if

$$\lim_{n\to\infty} E[|X_n - X_\infty|^p] = 0.$$

In the case $p = 2$, this type of convergence is called "convergence in quadratic mean." More generally, if $X_1, X_2, \ldots, X_\infty$ take values in a metric space (S, ρ) equipped with the Borel σ-field generated by the open sets of S and ρ is the metric, then we say that X_n converges in the pth mean to X_∞ if

$$E\{[\rho(X_n, X_\infty)]^p\} \to 0 \qquad \text{as } n \to \infty.$$

In general, convergence almost surely implies convergence in probability, which implies convergence in distribution. Also, convergence in the pth mean for any $p > 0$ implies convergence in probability. Many of the examples considered in this chapter show that no other simple implications from the preceding four types of convergence hold.

The laws of large numbers for independent and identically distributed random variables say that the average of a sample should, in some sense, converge to the true population mean of the sample as the sample size increases. Precisely, the weak law of large numbers (Khinchine's Theorem) says: If $X_1, X_2, \ldots$ is a sequence of independent and identically distributed random variables and $E[X_i] = \mu$ exists, then $\bar{X}_n = (X_1 + \cdots + X_n)/n$ converges to μ in probability. The strong law of large numbers (Kolmogorov's Version) asserts the following: If $X_1, X_2, \ldots$ is a sequence of independent and identically distributed random variables, then a necessary and sufficient condition that $\bar{X}_n$ converges to μ almost surely is that $E[X_i]$ exists and is equal to μ.

The laws of large numbers for independent and identically distributed random variables say that S_n/n converges almost surely and in probability to $E[X_1]$. However, this gives no information about the distribution of S_n. Various versions of the Central Limit Theorem are concerned with approximating the distribution of S_n for large n by describing the weak limit of $(S_n - a_n)/b_n$ for suitable sequences of constants a_n and b_n. Here we are concerned about convergence in law of the distributions of $(S_n - a_n)/b_n$.

A simple version of the Central Limit Theorem is the following: If X_1, $X_2, \ldots$ is a sequence of independent and identically distributed random variables with expectation $E[X_i] = \mu$ and finite variance $\mathrm{Var}[X_i] = \sigma^2 \neq 0$, then $\mathbf{L}[(S_n - n\mu)/(n^{-1/2}\sigma)] \to \mathbf{L}(Z)$ where Z has the standard Gaussian distribution.

More general theorems exist concerning the limiting behavior of S_n. The most satisfactory condition under which S_n converges to a Gaussian distribution is due to Lindeberg. We consider a triangular array of random variables:

$$\begin{array}{cccc} X_{11} & X_{12} & \cdots & X_{1n_1} \\ X_{21} & X_{22} & \cdots & X_{2n_2} \\ \vdots & \vdots & & \vdots \end{array}$$

We assume that for each i, the n_i random variables in the ith row, $X_{i1}, \ldots,$ X_{in_i}, are independent. We do not assume anything about independence between rows, nor do we assume that any of the random variables are identically distributed. Let

$$S_i = \sum_{j=1}^{n_i} X_{ij} \qquad \mu_{ij} = E[X_{ij}] \qquad \mu_i = \sum_{j=1}^{n_i} \mu_{ij}$$

$$\sigma_{ij}^2 = \mathrm{Var}[X_{ij}] \qquad \sigma_i^2 = \sum_{j=1}^{n_i} \sigma_{ij}^2.$$

We will assume that each σ_{ij}^2 is finite and nonzero. If, in addition, the Lindeberg condition holds, which requires that for every $\varepsilon > 0$,

$$\frac{1}{\sigma_i^2} \sum_{j=1}^{n_i} E\left[(X_{ij} - \mu_{ij})^2 I\left(\left|\frac{X_{ij} - \mu_{ij}}{\sigma_i}\right| > \varepsilon\right)\right] \to 0 \qquad \text{as } i \to \infty,$$

then $\mathbf{L}[(S_i - \mu_i)/\sigma_i] \to \mathbf{L}(Z)$ where Z has the standard Gaussian distribution. A useful sufficient condition for Lindeberg's condition to hold is Lyapounov's condition, which is often easier to verify than Lindeberg's condition. Lyapounov's condition holds if we can find a value of $\delta > 0$ for which

$$\frac{1}{\sigma_i^{2+\delta}} \sum_{j=1}^{n_i} E[|X_{ij} - \mu_{ij}|^{2+\delta}] \to 0 \qquad \text{as } i \to \infty.$$

Note that Lindeberg's condition implies that the triangular array satisfies the following condition for tail probabilities: For every $\varepsilon > 0$,

$$\max_{1 \le j \le n} P\left\{\frac{|X_{ij} - \mu_{ij}|}{\sigma_i} > \varepsilon\right\} \to 0 \qquad \text{as } i \to \infty.$$

A triangular array with this property is called "uniformly asymptotically negligible." This condition makes precise the intuitive notion that any one standardized variable $(X_{ij} - \mu_{ij})/\sigma_i$ must be small relative to the sum $(S_i - \mu_i)/\sigma_i$ in order for us to be able to say something definite about the asymptotic distribution. This forces some averaging of independent variables and prevents a single term from dominating the sum.

The following elegant converse to Lindeberg's condition shows why the Lindeberg condition is "satisfactory" for convergence to a Gaussian distribution. With the same notation as before, if a triangular array with independent random variables in each row is uniformly asymptotically negligible with nonzero finite σ_i, then $\mathbf{L}[(S_i - \mu_i)/\sigma_i] \to \mathbf{L}(Z)$, where Z has the standard Gaussian distribution, only if Lindeberg's condition is satisfied.

For uniformly asymptotically negligible arrays, there is a more general central limit theorem that characterizes the possible limit laws of S_i (properly normalized) as infinitely divisible probability laws. See Feller (1971) for details. Multivariate central limit theorems can be found in Serfling (1980).

The following convergence theorem, due to Slutsky, is often useful. Suppose X_n, Y_n, and Z_n are all defined on a probability space $(\Omega, \mathbf{F}, P)$. If $\mathbf{L}(X_n) \to \mathbf{L}(X)$, $Y_n \xrightarrow{\mathbf{P}} y$, and $Z_n \xrightarrow{\mathbf{P}} z$ where y and z are real numbers, then $\mathbf{L}(Y_n X_n + Z_n) \to \mathbf{L}(yX + z)$. A consequence of Slutsky's theorem is the following: If a_n is a sequence of numbers that tends to ∞ and b is any real number such that $\mathbf{L}[a_n(X_n - b)] \to \mathbf{L}(X)$, then for any function g from $\mathbf{R}$ to $\mathbf{R}$ that is differentiable at b, we have $\mathbf{L}\{a_n[g(X_n) - g(b)]\} \to \mathbf{L}[g'(b)X]$. In particular, if X has the Gaussian distribution $N(\mu, \sigma^2)$, then $g'(b)X$ has the Gaussian distribution $N(g'(b)\mu, [g'(b)]^2\sigma^2)$, which may possibly be a degenerate Gaussian distribution if $g'(b) = 0$.

The continuity theorem often provides a method of showing convergence in distribution relatively easily using characteristic functions. It states that if $X_1, X_2, \ldots$ is a sequence of real random variables with corresponding distribution functions $F_1(x), F_2(x), \ldots$ and corresponding characteristic functions $\phi_1(t), \phi_2(t), \ldots$, then X_n converges in distribution to a random variable X if and only if the sequence $\phi_1(t), \phi_2(t), \ldots$ converges pointwise for every real t to a function $\phi(t)$ that is continuous at $t = 0$. The function $\phi(t)$ is then the characteristic function of the limit X.

5.1. A sequence of random variables that converges in law to each of two limiting random variables that are almost surely different.

The point of this example is to show that convergence in law refers only to distributions of random variables and need not imply very much about their pointwise behavior. For example, consider a sequence of independent and identically distributed standard Gaussian random variables $X_1, X_2, \ldots \sim N(0, 1)$. Then, because all the distribution laws are identical, we will certainly have

$$X_n \xrightarrow{d} X_1 \quad \text{and} \quad X_n \xrightarrow{d} X_2$$

even though X_1 and X_2 are independent nondegenerate Gaussians and are therefore almost surely different; that is,

$$P(X_1 = X_2) = 0.$$

A more extreme class of examples may also be given. In order to have convergence in distribution, we need not have the random variables defined on the same probability spaces. All that matters are the probability distributions induced on the real numbers, not the underlying probability spaces. When the limiting random variables are defined on different probability spaces, convergence in distribution can still happen even though it would make no sense to talk about "almost sure" pointwise considerations.

5.2. Two independent sequences of random variables, each converging in law to a limit, such that the sequence of term-by-term sums does not converge in law to the sum of the limits.

This example again emphasizes that the distribution is only one of many properties of random variables. In particular, the distribution law of a sum depends not only on the marginal distributions of each summand but also on their joint distribution (which reflects any statistical independence or dependence in the situation). Note, however, that if X_n and Y_n are independent, then $\mathbf{L}(X_n + Y_n)$ converges to the convolution $\mathbf{L}(X) * \mathbf{L}(Y)$, as may be seen by considering characteristic functions. As in Example 5.1, X_n and Y_n may be defined on different probability spaces in order to exhibit a counterexample. Even if they are defined on the same space, however, it is still possible that $\mathbf{L}(X_n + Y_n)$ might not converge to $\mathbf{L}(X + Y)$.

Consider two independent sequences of standard Gaussian random variables $X_1, X_2, \ldots \sim N(0, 1)$ and $Y_1, Y_2, \ldots \sim N(0, 1)$. Consider two other random variables: Let $X \sim N(0, 1)$, and then define $Y = -X$ so that Y also has a standard Gaussian distribution but is not independent of X. Because all the distributions are identical, we will certainly have convergence in distribution:

$$X_n \xrightarrow{\text{d}} X \quad \text{and} \quad Y_n \xrightarrow{\text{d}} Y.$$

However, the sums behave differently because X_n and Y_n are independent of each other (for each n), whereas X and Y are not. We have

$$\mathbf{L}(X_n + Y_n) = N(0, 2)$$

whereas $X + Y$ is degenerate at zero (because we defined $Y = -X$), so that we may write, for example,

$$\mathbf{L}(X + Y) = N(0, 0).$$

In particular, because these last two normal distributions are different, it follows that $X_n + Y_n$ does *not* converge in law to $X + Y$.

5.3. Two sequences of random variables, each converging in law to a limit, such that the joint distribution does not converge in law to any limit.

Suppose $X_n \to X$ and $Y_n \to Y$, where X_n and Y_n are defined on the same probability space, so that (X_n, Y_n) has a joint distribution. The sequence of joint distributions is tight (i.e., no mass escapes to infinity) if and only if the marginal distributions are tight. This implies that a subsequence of (X_n, Y_n) converges in law to a limit. The sequence itself may not converge, however.

Let f be a nondegenerate univariate density. If n is odd, let (X_n, Y_n) be chosen from the bivariate density $f(x)f(y)$ so that X_n and Y_n are independent, each with marginal density f. If n is even, let (X_n, Y_n) be chosen so that $Y_n = X_n$ is chosen from f. Marginally, X_n converges in law to f and Y_n does also. However, the joint distribution (X_n, Y_n) is oscillating and does not converge in law to a limit.

5.4. A sequence of distribution functions that converges in distribution such that the sequence of expected values of a bounded function does not converge to the expectation of that same function under the limiting distribution.

This example shows that continuity may not be eliminated from the equivalence between convergence in distribution and convergence of the expectations of all bounded continuous functions. Example 5.5 will show that boundedness is also necessary.

Define the sequence of distribution functions by letting F_n represent the distribution that is concentrated (degenerate) at $1/n$. Let $X_n \sim F_n$, so that X_n is identically equal to $1/n$. Observe that $X_n \xrightarrow{d} X_\infty$ where X_∞ is identically zero. Consider the bounded (but not continuous) function defined by

$$f(x) = \begin{cases} 1 & \text{if } x > 0 \\ 0 & \text{if } x \leq 0. \end{cases}$$

Then we have the following expectations for each finite n:

$$E[f(X_n)] = f\left(\frac{1}{n}\right) = 1,$$

which do not converge to the expectation of that same function of the limiting distribution:

$$E[f(X_\infty)] = f(0) = 0.$$

5.5. A sequence of distribution functions that converges in distribution such that the sequence of expected values of a continuous function does not converge to the expectation of that same function under the limiting distribution.

This example shows that boundedness may not be eliminated from the equivalence between convergence in distribution and convergence of the expectations of all bounded continuous functions. Example 5.4 showed that continuity was also necessary.

Define the sequence of distribution functions F_n so that the corresponding random variables X_n are as follows:

$$X_n = \begin{cases} 0 & \text{with probability } 1 - \dfrac{1}{n} \\[2ex] n & \text{with probability } \dfrac{1}{n}. \end{cases}$$

Observe that $X_n \xrightarrow{d} X_\infty$ where X_∞ is identically equal to zero. Consider the function $f(x) = x$, which is continuous but not bounded. For any finite n, the expectations are:

$$E[f(X_n)] = 0\left(1 - \frac{1}{n}\right) + n\left(\frac{1}{n}\right) = 1,$$

which do not converge to the expectation of the function evaluated at the limit:

$$E[f(X_\infty)] = E[f(0)] = 0.$$

5.6. A sequence of random variables that converges almost surely (and hence in probability) to a random variable X but does not converge in mean to X.

Consider the sample space $\Omega = (0, 1)$ with uniform probability (Lebesgue) measure. Define

$$X_n(\omega) = nI_{(0,1/n)}(\omega).$$

Then, clearly, X_n converges almost surely to X_∞, which is a trivial random variable that is identically zero. However,

$$E[X_n] = \int_0^{1/n} n\, d\omega = 1.$$

Thus, $E[X_n]$ converges to one as $n \to \infty$, and so X_n does not converge to $X_\infty = 0$ in mean. Here, even though the first moment converges to a finite value, it converges to the wrong value. Note that for this example we have convergence in the pth mean provided $0 < p < 1$.

5.7. A sequence of random variables that converges in probability to a random variable X but does not converge almost surely to X.

Consider as sample space the unit interval (0, 1) with uniform probability (Lebesgue) measure. Define the function $f_{k,j}$ to be the indicator function of the interval $((j-1)/2^k, j/2^k)$ for $k \geq 1$ and for $1 \leq j \leq 2^k$; that is,

$$f_{k,j}(\omega) = \begin{cases} 1 & \text{if } \dfrac{j-1}{2^k} \leq \omega \leq \dfrac{j}{2^k} \\ 0 & \text{otherwise.} \end{cases}$$

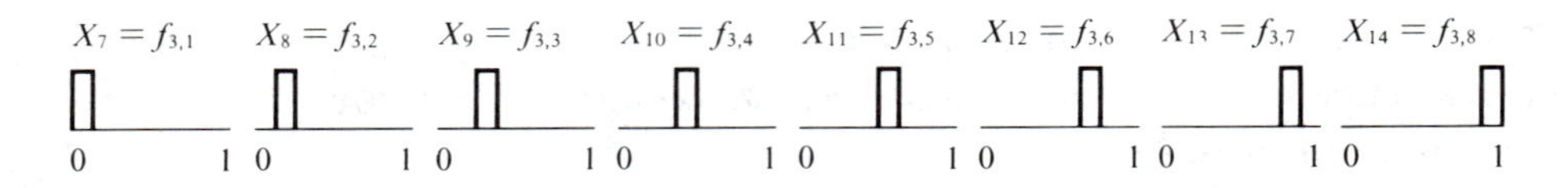

FIGURE 5.7.1.
Random variables that converge in probability but not almost surely.

Ordering these functions lexicographically first according to k increasing and then for each k according to increasing values of j, we obtain a sequence of random variables $X_1, X_2, \ldots$ that represent $f_{1,1}, f_{1,2}, f_{2,1}, f_{2,2}, f_{2,3}, f_{2,4}, f_{3,1}, \ldots$. Note that if $X_n = f_{k_n, j_n}$, then $k_n \to \infty$ as $n \to \infty$. Figure 5.7.1 shows the first several functions in this sequence and makes clear that the sequence consists of "boxcar-shaped" functions that get smaller and smaller but still always cover the whole interval when taken a whole row at a time.

It is clear from this construction that the sequence $X_1, X_2, \ldots$ does not converge almost surely to a random variable. In fact, for *no* element ω does $\lim_{n\to\infty} X_n(\omega)$ even exist, since for every element ω in $(0, 1)$, $X_n(\omega)$ takes on each of the values zero and one infinitely often.

However, the sequence $X_1, X_2, \ldots$ converges in probability to zero. This follows from the fact that for any ε in $(0, 1)$, $P\{|X_n| > \varepsilon\} = P\{X_n = 1\}$. But the probability $P\{X_n = 1\}$ decreases by $\frac{1}{2}$ in groups of 2^k; that is, $P\{X_n = 1\} = (\frac{1}{2})^{k_n} \to 0$ as $n \to \infty$. Thus, we have $X_n \overset{\mathrm{P}}{\to} 0$.

5.8. A sequence of independent random variables that converges in probability to a random variable X but does not converge almost surely to X.

Note that in Example 5.7 the sequence was not independent because within a row of functions (as was shown in Figure 5.7.1) some function takes the value one, whereas all others are zero (with probability one). In this example we show that an independent sequence can also be constructed.

Let $H_1, H_2, \ldots$ be a sequence of independent Bernoulli coin tosses, with probability $1/n$ so that

$$H_n = \begin{cases} 1 & \text{with probability } \dfrac{1}{n} \\ 0 & \text{with probability } 1 - \dfrac{1}{n}. \end{cases}$$

Then $P\{H_n \neq 0\} = 1/n \to 0$ as $n \to \infty$, and therefore H_n converges in probability to zero. Since $\Sigma_{n=1}^{\infty} P\{H_n = 1\} = \Sigma_{n=1}^{\infty} 1/n = \infty$, however, the second Borel-Cantelli Lemma implies that $P\{H_n = 1 \text{ infinitely often}\} = 1$. Thus, H_n does not converge almost surely to zero.

More generally, in place of the probabilities $1/n$ we may use probabilities $P\{H_n = 1\} = p_n$. We will have H_n converging in probability to zero but not converging almost surely provided $p_n \to 0$ as $n \to \infty$ and $\Sigma_{n=1}^{\infty} p_n = \infty$.

5.9. A sequence of independent random variables that converges in probability to a random variable, but the lim sup is almost surely ∞ and the lim inf is almost surely $-\infty$.

How pathological can the "almost sure" behavior be when we have convergence in probability? At least in Example 5.8 the lim sup of the sequence was bounded. It can happen, however, that even though an independent sequence converges in probability, the lim sup is almost surely infinite. Such an example may be constructed by defining $X_n = (-1)^n n H_n$ where H_n are independent Bernoulli random variables with $P\{H_n = 1\} = 1/n$. The sequence $X_1, X_2, \ldots$ is then independent and converges in probability to zero because $P\{X_n = 0\} = P\{H_n = 0\} = 1 - 1/n \to 1$ as $n \to \infty$. However, by the second Borel-Cantelli Lemma, we have

$$\sum_{n=1}^{\infty} P\{X_n = n\} = \sum_{k=1}^{\infty} \frac{1}{2k} = \infty$$

and so $P\{X_n = n \text{ infinitely often}\} = 1$, and therefore

$$\limsup_{n\to\infty} X_n = \infty \quad \text{almost surely.}$$

Similarly, $P\{X_n = -n \text{ infinitely often}\} = 1$, and so

$$\liminf X_n = -\infty \quad \text{almost surely.}$$

5.10. A sequence of random variables that converges almost surely to a limit, and a function such that the sequence of function values does not even converge in probability to the function evaluated at the limit.

This simple example reinforces the idea that convergence is not necessarily preserved for functions of sequences. Let X_n be a degenerate random variable concentrated at $1/n$; it may be directly verified that $X_n \to 0$ with probability one. Consider the function defined by

$$g(t) = \begin{cases} t & \text{if } t \neq 0 \\ 1 & \text{if } t = 0. \end{cases}$$

The function values are $g(X_n) = 1/n$ for all finite values of n, and so the function values converge to zero. This is not equal to the function evaluated at the limit, however, which is $g(0) = 1$.

5.11. Random variables X_n converging in law to X and a continuous measurable real-valued function g such that $g(X_n)$ does not converge in law to $g(X)$, even though the set of discontinuities of g has probability zero under the law of X.

The so-called Continuous Mapping Theorem is perhaps the most useful result in the theory of weak convergence. One version is the following (see Billingsley 1968, Theorem 5.1): Let $(S_i, \mathbf{S}_i)$, where $i = 1, 2$, be metric spaces equipped with their Borel σ-fields; let X_n and X be random variables on $(S_1, \mathbf{S}_1)$; let g be a measurable map from $(S_1, \mathbf{S}_1)$ to $(S_2, \mathbf{S}_2)$; and let Dg denote the set of discontinuities of g. If X_n converges in distribution to X and $P(X \in Dg) = 0$, then $g(X_n)$ converges in distribution to $g(X)$ on $(S_2, \mathbf{S}_2)$.

This counterexample, adapted from Pollard (1984, p. 66), shows that this theorem does not extend directly if $\mathbf{S}_i$ is not the Borel σ-field of S_i. However, Pollard (1984, Theorem 12, p. 70) provides a generalization under additional assumptions.

Let $S_1 = S_2 = \mathbf{R}$ be the real line equipped with the usual Euclidean metric. Let $\mathbf{S}_2 = \mathbf{B}$ be the usual Borel σ-field, but let $\mathbf{S}_1$ be

$$\{\phi, \mathbf{R}, (-\infty, 0], (0, \infty]\}.$$

Let X_n and X be random variables that take values in $(S_1, \mathbf{S}_1)$, with X_n identically equal to 3 and X identically equal to -1. Since any bounded continuous measurable function f from $(S_1, \mathbf{S}_1)$ to $(\mathbf{R}, \mathbf{B})$ must be constant, we trivially have $E[f(X_n)] \to E[f(X)]$, so that X_n converges in distribution to X. However, let g be the map from $\mathbf{R}$ to $\mathbf{R}$ defined by

$$g(x) = \begin{cases} 1 & \text{if } x > 0 \\ 0 & \text{otherwise.} \end{cases}$$

This map is measurable and continuous at the point $x = -1$, which has probability one under X. However, $g(X_n)$ does not converge in distribution to $g(X)$ because $g(X_n)$ has mass one at 1, but $g(X)$ has mass one at -1.

5.12. A determining class that is not a convergence-determining class.

Although a convergence-determining class is a determining class, the converse is false. Consider the class $\mathbf{C}$ of finite intervals (a, b) on the real number line that do not include the point zero. It is easy to see that $\mathbf{C}$ is a determining class. However, if Q_n is the probability measure that has mass one at n and Q is the probability measure that has mass one at 0, then for any $A \in \mathbf{C}$,

$$Q_n(A) \to Q(A) = 0.$$

However, Q_n does not converge weakly to Q because

$$0 = Q_n\{(-\infty, \tfrac{1}{2}]\} \quad \text{does not converge to} \quad 1 = Q\{(-\infty, \tfrac{1}{2}]\}.$$

For an additional example, see Billingsley (1968, pp. 19–20).

5.13. A sequence of random variables that converges in quadratic mean to a random variable X but does not converge almost surely to X.

Let the random variables $X_1, X_2, \ldots$ be defined as in Example 5.7, and recall that this sequence does not converge almost surely to any random variable. We will show that this sequence converges in quadratic mean to the random variable, which is identically zero; that is, $E[X_n^2] \to 0$ as $n \to \infty$. For any $n \geq 1$, X_n may be written as

$$X_n(\omega) = \begin{cases} 1 & \text{if } \dfrac{j_n - 1}{2^{k_n}} < \omega < \dfrac{j_n}{2^{k_n}} \\ 0 & \text{otherwise} \end{cases}$$

for some j_n and k_n. It follows that $E[X_n^2] = 1/(2^{k_n})$. In fact, $E[X_n^2]$ is simply the length of the subinterval where X_n takes the value one. Therefore, since these subintervals tend to zero as $n \to \infty$, it follows that $E[X_n^2] \to 0$ as $n \to \infty$. Thus, X_n tends to zero in quadratic mean (i.e., in L^2).

5.14. A sequence of random variables that converges in the pth mean to a random variable X for every positive value of p but does not converge almost surely to X.

This example shows a more general result than does Example 5.13. Consider the same sequence $X_1, X_2, \ldots$ as in Examples 5.13 and 5.7. We have shown that this sequence converges in quadratic mean (L^2) but that it does not converge almost surely to any random variable. However, because each X_n is simply an indicator of an interval and therefore takes on only the values zero and one, we have

$$E[|X_n - 0|^p] = E[X_n^2] \to 0 \qquad \text{as } n \to \infty.$$

It follows that X_n converges in L^p to zero.

5.15. A sequence of random variables that converges in probability to a random variable X but does not converge in the pth mean.

As in Example 5.7, consider the uniform (Lebesgue) probability measure over the class of Lebesgue-measurable sets in (0, 1), and define a related (but different) sequence of functions as follows:

$$f_{k,j}(\omega) = \begin{cases} k^{1/p} & \text{if } \dfrac{j-1}{k} < \omega < \dfrac{j}{k} \\ 0 & \text{otherwise} \end{cases}$$

for $k = 1, 2, \ldots$ and for $j = 1, \ldots, k$. Ordering these functions lexicographically first according to k increasing and then according to j increasing, we obtain a sequence of random variables $X_1, X_2, \ldots$ with $X_n = f_{k_n, j_n}$ for some k_n and j_n. As before, note that for any ω in (0, 1), $X_n(\omega)$ takes on the values zero and one infinitely often, so that the set on which X_n converges (to anything) is empty. In particular, X_n does not converge almost surely to any random variable.

However, X_n does converge in probability to zero. To see this, for any ε in

(0, 1),

$$P\{|X_n| > \varepsilon\} = P\{X_n \geq 1\} = \frac{1}{k_n} \to 0 \qquad \text{as } n \to \infty.$$

However, X_n does not converge in the pth mean to zero, since for all $n = 1, 2, \ldots$, we have:

$$E[|X_n - 0|^p] = E[X_n^p] = \int_{(j_n-1)/k_n}^{j_n/k_n} (k_n^{1/p})^p \, d\omega = 1.$$

Thus, $\lim_{n\to\infty} E[|X_n - 0|^p] = 1$, so that X_n does not converge to zero in L^p. Note, however, that $X_n \to 0$ in L^s whenever $0 < s < p$.

5.16. A sequence of random variables that converges almost surely to X but does not converge in the pth mean to X.

Again consider uniform probability on (0, 1). Define the sequence $X_1, X_2, \ldots$ of random variables as follows:

$$X_n(\omega) = 2^n I_{(0,\,1/n)}(\omega).$$

Then it is clear that $X_n \xrightarrow{\text{a.s.}} 0$; in fact, the complement of the set for which X_n converges not only has probability zero but also is empty. In addition, it follows that $X_n \xrightarrow{\text{P}} 0$. However,

$$E[|X_n - 0|^p] = E[X_n^p] = \int_0^{1/n} (2^n)^p \, d\omega = \frac{2^{np}}{n}$$

which tends to ∞ as $n \to \infty$ for every $p > 0$. Thus, X_n does not converge to zero in L^p.

5.17. A sequence of random variables that converges in distribution to X but not in probability to X (from Roussas 1973, p. 137).

Consider the discrete probability space with sample space $S = \{0, 1\}$ and equal probability mass $\frac{1}{2}$ at each point in S. Define the random variable X and the sequence of random variables $X_1, X_2, \ldots$ as follows:

$$X_n(1) = 1 \qquad X_n(0) = 0 \qquad \text{for all } n = 1, 2, \ldots$$

and

$$X(1) = 0 \qquad X(0) = 1.$$

Then the distribution function of X_n (for all n) and X is given by

$$F(x) = \tfrac{1}{2}I_{(0,1)}(x) + I_{[1,\infty)}(x).$$

Thus, trivially, X_n converges in distribution to X. For any s in S, however, we have $|X_n(s) - X(s)| = 1$. So X_n does not converge in probability to X.

In fact, any sequence $X_1, X_2, \ldots$ of independent and identically distributed nondegenerate random variables would also provide an example here.

5.18. A sequence of random variables that converges in distribution to X but does not converge in the pth mean to X.

Consider the sequence of random variables $X_1, X_2, \ldots$ and the limit X as defined in Example 5.17. Then $X_n \xrightarrow{\text{d}} X$. However,

$$E[|X_n - 0|^p] = E[1^p] = 1 \neq 0.$$

Thus, for any $p > 0$, X_n does not converge in L^p to X.

Other illustrations were provided by Examples 5.15 and 5.16, which concerned modes of convergence that imply convergence in distribution.

5.19. A sequence of random variables whose characteristic functions converge, but the random variables do not converge in distribution to any random variable.

The point of the following two examples is that the continuity requirement of the limiting characteristic function $\phi(t)$ at $t = 0$ cannot be eliminated in the Continuity Theorem. This is related to the requirement of tightness—to guarantee that mass cannot "escape" to infinity.

The first example defines (as in Lukacs 1970, p. 48) the sequence of random variables $X_1, X_2, \ldots$ to have corresponding cumulative distribution functions given by (for $n = 1, 2, \ldots$)

$$F_n(x) = \begin{cases} 0 & \text{if } x < -n \\ \dfrac{n + x}{2n} & \text{if } -n \leq x \leq n \\ 1 & \text{if } x > n. \end{cases}$$

The characteristic function of X_n is then given by

$$\phi_n(t) = \int_{-\infty}^{\infty} e^{itx}\, dF_n(x) = \int_{-n}^{n} e^{itx} \frac{n+x}{2n}\, dx.$$

Integrating by parts, we find that

$$\phi_n(t) = \begin{cases} \dfrac{\sin(nt)}{nt} & \text{if } t \neq 0 \\ 1 & \text{if } t = 0. \end{cases}$$

Thus, in the limit, we have

$$\lim_{n\to\infty} \phi_n(t) = \begin{cases} 0 & \text{if } t \neq 0 \\ 1 & \text{if } t = 0. \end{cases}$$

Note that the limit $\phi(t)$ exists but fails to be continuous at $t = 0$. To see that X_n does not converge in distribution to any random variable X, we note that

$$\lim_{n\to\infty} F_n(x) = \tfrac{1}{2} \qquad \text{for all } x.$$

Because this function, which is identically $\frac{1}{2}$, is not a cumulative distribution function, it follows that X_n does not converge in distribution.

The second example (from Roussas 1973, p. 138) lets X_n have the Gaussian distribution with mean zero and variance n. Its characteristic function is

$$\phi_n(t) = \exp\left(\frac{-nt^2}{2}\right).$$

This sequence of characteristic functions tends pointwise to the limiting function $\phi(t)$ as $n \to \infty$ where

$$\phi(t) = \begin{cases} 0 & \text{if } t \neq 0 \\ 1 & \text{if } t = 0. \end{cases}$$

However, if G is the cumulative distribution function of the standard Gaussian distribution with mean zero and variance one, then the distribution function of X_n is given by

$$F_{X_n}(x) = P\{X_n \le x\} = P\{n^{-1/2} X_n \le n^{-1/2} x\} = G(n^{-1/2} x).$$

In the limit, for every real number x, $G(n^{-1/2}x) \to \frac{1}{2}$ as $n \to \infty$. However, $F(x) = \frac{1}{2}$ is not a distribution function and therefore the sequence cannot converge in distribution.

5.20. A sequence of random variables that converges in distribution but whose cumulative distribution functions do not converge pointwise.

The point of this example is to show that the restriction to points of continuity of the limiting cumulative distribution function cannot be eliminated. Technicalities do matter: A sequence cannot be said to "converge pointwise" if it fails to converge at even a single point.

Let X_n be degenerate, alternating between 0 and $1/n$ according to the following definition:

$$X_n = \begin{cases} \dfrac{1}{n} & \text{if } n \text{ is odd} \\ 0 & \text{if } n \text{ is even.} \end{cases}$$

It can be verified that $X_n \xrightarrow{d} 0$ by showing pointwise convergence of the cumulative distribution functions everywhere except at $x = 0$ because the limiting cumulative distribution function, $I\{x \geq 0\}$, is continuous except at zero. The cumulative distribution functions of X_n do not even form a pointwise convergent sequence, however, because at zero we have

$$F_n(0) = P\{X_n \leq 0\} = \begin{cases} 0 & \text{if } n \text{ is odd} \\ 1 & \text{if } n \text{ is even.} \end{cases}$$

This oscillating behavior does not have a limit as $n \to \infty$.

It is interesting to note that if $F_n(x) \to F(x)$ for all x and F is continuous, then the stronger statement

$$\sup_x |F_n(x) - F(x)| \to 0$$

is also true. This example shows that we may have F_n converging weakly to F, but $\sup_x |F_n(x) - F(x)|$ does not converge to zero because F is not continuous.

5.21. A sequence of distributions F_n on the real line for which no subsequence converges weakly to a distribution, even though for every real number x, the sequence $F_n(x)$ converges.

Helly's selection theorem states that every sequence of cumulative distribution functions possesses a subsequence that converges weakly to a limiting function. However, this limiting function need not be a distribution function because mass might "escape" to ∞ or to $-\infty$. The condition of "tightness" would prevent this and ensure that the limiting function is indeed a distribution.

Let the distribution F_n correspond to placing unit mass at n. Then every subsequence converges pointwise to the zero function, which is not a distribution function.

5.22. A sequence of random variables that converges in distribution but whose sequence of corresponding probability measures does not converge for all Borel sets (from Billingsley 1979, p. 281).

Suppose X_n has distribution function F_n and probability measure P_n. Convergence in distribution to X is equivalent to $\lim_{n\to\infty} P_n(A) = P(A)$ for every A of the form $A = (-\infty, x]$ for which $P(\{x\}) = 0$. P is then the probability measure associated with the random variable X. The statement $\lim_{n\to\infty} P_n(B) = P(B)$ may not hold in general, however, if B is an arbitrary Borel set. This may seem somewhat surprising because the sets $(-\infty, x]$ generate the Borel sets.

An example is provided by letting P_n represent a probability mass of $1/n$ at j/n for $j = 0, 1, \ldots, n - 1$. Let P be the Lebesgue (uniform) probability measure in (0, 1). The corresponding distribution functions are given by

$$F_n(x) = \begin{cases} 0 & \text{if } x < 0 \\ \dfrac{[nx]}{n} & \text{if } 0 \leq x \leq 1 \\ 1 & \text{if } x > 1 \end{cases}$$

where $[t]$ denotes the greatest integer less than or equal to t; and

$$F(x) = \begin{cases} 0 & \text{if } x < 0 \\ x & \text{if } 0 \leq x \leq 1 \\ 1 & \text{if } x > 1. \end{cases}$$

Because $[nx]/n$ differs from x by at most $1/n$ in absolute value, it follows that $F_n(x) \to F(x)$ as $n \to \infty$ for all real x. However, consider the set Q of rational numbers. For every n, $P_n(Q) = 1$ because each P_n is concentrated on the rational numbers. The Lebesgue probability measure of Q is zero. Thus,

$$\lim_{n\to\infty} P_n(Q) = 1 \neq 0 = P(Q).$$

This example holds for all nonempty subsets of the rationals as well—for example, the singleton set $\{r\}$ where r is any rational number in (0, 1). Note that Q and its subsets are relatively simple types of Borel sets.

Another example, somewhat simpler than the earlier one, is provided by Example 5.20 if we look at the behavior of the probability of the singleton set $\{0\}$, which is not convergent.

5.23. A sequence of discrete random variables that converges in distribution to a continuous random variable.

As in Example 5.22, let X_n have the cumulative distribution $F_n(x)$ and let X have the cumulative distribution function $F(x)$. In this case, X_n is discrete and X is continuous. However, we still have $X_n \xrightarrow{d} X$.

In fact, it is possible to have almost sure convergence (not just convergence in probability) of a sequence of discrete random variables to a continuous random variable. Let $(\Omega, \mathbf{B}, P)$ denote the interval (0, 1) with the Borel σ-field and Lebesgue measure. If we define the sequence of random variables ($n = 1, 2, \ldots$) by

$$X_n(\omega) = \frac{[n\omega]}{n}$$

where $[t]$ denotes the greatest integer function, then each X_n is discrete because it can take on only the values $0, 1/n, 2/n, \ldots, (n-1)/n, 1$. However, for each ω we have

$$\lim_{n\to\infty} X_n(\omega) = \omega \qquad \text{for each } \omega \text{ in } \Omega$$

which defines a continuous random variable $X(\omega) = \omega$ in the limit, with almost sure convergence of X_n to X.

5.24. A sequence of continuous random variables that converges in distribution to a discrete random variable.

Let X_n have the cumulative distribution function given by

$$F_n(x) = \begin{cases} 0 & \text{if } x \leq 0 \\ x^n & \text{if } 0 < x < 1 \\ 1 & \text{if } x \geq 1. \end{cases}$$

Let X have the cumulative distribution function given by

$$F(x) = I_{[1,\infty)}(x).$$

Then $F(x)$ is continuous except at $x = 1$. For all $x \neq 1$, we have

$$\lim_{n\to\infty} F_n(x) = F(x).$$

Thus, $X_n \xrightarrow{d} X$ even though X_n is a continuous random variable for each n and X is a discrete random variable.

It is, in fact, possible to have almost sure convergence of a sequence of continuous random variables to a discrete random variable. For the interval (0, 1) with the Borel sets and Lebesgue measure, the sequence of continuous random variables defined by

$$X_n(\omega) = \omega^n$$

converges pointwise for each ω to the degenerate discrete random variable $X = 0$.

In fact, any discrete distribution is a weak limit of absolutely continuous distributions. To see this, suppose X has cumulative distribution function F, which is discrete (or singular). Then, if Y_n is independent of X and has distribution $\mathbf{L}(Y_n) = N(0, \sigma_n^2)$ and $\sigma_n^2 \to 0$, then $\mathbf{L}(X + Y_n)$ converges weakly to F. But note that $\mathbf{L}(X + Y_n)$ is absolutely continuous because the convolution of an absolutely continuous distribution with a discrete one must be absolutely continuous.

5.25. A sequence of absolutely continuous distributions with support equal to the entire plane that converges to a limit law that is degenerate at the origin.

Let X_n and Y_n be independent Gaussian random variables with mean zero and variance $1/n$. Then the joint characteristic function of (X_n, Y_n) is given by

$$\phi_{X_n, Y_n}(s, t) = \exp\left(-\frac{s^2 + t^2}{2n}\right) \to 1 \qquad \text{as } n \to \infty$$

for all s and t. But the function identically equal to one is the characteristic function of the degenerate probability distribution in the plane, placing unit mass at the origin, and (X_n, Y_n) converges in distribution to this limit.

5.26. A sequence of continuous random variables converging in distribution to a continuous random variable even though the sequence of densities does not converge (Billingsley 1979, p. 293).

Suppose the continuous random variable X_n has cumulative distribution function given by

$$F_n(x) = \begin{cases} 0 & \text{if } x < 0 \\ x + (2\pi n)^{-1} \sin(2\pi n x) & \text{if } 0 \le x \le 1 \\ 1 & \text{if } x > 1 \end{cases}$$

for $n = 1, 2, \ldots$. Then, for all x, $F_n(x) \to F(x)$ as $n \to \infty$, where $F(x)$ represents the uniform distribution on (0, 1). Thus, $X_n \xrightarrow{d} X$, where the random variable X has the uniform distribution on (0, 1). However, the sequence of density functions on (0, 1) is given by

$$f_n(x) = 1 + \cos(2\pi n x) \qquad \text{for } 0 < x < 1.$$

For large values of n, this density function oscillates many times between zero and two, and so f_n is not a convergent sequence of functions. In particular, it does not converge to $dF(x)/dx = 1$ on the interval of support, (0, 1).

5.27. A sequence of discrete random variables that converges in distribution to a discrete random variable even though the sequence of discrete probability functions does not converge.

Let the discrete random variable X_n have the discrete probability function given by

$$p_n(x) = \begin{cases} \frac{1}{2} & \text{if } x = 1 - \dfrac{1}{n} \quad \text{or} \quad x = 1 + \dfrac{1}{n} \\ 0 & \text{otherwise} \end{cases}$$

for $n = 1, 2, \ldots$ so that X_n takes on only the two values $1 - 1/n$ and $1 + 1/n$ with equal probabilities. In this case, $p_n(x) \to p(x)$ for all real x, where $p(x)$ is identically zero. But then $p(x)$ is not a discrete probability function (nor is it a density in any sense). Nonetheless, X_n does converge in distribution. The distribution function of X_n is given by

$$F_n(x) = \begin{cases} 0 & \text{if } x < 1 - \dfrac{1}{n} \\ \frac{1}{2} & \text{if } 1 - \dfrac{1}{n} \le x < 1 + \dfrac{1}{n} \\ 1 & \text{if } x \ge 1 + \dfrac{1}{n}. \end{cases}$$

Let the discrete random variable X correspond to a point mass at one, with cumulative distribution function given by

$$F(x) = I_{[1,\infty)}(x).$$

Then at all points of continuity of $F(x)$ (which excludes the value $x = 1$), we have

$$\lim_{n\to\infty} F_n(x) = F(x).$$

Thus X_n converges in distribution to X.

5.28. A sequence of dependent bivariate random variables that converges in distribution to an independent bivariate random variable.

In general, if X_n and Y_n are independent and (X_n, Y_n) converges in distribution to (X, Y), then (X, Y) are also independent. This property need not hold if we replace "independent" by "dependent."

Let (X_n, Y_n) have the bivariate Gaussian distribution so that each component has mean zero and variance one, and so that the correlation of X_n with Y_n is $1/n$. Then each term of the sequence is dependent, but the sequence converges in distribution to (X, Y) where X and Y are independent standard Gaussian random variables.

5.29. A sequence of discrete random variables that converges almost surely and in distribution but does not converge in the pth mean.

Define the random variable $X_n(\omega) = n^n I_{(0,1/n)}(\omega)$ with respect to the probability space consisting of (0, 1) with the Borel sets and Lebesgue measure. The random variable X_n is discrete and takes on the values n^n and zero with probabilities $1/n$ and $1 - 1/n$, respectively. X_n converges to zero almost surely, which implies that it also converges in probability and in distribution. If X is the (degenerate) zero random variable, then $E[X_n^p] =$

$n^{np-1} \to \infty$ as $n \to \infty$ for all $p > 0$. Hence, X_n does not converge in L^p for any positive p.

5.30. A sequence of continuous random variables with corresponding densities that converge pointwise but not in mean.

Convergence in mean of the density functions would hold if the limiting function were a density, owing to a theorem of Scheffe (Rao 1973, p. 124). However, the conclusion may be false if the limiting function exists but is not a density function.

Suppose X_n has a density function given by

$$f_n(x) = nI_{(0,\,1/n)}(x).$$

Then we have $f_n(x) \to 0$ as $n \to \infty$ for all real x. However, these densities do not converge to zero in L^1 because [using $f(x) = 0$ for the limiting function] we see that

$$\int_{\mathbf{R}} |f_n(x) - f(x)|\,dx = \int_{\mathbf{R}} f_n(x)\,dx = \int_0^{1/n} n\,dx = 1$$

which does not tend to zero as $n \to \infty$.

5.31. A sequence of nonsymmetric distributions that converges in distribution to a symmetric distribution.

Let X_n have a Poisson distribution with mean n, and define $Y_n = (X_n - n)/n^{1/2}$. Then by the Central Limit Theorem, Y_n converges in distribution to a standard Gaussian distribution, which is symmetric about zero. However, for each finite n, Y_n is not symmetrically distributed because, for example, its third moment is not zero.

5.32. An example where the sum of the expectations is not equal to the expectation of the sum.

Let $Y_n(\omega) = nI_{(0,\,1/n)}(\omega)$ for $\omega \in (0, 1)$ with the Borel subsets and Lebesgue measure. Define $X_n = Y_{n+1} - Y_n$ for $n = 1, 2, \ldots$. Then $E(X_n) = 0$ and so $\Sigma\, E(X_n) = 0$. However, $\Sigma\, X_n = -Y_1 = -1$, and therefore

$$E\left(\sum_{n=1}^{\infty} X_n\right) = -1 \neq 0 = \sum_{n=1}^{\infty} E(X_n).$$

5.33. A sequence of random variables that converges almost surely to a limit such that the expectations converge to the expectation of the limit, yet the variances do not converge to the variance of the limit.

Consider the uniform (Lebesgue) probability measure on (0, 1). Define the sequence of random variables:

$$X_n(\omega) = n^{1/2} I_{(0,\,1/n)}(\omega).$$

Then we clearly have

$$X_n \xrightarrow{\text{a.s.}} 0 \quad \text{and hence} \quad X_n \xrightarrow{\text{P}} 0 \quad \text{and} \quad X_n \xrightarrow{\text{d}} 0.$$

Also,

$$E[X_n] = \int_0^{1/n} n^{1/2}\, dx = n^{-1/2} \to 0 \qquad \text{as } n \to \infty.$$

Thus, if X is identically zero, we have $\lim_{n\to\infty} E[X_n] = E[X]$. However,

$$E[X_n^2] = \int_0^{1/n} [n^{1/2}]^2\, dx = 1.$$

Therefore, it follows that

$$\text{Var}[X_n] = E[X_n^2] - \{E[X_n]\}^2 = 1 - n^{-1/2}$$

so that

$$\lim_{n\to\infty} \text{Var}[X_n] = 1 \neq 0 = \text{Var}[X].$$

5.34. A sequence of random variables with finite moments of all orders that converges in distribution, yet the variances do not converge to any finite number.

Suppose X has a uniform distribution on (0, 1). Define

$$X_n = \begin{cases} X & \text{if } X \leq 1 - \dfrac{1}{n} \\ n & \text{if } X > 1 - \dfrac{1}{n}. \end{cases}$$

Then $X_n \xrightarrow{d} X$ may be verified directly. However, note that

$$E[X_n] = \int_0^{1-1/n} x\,dx + \int_{1-1/n}^{1} n\,dx = \frac{(1-1/n)^2}{2} + 1$$

and

$$E[X_n^2] = \int_0^{1-1/n} x^2\,dx + \int_{1-1/n}^{1} n^2\,dx = \frac{(1-1/n)^3}{3} + n.$$

Hence, it follows that $\mathrm{Var}[X_n] = E[X_n^2] - \{E[X_n]\}^2 \to \infty$ as $n \to \infty$, even though $\mathrm{Var}[X]$ exists. In general, if $X_n \xrightarrow{d} X$ and variances exist, then $\liminf_{n\to\infty} \mathrm{Var}[X_n] \geq \mathrm{Var}[X]$ (see Bickel and Doksum 1977, p. 150). Thus, this example also shows that this inequality can be strict.

5.35. A sequence X_n of real random variables such that $\Sigma_{n=1}^{\infty} P(|X_n| > \varepsilon) = \infty$, but X_n converges almost surely to zero.

A sufficient condition for a sequence X_n of real random variables to converge to zero almost surely is that, for every $\varepsilon > 0$, $\Sigma_{n=1}^{\infty} P(|X_n| > \varepsilon) < \infty$. This condition is also necessary when the X_n sequence is independent, but this example shows that it is not necessary in general.

Let U be uniformly distributed in [0, 1]. Define

$$X_n = \begin{cases} 1 & \text{if } U < \dfrac{1}{n} \\ 0 & \text{otherwise.} \end{cases}$$

It can easily be verified that, for $0 < \varepsilon < 1$, $\Sigma_{n=1}^{\infty} P(|X_n| > \varepsilon) = \infty$, but clearly X_n converges almost surely to zero.

5.36. A sequence of independent and identically distributed continuous random variables where the laws of large numbers do not hold.

As in Example 4.6, let X_i (for $i = 1, 2, \ldots$) be independent and identically distributed with the following density function:

$$f(t) = (2\pi)^{-1/2} t^{-3/2} \exp\{-\tfrac{1}{2t}\} I_{(0,\infty)}(t).$$

As was shown in Example 4.6, the sample average $\bar{X}_n$ has the same distribution as nX_i for any i. Indeed, the sample average increases linearly with n, rather than converging as we would usually expect. It is generally believed, and is reinforced by the laws of large numbers, that the average $\bar{X}_n$ will tend to some "true" value. The problem arises here because the X_i themselves do not have an expectation. Note also that the distribution of $\bar{X}_n$ is not asymptotically Gaussian, so that the Central Limit Theorem does not always hold when the random variables do not have (finite) means.

5.37. A sequence of independent and identically distributed discrete random variables where the laws of large numbers do not hold (Feller 1968, p. 246).

Consider a symmetric random walk on the nonnegative integers; that is, let Y_i take on values 1 and -1, each with probability $\frac{1}{2}$. Let X_k be the waiting time from the $(k-1)$st return to the origin to the kth return. (A "return to the origin" is said to happen at time t if $Y_1 + \cdots + Y_t = 0$.) The sum $S_n = X_1 + \cdots + X_n$ is the time for the nth return to the origin in a symmetric random walk. It can be shown that $P\{S_n < n^2 t\} \to 2\{1 - G(t^{-1/2})\}$ where G is the cumulative distribution function of the standard Gaussian distribution. In this case, it is $\bar{X}_n/n$ that has a limiting distribution rather than $n^{1/2}\bar{X}_n$. Again, the problem is because of the surprising fact that the X_k have infinite means.

5.38. Independent and identically distributed random variables whose partial averages converge in probability to zero even though the expectation of each term does not exist.

Khinchine's form of the weak law of large numbers states that if $X_1, X_2, \ldots$ is a sequence of independent and identically distributed random variables and if $E[|X_i|] < \infty$, then

$$\bar{X}_n = \frac{X_1 + \cdots + X_n}{n} \xrightarrow{\text{P}} E[X_i].$$

This example shows that $S_n/n \xrightarrow{\text{P}} \mu$ need not imply that $E[X_i] = \mu$. In fact, if $X_1, X_2, \ldots$ are independent and identically distributed, then the necessary and sufficient condition that there exist constants μ_n such that $\bar{X}_n - \mu_n \xrightarrow{\text{P}} 0$ is that $tP\{|X_i| > t\} \to 0$ as $t \to \infty$, in which case we may take $\mu_n = E[X_i I(|X_i| \leq n)]$. (See Feller 1971, VII. 7, Theorem 1.)

Let $X_1, X_2, \ldots$ be independent, symmetric random variables with common distribution given by

$$P\{|X_i| > t\} = \frac{e}{t \log t} \qquad \text{for } t \geq e.$$

Note that $E[|X_i|] = \infty$; however, we will show that $\bar{X}_n \xrightarrow{P} 0$. This follows from the earlier theorem; however, we will show it directly without appealing to the general result. To this end, define truncated random variables $X_j' = X_j I(|X_n \leq n|)$, where $1 \leq j \leq n$, and set $S_n' = \Sigma_{j=1}^n X_j'$. For each n, $X_1', \ldots, X_n'$ are independent and identically distributed, and if $\varepsilon > 0$, then we have

$$P\left\{\left|\frac{S_n}{n} - \frac{S_n'}{n}\right| > \varepsilon\right\} \leq P\{S_n \neq S_n'\} \leq P\{X_j \neq X_j' \quad \text{for some } j\}$$

$$\leq \sum_{j=1}^{n} P\{X_j \neq X_j'\} = nP\{|X_1| > n\}$$

$$= \frac{en}{n \log n} \to 0 \qquad \text{as } n \to \infty.$$

Thus $(S_n - S_n')/n \xrightarrow{P} 0$. So, to show that $S_n/n \xrightarrow{P} 0$, it suffices to show that $S_n'/n \xrightarrow{P} 0$ (by Slutsky's Theorem). If $\varepsilon > 0$, since $E[X_i'] = 0$, then by Chebyshev's Inequality,

$$P\left\{\left|\frac{S_n'}{n}\right| > \varepsilon\right\} \leq \frac{E[|S_n'|^2]}{\varepsilon^2 n^2} = \frac{E[(X_1')^2]}{\varepsilon^2 n}.$$

Therefore, we must show that $E[X_1 I(|X_1| \leq n)]^2/n \to 0$ as $n \to \infty$. But,

$$\frac{E[X_1 I(|X_1| \leq n)]^2}{n} = \frac{2}{n}\int_e^n t^2 \, dP(X \leq t)$$

$$= \frac{2}{n}\int_e^n t^2 \left[\frac{e}{2(t \log t)^2} + \frac{e}{2t^2 \log t}\right] dt$$

$$= \frac{e}{n}\left[\int_e^n \frac{dt}{[\log t]^2} + \int_e^n \frac{dt}{\log t}\right]$$

$$\leq \frac{2e}{n}\int_e^n \frac{dt}{\log t} = \frac{2e}{n}\left[\int_e^{n^{1/2}} \frac{dt}{\log t} + \int_{n^{1/2}}^n \frac{dt}{\log t}\right]$$

$$\leq \frac{2e}{n}\left[(n^{1/2} - e) + \frac{(n - n^{1/2})}{\log(n^{1/2})}\right] \to 0 \qquad \text{as } n \to \infty$$

and therefore $S_n/n \xrightarrow{P} 0$.

5.39. Independent and identically distributed random variables such that the partial averages converge in probability to zero but do not converge almost surely.

As in Example 5.38, let $X_1, X_2, \ldots$ be independent, symmetric random variables with common distribution determined by

$$P\{|X_i| > t\} = \frac{e}{t \log t} \qquad \text{for } t \geq e.$$

It was shown in Example 5.38 that $S_n/n = (X_1 + \cdots + X_n)/n \xrightarrow{P} 0$. To show that S_n/n does not converge almost surely to zero, we will actually show that the $\limsup_{n\to\infty} |S_n/n|$ is almost surely infinite.

Note that the event $A = \{\limsup_{n\to\infty} |S_n/n| = \infty\}$ is a symmetric or permutable event. According to the Hewitt-Savage Zero-One Law, the event A must have probability either zero or one (see Chung 1974, Theorem 8.1.4). To argue by contradiction, if $P(A) = 0$, then there exists a number K such that

$$P\left\{\limsup_{n\to\infty} \left|\frac{S_n}{n}\right| \leq K\right\} > 0 \quad \text{or} \quad P\left\{\limsup_{n\to\infty} |S_n| \leq nK\right\} > 0.$$

But

$$\sum_{n=1}^{\infty} P\{|X_n| \geq 2nK\} = \sum_{n=1}^{\infty} \frac{e}{2nK \log(2nK)} = \infty$$

and so by the second Borel-Cantelli Lemma,

$$P\{|X_n| \geq 2nK \text{ infinitely often}\} = 1,$$

which contradicts $P\{\limsup_{n\to\infty} |S_n| \leq nK\} > 0$. Hence it follows by contradiction that $P\{A\} = 1$. Since the random variables are symmetric, it follows immediately that

$$\limsup_{n\to\infty} \frac{S_n}{n} \overset{\text{a.s.}}{=} \infty \quad \text{and} \quad \liminf_{n\to\infty} \frac{S_n}{n} \overset{\text{a.s.}}{=} -\infty.$$

Therefore, although S_n/n converges in probability to zero, it does not converge almost surely in this example.

5.40. A sequence of independent random variables that converges in probability to zero but whose partial averages do not converge in probability to zero (Chung 1974, Problem 5.2.2, p. 114).

Consider the sequence of independent random variables defined so that

$$X_n = \begin{cases} 2^n & \text{with probability } \dfrac{1}{n} \\ 0 & \text{with probability } 1 - \dfrac{1}{n}. \end{cases}$$

We have the convergence of X_n to zero in probability because, in particular, $P\{X_n = 0\} \to 1$ as $n \to \infty$. To show that the sequence of partial averages, $\bar{X}_n$, does not converge in probability to zero, we will show that $P\{|\bar{X}_n| < 1\}$ does not converge to one. Take N so that $2^{N/2} > N$ (any $N \geq 4$ will suffice). For $n \geq N$, we have

$$P\{|\bar{X}_n| \leq 1\} \leq P\left\{X_k = 0 \quad \text{whenever } \frac{n+1}{2} < k \leq n\right\}$$

$$\leq \left(1 - \frac{1}{n}\right)^{n/2} \to e^{-1/2} \qquad \text{as } n \to \infty.$$

Because this does not converge to zero, the partial averages do not converge in probability to zero.

5.41. The strong law of large numbers does not extend directly to triangular arrays.

Given a triangular array of random variables (all defined on a common probability space) such that the nth row consists of n independent and identically distributed random variables $X_{n1}, \ldots, X_{nn}$ according to a distribution F_n with mean zero, we might ask whether or not

$$\frac{S_n}{n} = \frac{1}{n}\sum_{j=1}^{n} X_{nj} \to 0 \quad \text{a.s.}$$

We give three examples to show the problems that arise.

(i) First, consider the case $F_n = F$, independent of n. If $X_{nj} = X_{mj}$ whenever $n, m \geq j$ (so that all random variables in any given column are identical), then by the usual strong law of large numbers, $S_n/n \to \mu$ a.s. if the mean of F is μ, but S_n/n diverges a.s. if the mean of F does not exist. Without this special row structure, S_n/n may diverge a.s. when F has a mean. Before we give an example, note that if F has mean zero and a fourth finite moment μ_4, then we easily see that $S_n/n \to 0$ a.s. regardless of the row structure. Indeed,

$$\sum_{n=1}^{\infty} P\left(\left|\frac{S_n}{n}\right| > \varepsilon\right) \leq \sum_{n=1}^{\infty} \frac{E|S_n/n|^4}{\varepsilon^4} < \infty$$

so by the Borel-Cantelli Lemma the result follows.

For the counterexample, let F be the symmetric distribution such that $P(|X| > t) = 1/t^2$ if $t > 1$. Note that F has mean zero. Let all variables X_{nj} be independent and identically distributed from F so that the rows are independent. We make use of the following fact: Whenever $Y_1, \ldots, Y_n$ are independent and identically distributed according to a symmetric distribution, then

$$P\left(\sum_{j=1}^{n} |Y_j| > t\right) \geq P\left(\max_{1\leq j\leq n} |Y_j| > t\right).$$

This may be found in Feller (1971, Section V.5). Hence,

$$\sum_{n=1}^{\infty} P\left(\frac{|S_n|}{n} > 1\right) \geq \sum_{n=1}^{\infty} P\left(\max_{1\leq j\leq n} |X_{nj}| > n\right)$$

$$= \sum_{n=1}^{\infty}\left[1 - \left(1 - \frac{1}{n^2}\right)^n\right]$$

$$= \sum_{n=1}^{\infty}\left\{1 - \exp\left[n \log\left(1 - \frac{1}{n^2}\right)\right]\right\}$$

$$\geq \sum_{n=1}^{\infty}\left[-n \log\left(1 - \frac{1}{n^2}\right) + n^2 \log^2\left(1 - \frac{1}{n^2}\right)\right]$$

$$\geq \sum_{n=1}^{\infty}\left[-n\left(-\frac{1}{n^2}+\frac{1}{n^4}\right)+n^2\left(-\frac{1}{n^2}+\frac{1}{n^4}\right)^2\right]$$

$$=\sum_{n=1}^{\infty}\frac{1}{n}+0\left(\frac{1}{n^2}\right)=\infty.$$

By the Borel-Cantelli Lemma, $P(|S_n|/n$ infinitely often$) = 1$. However, since F has mean zero, by the usual weak law of large numbers, $S_n/n \xrightarrow{P} 0$, so that S_n/n diverges almost surely.

(ii) In the preceding example, F had moments of order p whenever $p < 2$. Here we give an example where F_n has mean zero and variance one. Let the nth row consist of n independent and identically distributed variables that take on the values n, 0, and $-n$ with probabilities $1/(2n^2)$, $1 - 1/n^2$, and $1/(2n^2)$, respectively, and again let the rows be independent. Note by Chebyshev's Inequality that $P(|S_n|/n > \varepsilon) \leq 1/(n\varepsilon^2) \to 0$ so that $S_n/n \xrightarrow{P} 0$. However,

$$\sum_n P\left(\frac{|S_n|}{n} > \frac{1}{2}\right) \geq \sum P\begin{bmatrix}\text{exactly one } X_{nj} \neq 0 \text{ and the} \\ \text{other } n-1 \text{ variables are zero}\end{bmatrix}$$

$$=\sum_n n\left(\frac{1}{n^2}\right)\left(1-\frac{1}{n^2}\right)^{n-1}$$

$$=\sum_n\left(\frac{1}{n}\right)\left(1-\frac{1}{n^2}\right)^{n-1}=\infty.$$

Thus, $P(|S_n|/n > \frac{1}{2}$ infinitely often$) = 1$ and so S_n/n does not converge almost surely to zero. Note that F_n converges weakly to the degenerate distribution, δ_0, with all mass at zero.

(iii) Without the existence of a bounded second moment, we may not even have $S_n/n \xrightarrow{P} 0$. Indeed, suppose the nth row consists of n independent Gaussian random variables with mean zero and variance n. Then $P(|S_n|/n > \varepsilon) = P(N(0, 1) > \varepsilon)$, which does not tend to zero.

5.42. A limit in distribution of partial sums of independent random variables that is not infinitely divisible.

If the sequence of partial sums of a sequence of independent random variables has a limit in distribution, there is no reason to expect that limit to be infinitely divisible. For example, let X_1 be Bernoulli with probability $\frac{1}{2}$ (so that X_1 takes on the values zero and one with equal probability); in fact, any distribution that is not infinitely divisible would suffice for X_1. Then choose all the others to be degenerate at zero; that is, $X_2 = X_3 = \cdots = 0$. Then the partial sums are always X_1 and so converge in distribution to X_1, which is not infinitely divisible.

5.43. A sequence of independent and identically distributed random variables where the Central Limit Theorem does not hold.

Let the independent random variables $X_1, X_2, \ldots$ each have the Cauchy distribution. In this case, the sample average is not asymptotically Gaussian, even though the limiting distribution exists, because it was shown in Example 4.5 that the sample average $\bar{X}_n$ actually has the same Cauchy distribution as any of the X_i. Thus, $\bar{X}_n$ is asymptotically Cauchy distributed. The problem, in this case, is that Cauchy-distributed random variables do not have a mean value. Thus, instead of $n^{1/2}\bar{X}_n$ converging in distribution to a Gaussian distribution, it is $\bar{X}_n$ itself that has a limiting Cauchy distribution.

5.44. A sequence of independent and identically distributed random variables with finite mean and finite variance whose sequence of sample averages, standardized by subtracting the mean and dividing by the standard deviation of the sample average, is not asymptotically Gaussian.

The point of this surprising but trivial example is that we cannot ignore the requirement that $\mathrm{Var}[X_i] \neq 0$, although some very respectable texts fail to include this requirement in their statements of the Central Limit Theorem. Let X_i be identically zero for each i. Then the limiting distribution of $\bar{X}_n$ is the zero distribution, and standardization cannot be performed properly. In particular, with $\mu = \sigma = 0$, we have a finite mean and variance but

$$n^{1/2}\frac{\bar{X}_n - \mu}{\sigma} \quad \text{does not converge in law to} \quad N(0, 1)$$

because this expression involves division by zero and is therefore undefined.

5.45. A sequence of identically distributed, pairwise independent random variables with finite nonzero variance for which the Central Limit Theorem does not hold.*

This example shows that the condition of independence cannot be relaxed to pairwise independence (even over all possible pairs from the sequence) in the Central Limit Theorem. Begin with the sequence $U_1, U_2, \ldots,$ which is

* The possibility of such an example was suggested by P. Diaconis.

independent and identically distributed, with each term taking the values -1 and 1 with probability $\frac{1}{2}$. Then construct a new sequence as follows:

$$X_1 = U_1$$
$$X_2 = U_2 X_1$$
$$X_3 = U_3 X_1, \qquad X_4 = U_3 X_2$$
$$X_5 = U_4 X_1, \qquad X_6 = U_4 X_2, \qquad X_7 = U_4 X_3, \qquad X_8 = U_4 X_4$$
$$\vdots$$

Note that this sequence is pairwise independent because each term is a distinct product of the form $U_{i_1} \cdot \cdots \cdot U_{i_k}$ and may also be seen to be identically distributed from this fact. The sequence of partial sums behaves in a peculiar way, however. The 2^nth partial sum is

$$S_{2^n} = X_1 + \cdots + X_{2^n} = U_1(1 + U_2)(1 + U_3) \cdots (1 + U_{n+1})$$
$$= \begin{cases} 0 & \text{with probability } 1 - 2^{-n} \\ 2^n & \text{with probability } 2^{-(n+1)} \\ -2^n & \text{with probability } 2^{-(n+1)}. \end{cases}$$

This follows because each independent term of the form $(1 + U_i)$ is zero with probability $\frac{1}{2}$. Thus a subsequence of the normalized partial sums (which would tend to a standard Gaussian if the Central Limit Theorem holds) tends instead in distribution to zero:

$$\frac{S_{2^n}}{\sigma_{(S_{2^n})}} = \frac{S_{2^n}}{2^{n/2}} \to 0 \qquad \text{in distribution.}$$

5.46. A triangular array that satisfies the Lindeberg condition but does not satisfy any Lyapounov condition.

Partial sums of an independent and identically distributed sequence of random variables represent a special case of a triangular array. Consider the sequence of random variables $X_1, X_2, \ldots$ that are independent and each has density function

$$f(x) = \frac{c}{x^3(\log x)^2} I_{(e,\infty)}(x)$$

where c is the constant that causes f to integrate to one and be a density.

It may be verified directly that the first two moments are nonzero and finite.

To see that the Lyapounov condition is not satisfied for any $\delta > 0$, observe that

$$E[|X_1|^{2+\delta}] = \int_e^{\infty} \frac{c}{x^{1-\delta}(\log x)^2}\,dx.$$

Changing variables to $u = \log x$, we find that this becomes

$$\int_e^{\infty} \frac{cx^{\delta}}{(\log x)^2}\,d(\log x) = \int_0^{\infty} \frac{ce^{\delta u}}{u^2}\,du = \infty.$$

Because this is infinite, the Lyapounov condition cannot hold in this case.

To verify that the Lindeberg condition is indeed satisfied in this example, we need only observe that it is always satisfied in an independent and identically distributed situation with finite mean and finite nonzero variance such as we have here (Billingsley 1979, p. 310).

5.47. A sequence of independent and identically distributed random variables with finite moments of all positive real orders less than two, but whose partial sums do not converge to a Gaussian distribution when normalized by subtracting the expectation and dividing by the square root of the sample size.

The point of this example is that we cannot in general extend the Central Limit Theorem to independent and identically distributed random variables without a second moment (but see Example 5.48).

Suppose $X_1, X_2, \ldots$ are independent with a common symmetric distribution whose density is

$$f(x) = \frac{1}{|x|^3} I(|x| \geq 1).$$

Note that the absolute moment of order $1 + \delta$, where $\delta < 1$, is finite because

$$E[|X|^{1+\delta}] = 2\int_1^{\infty} |x|^{\delta-2}\,dx$$

which is finite if and only if $\delta < 1$. Because $f(x)$ is an even function,

$E[X_i] = 0$. We will actually show that

$$\frac{X_1 + \cdots + X_n}{(n \log(n))^{1/2}} \xrightarrow{d} N(0, 1).$$

To this end, let $Y_{nk} = X_k I(|X_k| \le n^{1/2})$. Also let $S_n = X_1 + \cdots + X_n$ and $S'_n = Y_{n1} + \cdots + Y_{nn}$. First, we will find the asymptotic distribution of S'_n, which will require a simple application of Lindeberg's Central Limit Theorem. Note that $E[Y_{nk}] = 0$ and

$$\operatorname{Var}[Y_{nk}] = 2\int_1^{n^{1/2}} y^2 f(y)\,dy = 2\log(n^{1/2}) = \log n.$$

Therefore, $\operatorname{Var}[S'_n] = n \log n$. To verify Lindeberg's condition for the triangular array Y_{nk}, we must show that for any $\varepsilon > 0$,

$$\sum_{k=1}^{n} \frac{1}{\operatorname{Var}[S'_n]} E\{Y_{nk}^2 I(Y_{nk}^2 > \varepsilon \operatorname{Var}[S'_n])\} \to 0 \qquad \text{as } n \to \infty.$$

To see this, the left-hand side becomes

$$\frac{n}{\operatorname{Var}[S'_n]} E\{Y_{nk}^2 I(Y_{nk}^2 > \varepsilon \operatorname{Var}[S'_n]\} = \frac{1}{\log n} \int_{\{y\,:\,y^2 > \varepsilon n \log n \text{ and } |y| \le n^{1/2}\}} y^2 f(y)\,dy.$$

But for any $\varepsilon > 0$, for all large enough n, the set of y such that $y^2 > \varepsilon n \log n$ and $|y| \le n^{1/2}$ is empty! Therefore, for large enough n, the integral vanishes, verifying Lindeberg's condition. It follows that

$$\frac{S'_n}{(n \log n)^{1/2}} \xrightarrow{d} N(0, 1).$$

Now to translate this result back to the original sequence, note that

$$\frac{S_n}{(n \log n)^{1/2}} = \frac{S'_n}{(n \log n)^{1/2}} + \frac{S_n - S'_n}{(n \log n)^{1/2}},$$

so to show that $S_n/(n \log n)^{1/2} \xrightarrow{d} N(0, 1)$, by Slutsky's Theorem it suffices to show that $(S_n - S'_n)/(n \log n)^{1/2} \xrightarrow{P} 0$. But

$$\frac{S_n - S'_n}{(n \log n)^{1/2}} = \frac{\sum_{k=1}^{n} X_k I(|X_k| > n^{1/2})}{(n \log n)^{1/2}}.$$

When Markov's Inequality is applied, $P\{|X| > a\} \leq E[|X|]/a$ results in

$$P\left\{\frac{|S_n - S_n'|}{(n \log n)^{1/2}} > \varepsilon\right\} = P\left\{\left|\sum_{k=1}^{n} X_k I(|X_k| > n^{1/2})\right| > \varepsilon(n \log n)^{1/2}\right\}$$

$$\leq \frac{E\left[\left|\sum_{k=1}^{n} X_k I(|X_k| > n^{1/2})\right|\right]}{\varepsilon(n \log n)^{1/2}}$$

$$\leq \frac{\sum_{k=1}^{n} E[|X_k| I(|X_k| > n^{1/2})]}{\varepsilon(n \log n)^{1/2}}$$

$$= \frac{2n^{1/2}}{\varepsilon(\log n)^{1/2}} \int_{n^{1/2}}^{\infty} |x| f(x)\, dx$$

$$= \frac{2n^{1/2}}{\varepsilon(\log n)^{1/2}} \int_{n^{1/2}}^{\infty} \frac{1}{x^2}\, dx$$

$$= \frac{2}{\varepsilon(\log n)^{1/2}} \to 0 \qquad \text{as } n \to \infty.$$

Therefore $(S_n - S_n') \xrightarrow{P} 0$, which yields the result:

$$\frac{S_n}{(n \log n)^{1/2}} \xrightarrow{d} N(0, 1).$$

5.48. A sequence of independent and identically distributed random variables with infinite variance whose sequence of partial averages is nonetheless asymptotically Gaussian.

The point of this example is that when second moments do not exist, we cannot get a Gaussian limit for the asymptotic distribution of partial averages of independent and identically distributed random variables with normalization by the square root of n. However, we may be able to obtain a Gaussian limit with a different normalizing factor.

In fact, Example 5.47 exhibited an independent and identically distributed sequence $X_1, X_2, \ldots$ where their density was

$$f(x) = \frac{1}{|x|^3} I(|x| \geq 1).$$

The variance of X_i is infinite, but nonetheless it was shown that

$$\frac{\sum_{k=1}^{n} X_k}{(n \log n)^{1/2}} \xrightarrow{d} N(0, 1).$$

5.49. A triangular array, where in each row the variables are independent and identically distributed with mean zero and variance one, but whose normalized partial sums do not converge to a Gaussian distribution.

To construct such an example, the distributions in each row must be different. For each n, let X_{nk} (where $k = 1, 2, \ldots, n$) be independent random variables such that

$$X_{nk} = \begin{cases} n & \text{with probability } \dfrac{1}{2n^2} \\ -n & \text{with probability } \dfrac{1}{2n^2} \\ 0 & \text{with probability } 1 - \dfrac{1}{n^2}. \end{cases}$$

Then $E(X_{nk}) = 0$ and $\mathrm{Var}(X_{nk}) = 1$. To see that $S_n = \Sigma_{k=1}^{n} X_{nk}$ cannot converge to a Gaussian distribution, it is enough to observe that

$$\begin{aligned} P(S_n = 0) &\geq P(X_{nk} = 0 \quad \text{for } k = 1, \ldots, n) \\ &= \left(1 - \frac{1}{n^2}\right)^n \to 1 \qquad \text{as } n \to \infty. \end{aligned}$$

Note that, although $F_n = \mathbf{L}(X_{nk})$ has mean zero and variance one and F_n converges weakly to F_∞, the distribution degenerate at one, $\mathrm{Var}(F_n) = 1$ does not converge to $\mathrm{Var}(F_\infty) = 0$. In fact, if (for each n) X_{nk} are independent and identically distributed from F_n (for $k = 1, \ldots, n$) and $\mathbf{L}(X_\infty) = F_\infty$ such that $E(X_{n1}) \to E(X_\infty)$, $F_n = F_\infty$, *and* $\mathrm{Var}(X_{n1}) \to \mathrm{Var}(X_\infty)$, then Lindeberg's condition is satisfied so that S_n normalized will have a Gaussian limiting distribution.

A triangular array of binomials suitably normalized so that the sequence of partial sums converges to the Poisson distribution also provides a counterexample.

5.50. A triangular array, each row consisting of pairwise independent identically distributed Bernoulli random variables such that the expected number of successes in each row is one, but the limiting distribution is not the Poisson distribution with mean one.

In general, if X_{nj} is a Bernoulli random variable that takes the values one and zero with probabilities p_{nj} and $1 - p_{nj}$, respectively, such that $X_{n1}, \ldots, X_{nk_n}$ are independent for each fixed n, then $S_n = X_{n1} + \cdots + X_{nk_n}$ converges weakly to a Poisson distribution with mean λ if and only if (i) $\max_{1 \le j \le k_n} p_{nj} \to 0$ as $n \to \infty$, and (ii) $p_{n1} + \cdots + p_{nk_n} \to \lambda$ as $n \to \infty$. This is a well-known theorem from Koopman (1950). Many authors have shown that a limiting Poisson distribution persists under mild dependency. This counterexample shows that no extension is possible if we assume the variables are pairwise independent with conditions (i) and (ii) holding as well.

To construct the variables for the nth row, we use Feller's (1959) construction of a sequence of n random variables that are pairwise independent but not mutually independent. Fix n. Let T_1 be the sample space consisting of the $n!$ permutations of $(1, 2, \ldots, n)$, and let T_2 be the sample space consisting of the n points $(j, \ldots, j)$ for $1 \le j \le n$. Now consider T, the mixture of T_1 with T_2 mixed with weights $1 - 1/n$ for T_1 and $1/n$ for T_2. Draw a random n-tuple $X = (X_1, \ldots, X_n)$ from T with these weights. Define, for $1 \le j \le n$,

$$X_{nj} = \begin{cases} 1 & \text{if } X_j = 1 \\ 0 & \text{otherwise.} \end{cases}$$

Then $p_{nj} = P(X_{nj} = 1) = 1/n$. So conditions (i) and (ii) are clearly satisfied with $\lambda = 1$. A simple calculation verifies that the X_{nj} for $1 \le j \le n$ are pairwise independent. However, the sum S_n does not converge to a Poisson distribution with mean one. Indeed, if $k > 1$, then

$$P(S_n = k) \le P\{X = (1, 1, \ldots, 1)\}$$

$$= \frac{1}{n^2} \to 0.$$

5.51. A sequence of independent random variables whose partial sums, when standardized to have mean zero and variance one, converge to a Gaussian distribution but with variance less than one.

Let $Y_1, Y_2, \ldots$ be a sequence of independent and identically distributed standard Gaussian random variables. Independent of these, let $Z_1, Z_2, \ldots$ be a sequence of independent random variables given by

$$Z_i = \begin{cases} c & \text{with probability } \dfrac{1}{2i^2} \\ 0 & \text{with probability } 1 - \dfrac{1}{i^2} \\ -c & \text{with probability } \dfrac{1}{2i^2}. \end{cases}$$

Define $X_i = Y_i + Z_i$ and $S_n = X_1 + \cdots + X_n$. Note that $E[X_i] = 0$ and so $E[S_n] = 0$. Also note that $\text{Var}[X_i] = 1 + \text{Var}[Z_i] = 1 + c^2/i^2$ so that $\text{Var}[S_n] = n + \Sigma_{i=1}^n c^2/i^2$. Because $\Sigma_{i=1}^\infty P\{Z_i \neq 0\} = \Sigma_{i=1}^\infty 1/i^2 < \infty$, by the Borel-Cantelli Lemma $\Sigma_{i=1}^\infty Z_i < \infty$ almost surely. Because $\text{Var}[S_n] \to \infty$, we have $\Sigma_{i=1}^n Z_i/\{\text{Var}[S_n]\}^{1/2} \xrightarrow{P} 0$. By Slutsky's Theorem, the asymptotic distribution of $S_n/\{\text{Var}[S_n]\}^{1/2}$ is therefore the same as the asymptotic distribution of $(Y_1 + \cdots + Y_n)/\{\text{Var}[S_n]\}^{1/2}$. But $(Y_1 + \cdots + Y_n)/\{\text{Var}[S_n]\}^{1/2} \sim N(0, n/\text{Var}[S_n])$. Define σ to be the limit of this variance, so that $n/\text{Var}[S_n] \to 1/\{1 + c^2\Sigma_{i=1}^\infty 1/i^2\} = \sigma$. Note that σ may take on any value between zero and one, depending on the value of c. Choosing a positive value of c, we see that

$$\frac{S_n - E[S_n]}{\{\text{Var}[S_n]\}^{1/2}} \xrightarrow{d} N(0, \sigma^2) \qquad \text{where } \sigma^2 < 1.$$

5.52. A precise formulation of the uniformly asymptotically negligible condition is important (from Chung 1974, problem 7.1.1).

Given a triangular array $\{X_{nj} : 1 \leq j \leq K_n\}$ for which $K_n \to \infty$ as $n \to \infty$, four conditions might hold for every $\varepsilon > 0$:

(a) $\lim_{n\to\infty} P\{|X_{nj}| > \varepsilon\} = 0$ for all j.

(b) $\lim_{n\to\infty} \max_{1\leq j\leq K_n} P\{|X_{nj}| > \varepsilon\} = 0,$ the UAN condition.

(c) $\lim_{n\to\infty} P\left\{\max_{1\leq j\leq K_n} |X_{nj}| > \varepsilon\right\} = 0.$

(d) $\lim_{n\to\infty} \sum_{j=1}^{K_n} P\{|X_{nj}| > \varepsilon\} = 0.$

In fact, **(d)** implies **(c)** implies **(b)** implies **(a)**. All of these seem to represent the idea that no X_{nj} in the sum of its row can contribute too much toward the limiting distribution of S_n, although only condition **(b)** is "right."

We will consider the case $K_n = n$ in the following examples. To see that **(a)** does not imply **(b)**, let X_{nj} be independent degenerate random variables, with $X_{nj} = j/n$. Then **(a)** holds because for each fixed j the terms in the limit are eventually zero. However, **(b)** fails because the maximum over all j is one if $\varepsilon < 1$.

To see that **(b)** does not imply **(c)**, take X_{nj} independent as one with probability p_n and zero with probability $1 - p_n$. Then **(b)** holds if $p_n \to 0$ as $n \to \infty$—say, $p_n = 1/n$. Then **(c)** fails because if $0 < \varepsilon < 1$, then

$$P\{\max|X_{nk}| > \varepsilon\} = 1 - \left(1 - \frac{1}{n}\right)^n \to 1 - \frac{1}{e} \neq 0.$$

To see that **(c)** does not imply **(d)**, let $X_{nj} = X_{n1}$ $(j = 1, \ldots, n)$ where X_{n1} is one with probability p_n and zero with probability $1 - p_n$. Then **(c)** holds if $p_n \to 0$, but **(d)** fails if np_n does not tend to zero—for example, if $p_n = 1/n^{1/2}$ so that $np_n \to \infty$. However, Chung states that **(c)** and **(d)** are equivalent if the X_{nj} are independent in each row.

5.53. A uniformly asymptotically negligible triangular array of independent random variables whose normalized row sums converge weakly to a Gaussian distribution, but for which the Lindeberg condition fails to hold.

The point of this example is that in the "converse" to Lindeberg's Central Limit Theorem, we must assume that the variances of the row sums are finite and nonzero.

As in Example 5.52, let $X_1, X_2, \ldots$ be independent with common density $f(x) = 1/|x|^3 I(|x| \geq 1)$. Consider the triangular array $X_{ij} = X_j$ (where $1 \leq j \leq i$). Then, as we saw in Example 5.48,

$$\mathbf{L}\left[\frac{S_n}{(n \log n)^{1/2}}\right] \to N(0, 1) \qquad \text{as } n \to \infty,$$

although Lindeberg's condition fails because second moments do not even exist. The triangular array X_{ij} is uniformly asymptotically negligible, however. To see this, observe that

$$\max_{1 \leq j \leq n} P\left\{\frac{|X_j|}{(n \log n)^{1/2}} > \varepsilon\right\} = P\left\{\frac{|X_1|}{(n \log n)^{1/2}} > \varepsilon\right\}$$

because the random variables are identically distributed. This last probability, for large n, is

$$\frac{1}{\varepsilon^2 n \log n} \to 0 \qquad \text{as } n \to \infty,$$

which shows that the triangular array is uniformly asymptotically negligible.

5.54. A sequence of random variables, a sequence of normalizing constants, and a continuous function, such that the normalized sequence is asymptotically standard Gaussian, but applying the same normalization to the function values does not result in convergence to a Gaussian distribution.

As mentioned in the introduction to this chapter, if $a_n \to \infty$ and $\mathbf{L}[a_n(X_n - b)] \to \mathbf{L}(X)$, then $\mathbf{L}\{a_n[g(X_n) - g(b)]\} \to \mathbf{L}[g'(b)X]$, provided g is differentiable at b. This example shows that we cannot remove the hypothesis of differentiability.

Let X_n have a Gaussian distribution $N(0, 1/n)$. Set $b = 0$ and $a_n = n^{1/2}$. Then $\mathbf{L}[a_n(X_n - b)] = N(0, 1)$, the standard Gaussian distribution. If $g(t) = |t|$, then g is continuous. To see that $\mathbf{L}\{a_n[g(X_n) - g(b)]\}$ does not have a limiting Gaussian distribution, observe that

$$a_n[g(X_n) - g(b)] = n^{1/2}|X_n|,$$

which has the same distribution for all values of n and therefore converges in distribution to $|X_1|$ where $X_1 \sim N(0, 1)$. This limiting distribution cannot be Gaussian because, in particular, it can never be negative. It is, in fact, the square root of a chi-squared distribution with one degree of freedom and is therefore a chi distribution, and we have

$$a_n[g(X_n) - g(b)] \xrightarrow{\text{d}} \chi_1$$

instead of convergence to a Gaussian distribution.

5.55. A triangular array of random variables with independent random variables in each row that is not uniformly asymptotically negligible, but whose asymptotic distribution of the row sums is nonetheless standard Gaussian.

This example shows that we can get a Gaussian distribution in the limit for the row sums of independent random variables without assuming the uniformly asymptotically negligible condition. Of course, if uniformly asymptotic negligibility does not hold, then neither does the Lindeberg condition.

Consider a triangular array $X_{i,j}$, where $1 \leq j \leq n_i$. In the ith row, let the $n_i = i + 1$ random variables be independent with $X_{i,j} \sim N(0, 1)$ if $1 \leq j \leq i$ and $X_{i,i+1} \sim N(0, i)$. If we define $S_i = X_{i,1} + \cdots + X_{i,i+1}$, then $S_i \sim N(0, 2i)$. Therefore, $S_i/(2i)^{1/2} \sim N(0, 1)$ and so we have

$$\frac{S_i}{(2i)^{1/2}} \xrightarrow{\text{d}} N(0, 1).$$

To see that the array is not uniformly asymptotically negligible, however, note that if $\varepsilon > 0$, then

$$\begin{aligned}
\max_{1 \leq j \leq n_i} P\left\{\frac{|X_{i,j}|}{(\text{Var}[S_i])^{1/2}} > \varepsilon\right\} &\geq P\left\{\frac{|X_{i,i+1}|}{(\text{Var}[S_i])^{1/2}} > \varepsilon\right\} \\
&= P\left\{\frac{|X_{i,i+1}|}{(2i)^{1/2}} > \varepsilon\right\} \\
&= P\left\{\frac{|Z|}{2^{1/2}} > \varepsilon\right\} > 0
\end{aligned}$$

where $Z \sim N(0, 1)$, and note that this result holds for all values of i. In particular, this maximum probability does not tend to zero, and the triangular array is therefore not uniformly asymptotically negligible.

CHAPTER SIX
CONDITIONAL EXPECTATION, MARTINGALES, AND ALMOST SURE CONVERGENCE OF SUMS OF INDEPENDENT RANDOM VARIABLES

Introduction

Given a probability space $(\Omega, \mathbf{F}, P)$, let $\mathbf{G}$ be a sub-σ-field of $\mathbf{F}$ and let X be a random variable whose expectation exists (possibly ∞ or $-\infty$). The conditional expectation of X given $\mathbf{G}$, denoted $E[X \mid \mathbf{G}]$, is any $\mathbf{G}$-measurable random variable that satisfies the functional equation

$$\int_G E[X \mid \mathbf{G}](\omega)\, dP(\omega) = \int_G X(\omega)\, dP(\omega) \qquad \text{for all } G \in \mathbf{G}.$$

Conditional expectations exist by the Radon-Nikodym Theorem and are unique up to sets of P measure zero. A particular random variable $E[X \mid \mathbf{G}]$ is called a "version" of the conditional expectation. For a construction, see (for example) Billingsley (1979, Section 34). When we write $E[X \mid Y]$, we mean $E[X \mid \mathbf{G}]$ where $\mathbf{G}$ is the σ-field generated by Y.

The conditional probability $P[B \mid \mathbf{G}]$ of an event B in $\mathbf{F}$ given a sub-σ-field $\mathbf{G}$ is defined to be the conditional expectation of the corresponding indicator random variable, so that $P[B \mid \mathbf{G}] = E[I_B \mid \mathbf{G}]$ by definition.

The following properties are fundamental; equality and inequality are asserted almost surely.

(i) If $c \in \mathbf{R}$, then $E[cX + Y \mid \mathbf{G}] = cE[X \mid \mathbf{G}] + E[Y \mid \mathbf{G}]$.

(ii) If $X \le Y$, then $E[X \mid \mathbf{G}] \le E[Y \mid \mathbf{G}]$.

(iii) If $X_n \to X$ a.s. and $|X_n| \le Y$, where Y is integrable, then $E[X_n \mid \mathbf{G}] \to E[X \mid \mathbf{G}]$ a.s.

(iv) If X is $\mathbf{G}$-measurable, then $E[X \mid \mathbf{G}] = X$.

(v) If X is independent of $\mathbf{G}$, then $E[X \mid \mathbf{G}] = E(X)$.

(vi) If X is $\mathbf{G}$-measurable and both XY and Y are integrable, then $E[XY \mid \mathbf{G}] = XE[Y \mid \mathbf{G}]$.

(vii) If X is integrable and $\mathbf{G}$ is a sub-σ-field of $\mathbf{F}$, then $E[X \mid \mathbf{G}] = E\{E[X \mid \mathbf{F}] \mid \mathbf{G}\}$.

Let $X: (\Omega_1, \mathbf{F}_1, P) \to (\Omega_2, \mathbf{F}_2)$ be a random object and let $\mathbf{G}$ be a sub-σ-field of $\mathbf{F}_1$. A functional Q on $\Omega_1 \times \mathbf{F}_2$ is called a "regular conditional probability" for X given $\mathbf{G}$ if

(i) $Q(\omega, B)$ is a probability measure in B for each fixed $\omega \in \Omega_1$, and

(ii) for each fixed $B \in \mathbf{F}_2$, $Q(\omega, B) = P[X \in B \mid \mathbf{G}](\omega)$ for almost all ω.

One important fact is that if X is a real random variable on $(\Omega_1, \mathbf{F}_1, P)$, so that Ω_2 is the real line and $\mathbf{F}_2$ is the class of Borel sets, and $\mathbf{G}$ is a sub-σ-field of $\mathbf{F}_1$, then a regular conditional probability for X given $\mathbf{G}$ always exists. For a proof, see Ash (1972, Section 6.6).

Let $(\Omega, \mathbf{F}, P)$ be a probability space and let N be a subset of $\{-\infty, \ldots, -2, -1, 0, 1, 2, \ldots, \infty\}$. Let $\{\mathbf{F}_n, n \in N\}$ be a collection of sub-σ-fields of $\mathbf{F}$ with indices in N such that $\mathbf{F}_m$ is a subset of $\mathbf{F}_n$ whenever $m < n$, where $m \in N$ and $n \in N$. If $n \in N$, let X_n be a random variable defined on $(\Omega, \mathbf{F}, P)$ that is measurable $\mathbf{F}_n$. Then the sequence $\{X_n, \mathbf{F}_n, n \in N\}$ is called a "martingale" if for all $m, n \in N$ with $m < n$, we have

$$E[X_n \mid \mathbf{F}_m] = X_m \quad \text{a.s.} \qquad \text{(martingale).}$$

We have a "submartingale" if for all $m, n \in N$ with $m < n$,

$$E[X_n \mid \mathbf{F}_m] \geq X_m \quad \text{a.s.} \qquad \text{(submartingale)}$$

or a "supermartingale" if for all $m, n \in N$ with $m < n$,

$$E[X_n \mid \mathbf{F}_m] \leq X_m \quad \text{a.s.} \qquad \text{(supermartingale).}$$

We will often omit reference to the σ-fields $\mathbf{F}_n$ because if $\{X_n, \mathbf{F}_n, n \in N\}$ is a sub- or supermartingale, then so is $\{X_n, \mathbf{G}_n, n \in N\}$ where $\mathbf{G}_n = \sigma(X_j, j \in N, j \leq n)$ if $n \in N$. We will often consider the case when $N = \{1, 2, \ldots\}$.

The following are standard results about martingales. If $\{X_n, \mathbf{F}_n\}$ is a submartingale and g is an increasing convex function from $\mathbf{R}$ to $\mathbf{R}$, with $g(X_n)$ integrable for all n, then $\{g(X_n), \mathbf{F}_n\}$ is also a submartingale. If $\{X_n, \mathbf{F}_n\}$ is assumed to be a martingale with g a convex function from $\mathbf{R}$ to $\mathbf{R}$ and $g(X_n)$ is integrable for all n, then $\{g(X_n), \mathbf{F}_n\}$ is a submartingale.

If $\mathbf{F}_1, \mathbf{F}_2, \ldots$ is an increasing sequence of σ-fields, then a random variable T with range $\{1, 2, \ldots, \infty\}$ is called a "stopping time" for $\{\mathbf{F}_n\}$ if $\{T \leq n\} \in \mathbf{F}_n$ for all n and $P(T < \infty) = 1$. Intuitively, because a sub-σ-field may represent the information currently available, this measurability criterion says that the decision to stop at time T can be based on only past information and not on unavailable future information contained in the larger σ-fields farther out in the sequence. Let $T_1, T_2, \ldots$ be an increasing sequence of stopping times for $\{\mathbf{F}_n\}$ [so that $T_1(\omega) \leq T_2(\omega) \leq \cdots$ for all $w \in \Omega$], and let $\{X_n, \mathbf{F}_n\}$ be a (sub)martingale. Set $Y_n(\omega) = X_{T_n(\omega)}(\omega)$. The Optional Sampling Theorem gives sufficient conditions for $\{Y_n\}$ to be a (sub)martingale. If

(1) $E(|Y_n|) < \infty$ for all n, and

(2) $\underline{\lim}_{k\to\infty} \int_{\{T_n > k\}} |X_k| \, dP = 0$ for all n,

then $\{Y_n\}$ is a (sub)martingale. Furthermore, $E(X_1) = E(Y_n)$ for all n in the martingale case.

If $\{(X_n, \mathbf{F}_n) : n = 1, 2, \ldots, \infty\}$ is a (sub)martingale, where $\mathbf{F}_\infty$ is any sub-σ-field of $\mathbf{F}$ that contains all the $\mathbf{F}_n$, then X_∞ is called a "last element." Let $(X_1, \mathbf{F}_1), (X_2, \mathbf{F}_2), \ldots$ be a submartingale. If $\sup_n E[|X_n|] < \infty$, then there is an integrable random variable X_∞ such that $X_n \to X_\infty$ a.s. (Ash 1972, Theorem 7.4.3, p. 292). The convergence is in L^1 if and only if the X_n are uniformly integrable—that is, if

$$\sup_n \int_{\{|X_n| > c\}} |X_n| \, dP \to 0 \qquad \text{as } c \to \infty.$$

In this case, $X_n = E[X_\infty \mid \mathbf{F}_n]$, and if we define $\mathbf{F}_\infty = \sigma(\mathbf{F}_1 \cup \mathbf{F}_2 \cup \cdots)$, then X_∞ is a last element (Ash 1972, Theorem 7.6.6, p. 300).

An important special case of a martingale arises when $Y_1, Y_2, \ldots$ is a sequence of independent random variables with mean zero. Setting $S_n = Y_1 + \cdots + Y_n$, we find that the sequence of partial sums $\{S_n\}$ is a martingale.

A general criterion for the convergence of sums of independent random variables is given by the Kolmogorov Three Series Theorem. Let $Y_1, Y_2, \ldots$ be independent random variables. The sequence of partial sums $S_n = Y_1 + \cdots + Y_n$ converges almost surely if and only if

(a) $\sum_{n=1}^{\infty} P\{|Y_n| > 1\} < \infty.$

(b) $\sum_{n=1}^{\infty} E[Y_n I(|Y_n| \leq 1)]$ converges.

(c) $\sum_{n=1}^{\infty} \operatorname{Var}[Y_n I(|Y_n| \leq 1)] < \infty.$

For a proof, see Chow and Teicher (1978, p. 114).

Kolmogorov's strong law of large numbers gives a sufficient criterion for the convergence of S_n, suitably normalized, to zero almost surely: Let Y_1, $Y_2, \ldots$ be independent random variables with finite means and finite variances; let $S_n = Y_1 + \cdots + Y_n$ be the partial sums; and let $b_1, b_2, \ldots$ be an increasing sequence of positive constants that tends to infinity. The result is:

$$\text{If} \quad \sum \frac{\operatorname{Var}(Y_n)}{b_n^2} < \infty \quad \text{then} \quad \frac{S_n - E(S_n)}{b_n} \to 0 \quad \text{a.s.}$$

For a proof, see Ash (1972, p. 274).

6.1. Random variables X, Y, and Z such that X is independent of Y and of Z, $E(X \mid Y) = E(X \mid Z) = \frac{1}{2}$ is degenerate and hence independent of Y and of Z, but $E[X \mid (Y, Z)]$ is not independent of (Y, Z).

Let Y and Z be independent random variables that each take on the values 1 and -1 with probability $\frac{1}{2}$. Let X be defined by

$$X = \begin{cases} 1 & \text{if } Y + Z = 0 \\ 0 & \text{otherwise.} \end{cases}$$

Then X is independent of Y. Also, X is independent of Z. Because $P(X = 0) = P(X = 1) = \frac{1}{2}$, it follows that

$$E(X \mid Y) = E(X \mid Z) = \tfrac{1}{2}.$$

On the other hand, we find that

$$E[X \mid (Y, Z)] = I(Y + Z = 0) = X,$$

which is not degenerate.

6.2. Random variables such that Z is independent of X, Z is independent of Y, but $E[X \mid (Y, Z)] \neq E(X \mid Y)$ almost surely.

The conclusion would turn into an equality if we could assume that Z is independent of (X, Y) (see Breiman 1968, Problem 14, p. 76). However, it is false under the weaker assumption.

Take X, Y, and Z as in Example 6.1. It was shown that $E(X \mid Y) = \frac{1}{2}$

almost surely and $E[X \mid (Y, Z)] = X$ almost surely. Since X is not degenerate, the result follows.

6.3. $E(Y \mid X) = E(Y)$ need not imply that X and Y are independent, and $\operatorname{Cov}(X, Y) = 0$ need not imply that $E(Y \mid X) = E(Y)$ almost surely.

Consider the following three statements for integrable random variables X and Y defined on a common probability space:

1. X and Y are independent.
2. $E(Y \mid X) = E(Y)$ almost surely.
3. $\operatorname{Cov}(X, Y) = 0$.

In general, statement 1 implies 2, which implies 3. Example 4.11 shows that statement 3 need not imply 1. Here we show that statement 3 need not imply 2, and that 2 need not imply 1.

To see that statement 3 need not imply 2, let X be uniformly distributed on the interval $(-1, 1)$ and set $Y = X^2$. Then

$$\operatorname{Cov}(X, Y) = E(X^3) - E(X) \cdot E(X^2) = 0.$$

However, $E(Y \mid X) = E(X^2 \mid X) = X^2 = Y$ almost surely. Since Y is not degenerate, $E(Y)$ is not a version of the conditional expectation of Y given X.

To see that statement 2 need not imply statement 1, suppose (X, Y) is supported on the three points in the plane $(0, 0)$, $(0, 2)$, and $(1, 1)$, each with equal probability $\frac{1}{3}$. Then the marginal probabilities for Y are $P(Y = y) = \frac{1}{3}$ for $y = 0$, 1, and 2. Therefore, $E(Y) = 1$; however,

$$P(Y = y \mid X = 0) = \begin{cases} \frac{1}{2} & \text{if } y = 0 \text{ or } y = 2 \\ 0 & \text{otherwise.} \end{cases}$$

Therefore it follows that $E(Y \mid X = 0) = 1$. Because $X = 1$ implies $Y = 1$, we also have $E(Y \mid X = 1) = 1$. Since X is supported on the points zero and one, we have

$$E(Y \mid X) = 1 = E(Y).$$

However, X and Y are clearly not independent because, for example,

$$P(X = 0 \text{ and } Y = 1) = 0 \neq \tfrac{2}{9} = (\tfrac{2}{3})(\tfrac{1}{3}) = P(X = 0)P(Y = 1).$$

Indeed, whenever the support of (X, Y) is not a rectangle, X and Y are independent.

6.4. Random variables X and Y such that $E(Y \mid X)$ exists, but Y is not integrable.

The point of this example is that we do not need Y to be integrable in order to define $E(Y \mid \mathbf{G})$. In our definition, we assumed only that the expectation of Y exists, even though many texts do not point out this extension.

First, we have some standard terminology. If X and Y have a joint discrete probability function $f(x, y)$ on the integer coordinates in the plane so that

$$P(X = x \text{ and } Y = y) = f(x, y),$$

then naturally we may define the conditional discrete probability function of Y given $X = x$ as follows:

$$f(y \mid x) = P(Y = y \mid X = x) = \frac{f(x, y)}{f(x)}, \qquad \text{if } f(x) \neq 0,$$

where $f(x)$ is the marginal discrete probability function of X given by $f(x) = \Sigma_y f(x, y)$. If $f(x) = 0$, then we can define $f(y \mid x)$ so that it takes on any arbitrary value. For any fixed x with $f(x) \neq 0$, we can treat $P(Y = y \mid X = x)$ as a discrete probability function and take expectations with respect to this conditional distribution. In other words, it is natural to define the expectation of a function $g(Y)$ conditional on $X = x$ by

$$E[g(Y) \mid X = x] = \sum_y g(y) f(y \mid x)$$

if $f(x) \neq 0$ and arbitrarily otherwise. Of course, this cannot always be done in general. We have avoided any measurability difficulties by considering discrete X and Y, but it suffices for this example. Indeed, the preceding function is a version of $E[g(Y) \mid X]$ consistent with the definition in the beginning of this chapter.

Let (X, Y) have the joint discrete probability function given by

$$f(x, y) = \begin{cases} \dfrac{1}{x(x+1)} & \text{if } x = y = 1, 2, \ldots \\ 0 & \text{otherwise.} \end{cases}$$

Thus, $X = Y$ with probability one. Conditional on $X = x$, with x a positive integer, we find that $f(y \mid x) = 1$ if $y = x$ and $f(y \mid x) = 0$ otherwise. Thus the conditional expectation may be computed as

$$E(Y \mid X = x) = \sum_y y f(y \mid x) = x < \infty.$$

However, since the marginal discrete probability function of Y is $P(Y = y) = 1/[y(y + 1)]$ for $y = 1, 2, \ldots$, we find that

$$E(Y) = \sum_{y=1}^{\infty} y \frac{1}{y(y + 1)} = \infty.$$

Hence, $E(Y \mid X = x)$ exists and is finite for all x that occur with positive probability, but $E(Y)$ is not finite.

It is worthwhile to note that we can always define a regular conditional distribution function for Y given a random variable X. For a proof, see Ash (1972, p. 263). We then may wish to define the conditional expectation of $g(Y)$ given X as the integral of $g(Y)$ with respect to the regular conditional probability distribution. Thus, the preceding example may be easily generalized even if X and Y are not discrete. Indeed, suppose X is uniform on $(0, 1)$ and conditional on $X = x$, let $Y = 1/x$ with probability one. Then

$$E(Y \mid X) = \frac{1}{X} < \infty$$

but

$$E(Y) = \int_0^1 \frac{1}{x} dx = \infty.$$

6.5. A Borel-measurable function f and random variables U, V such that $E[f(U, V) \mid U]$ is almost surely equal to one, but for any fixed u we have $E[f(u, V) \mid U]$ is almost surely equal to zero.

This example is from Freedman (1971, p. 350). Let $U = V$ be uniformly distributed on $[0, 1]$. Set

$$f(u, v) = \begin{cases} 1 & \text{if } u = v \\ 0 & \text{if } u \neq v. \end{cases}$$

Because $f(U, V) = 1$ almost surely, it follows that $E[f(U, V) \mid U] = E[1 \mid U] = 1$ almost surely. However, for any fixed u, we have $f(u, V) = 0$

almost surely, so that

$$E[f(u, V) \mid U] = 0 \quad \text{almost surely.}$$

6.6. Random variables X and Y defined on a probability space $(\Omega, \mathbf{F}, P)$ with $\mathbf{G}$ a sub-σ-field of $\mathbf{F}$ and $X = E(Y \mid \mathbf{G})$ almost surely, but X is not a version of the conditional expectation of Y given $\mathbf{G}$.*

The point of this example is that X need not be $\mathbf{G}$-measurable. Let $\Omega = \{0, 1\}$ and $\mathbf{F} = \{\phi, \Omega, \{0\}, \{1\}\}$ with $P(\{1\}) = 1$. Let $X(\omega) = Y(\omega) = \omega$ and $\mathbf{G} = \{\phi, \Omega\}$. Since $X = Y = 1$ almost surely, we have

$$E(Y \mid \mathbf{G}) = E(1 \mid \mathbf{G}) = 1 = X \quad \text{almost surely.}$$

However, X is not $\mathbf{G}$-measurable because $\{\omega : X(\omega) \leq 0\} = \{0\}$ is not in $\mathbf{G}$. Thus X fails to be a version of the conditional expectation of Y given $\mathbf{G}$.

6.7. Independent random variables X and Y defined on $(\Omega, \mathbf{F}, P)$ and a sub-σ-field $\mathbf{G}$ of $\mathbf{F}$, such that $E(X \mid \mathbf{G})$ and $E(Y \mid \mathbf{G})$ are not independent for any versions of these conditional expectations.**

Let U and V be independent standard Gaussian distributions. Set $X = U + V$ and $Y = U - V$. Then X and Y are Gaussian with mean zero and variance two. Let $\mathbf{G} = \sigma(U)$, the σ-field generated by U. Then $E(X \mid \mathbf{G}) = E(U + V \mid \mathbf{G}) = U + E(V \mid \mathbf{G}) = U$ almost surely and $E(Y \mid \mathbf{G}) = U$ almost surely as well. Thus, in fact, any two versions of $E(X \mid \mathbf{G})$ and $E(Y \mid \mathbf{G})$ are almost surely equal and so cannot be independent (because they are not degenerate).

6.8. Random variables U, V, and W such that there exist versions of $E(W \mid U)$ and $E(W \mid V)$ that are continuous functions of U and V, respectively, the events $\{U = 1\}$ and $\{V = 0\}$ are the same, yet the conditional expectations differ when $U = 1$ and $V = 0$.

This example shows that conditioning with respect to an event of probability zero may lead to an apparent contradiction. It is important to note that a conditional expectation with respect to a sub-σ-field behaves the way we

* This example was suggested by C. Sugahara.

** This example was communicated by C. Sugahara.

expect only when integrated over a set in the sub-σ-field. This example is a slight modification of Billingsley (1979, Problem 33.2). Note that this example can be reformulated in terms of conditional densities.

Let X and Y be independent standard Gaussian random variables and let (R, θ) be their polar coordinate representation:

$$R^2 = X^2 + Y^2 = \left[\frac{(X + Y)^2}{2} + \frac{(X - Y)^2}{2}\right]$$

$$\theta = \tan^{-1}\left(\frac{Y}{X}\right).$$

Set $W = R^2$, $U = Y/X$ (which may possibly be ∞ or $-\infty$), and $V = X - Y$. Note that $X + Y$ and V are independent. Also, $(X + Y)^2/2$ has a chi-squared distribution with one degree of freedom. We compute

$$E(W \mid V) = E\left[\frac{(X + Y)^2}{2}\right] + \frac{V^2}{2} = 1 + \frac{V^2}{2}.$$

On the other hand, R and θ are independent so that R and $\tan(\theta)$ are independent. Also, R^2 is chi-squared with two degrees of freedom. Hence,

$$E(W \mid U) = E(R^2) = 2.$$

In summary, $E(W \mid V) = 1 + V^2/2$, and $E(W \mid U) = 2$. But the events $\{V = 0\}$ and $\{U = 1\}$ are the same, even though $1 \neq 2$.

Note that regular conditional probabilities exist for W given U and for W given V. Indeed, the conditional distribution of W given U is chi-squared with two degrees of freedom and the conditional distribution of W given V is chi-squared with one degree of freedom translated by $V^2/2$. These conditional distributions are different when we naively plug in the values $U = 1$ and $V = 0$.

6.9. A random variable X and a sub-σ-field G containing all one-point sets, yet $E(X \mid \mathrm{G})$ is not almost surely equal to X.

The point of this example is that, although our intuition leads us to interpret sub-σ-fields as representing partial information about a random experiment, such an informal interpretation may break down in certain cases. This example is adapted from Billingsley (1979, pp. 388–89).

Let $(\Omega, \mathbf{F}, P)$ be $[0, 1]$ with the Borel σ-field and uniform (Lebesgue) measure. Let $\mathbf{G}$ be the sub-σ-field consisting of sets that are either countable sets or complements of countable sets. Set $X(\omega) = I_{[0, 1/2]}(\omega)$. Then

$$E(X \mid \mathbf{G}) = E(X) = \tfrac{1}{2} \quad \text{almost surely}$$

because X is independent of $\mathbf{G}$. (This follows because each subset G of $\mathbf{G}$ has probability either zero or one.)

For any $\omega \in \Omega$, however, we have $\{\omega\} \in \mathbf{G}$. Informally, because $\mathbf{G}$ contains all one-point sets, the information $\mathbf{G}$ contains should be enough to know $\{\omega\}$ itself, and so we might expect that $E(X \mid \mathbf{G})$ would be

$$\begin{cases} 1 & \text{if } \omega \text{ is in } A \text{ (i.e., if } 0 \leq \omega \leq \tfrac{1}{2}) \\ 0 & \text{if } \omega \text{ is not in } A \text{ (i.e., if } \tfrac{1}{2} < \omega < 1) \end{cases} = X \quad \text{almost surely.}$$

However, X is not almost surely equal to $\tfrac{1}{2}$. Moreover, X is not $\mathbf{G}$-measurable. Of course, $E(X \mid \mathbf{G}) = \tfrac{1}{2}$ (almost surely) is the correct choice.

6.10. For every z, $T(X, z)$ has a distribution Q and is independent of Y, but $T(X, Z(Y))$ does not have distribution Q and is not independent of Y.

As pointed out in Perlman and Wichura (1975), the following statement is false:

> If for each $z \in \mathbf{Z}$ the random object $T(X, z)$ has distribution Q and is independent of Y, then the random object $T(X, Z(Y))$ also has distribution Q and is independent of Y.

Such statements have been used in multivariate distribution theory—for example, in some derivations of the distribution of Hotelling's T-squared statistic. This counterexample shows that the claim is false in general. It is from Perlman and Wichura (1975), who also give additional assumptions under which the statement is true.

Let $X = Y = Z(Y)$ have a uniform distribution on $[0, 1]$, and set

$$T(x, z) = zI_{\{z\}}(x).$$

Since $T(X, z) = 0$ almost surely, $T(X, z)$ is independent of Y for each fixed z. However, $T(X, Z(Y)) = X$, which is not degenerate at zero but is uniformly distributed on $[0, 1]$ and is not independent of Y but is equal to Y with probability one.

6.11. A sequence $X_1, X_2, \ldots$ of random variables and a sub-σ-field B such that $E(X_n \mid \mathbf{B}) \to E(X \mid \mathbf{B})$ in pth mean for all $p \geq 1$, but X_n does not converge in pth mean to X for any $p \geq 1$. Also, a sequence $X_1, X_2, \ldots$ of random variables such that X_n *does* converge in pth mean to X for all $p > 0$, but $E(X_n \mid \mathbf{B})$ does not converge to $E(X \mid \mathbf{B})$ almost surely.

In general, if $X_n \to X$ in L^p for some $p \geq 1$, then $E(X_n \mid \mathbf{B}) \to E(X \mid \mathbf{B})$ in L^p. This follows easily from Jensen's Inequality for conditional expectations. The converse is false. Take $X_1, X_2, \ldots$ independent with $P(X_n = 1) = P(X_n = 0) = \frac{1}{2}$. Set $X = \frac{1}{2}$ (degenerate) and $\mathbf{B} = \{\phi, \Omega\}$. Because $E(X_n \mid \mathbf{B}) = E(X_n) = \frac{1}{2} = E(X \mid \mathbf{B})$ for all n, it follows that $E(X_n \mid \mathbf{B}) \to E(X \mid \mathbf{B})$ in L^p for all $p \geq 1$. However, $E(|X_n - X|^p) = E(|X_n - \frac{1}{2}|^p) = (\frac{1}{2})^p$ for all n. Thus X_n does not converge to X in L^p for any $p \geq 1$.

This example illustrates two points. First, we cannot weaken the assumption of property (3) in the introduction to this chapter from $X_n \to X$ almost surely to $X_n \to X$ in pth mean and $X_n \to X$ in probability. Second, if we are considering the case $\mathbf{B} = \{\phi, \Omega\}$, then the result $E(X_n \mid \mathbf{B}) \to E(X \mid \mathbf{B})$ almost surely follows necessarily under the stated assumptions. Thus, this example shows that the stated assumptions do not apply to an arbitrary σ-field. (Also see Example 6.12.)

For this example, let $X_1, X_2, \ldots$ be independent with $P(X_n = 1) = 1/n$ and $P(X_n = 0) = 1 - 1/n$. Then $|X_n| \leq Y$ where $Y = 1$, and $E(|X_n|^p) = (1/n)^p \to 0$ as $n \to \infty$. Hence $X_n \to 0$ in pth mean for all $p > 0$, and $X_n \to 0$ in probability as well. If $\mathbf{B} = \sigma(X_1, X_2, \ldots)$, however, then $E(X_n \mid \mathbf{B}) = X_n$, which does not converge to zero almost surely. Indeed, $\Sigma\, P(X_n = 1) = \infty$ implies that $P(X_n = 1 \text{ i.o.}) = 1$.

6.12. A sequence $X_1, X_2, \ldots$ of uniformly integrable random variables and a σ-field B such that X_n converges almost surely to X, but $E(X_n \mid \mathbf{B})$ does not converge almost surely to $E(X \mid \mathbf{B})$.

Because the sequence is uniformly integrable, almost sure convergence implies convergence in expectation (see, for example, Ash 1972, Theorem 7.5.2). Convergence of conditional expectations need not follow, however; that is, we may have $X_n \to X$ almost surely and $E(X_n) \to E(X)$, but for some σ-field $\mathbf{B}$, $E(X_n \mid \mathbf{B})$ does not converge almost surely to $E(X \mid \mathbf{B})$. The result is true under slightly stronger assumptions. If, for example, the σ-field (with respect to which the conditional expectations are taken) is generated by a countable partition of the probability space, then the result necessarily follows. For an arbitrary σ-field, the result also follows provided the terms of the original sequence are uniformly bounded in absolute value by an integrable random variable (see, for example, Billingsley 1979, Theorem 34.2). This example shows that the conclusion is false in general. It is interesting to note that $X_n \to X$ in L^1 necessarily implies that $E(X_n \mid \mathbf{B}) \to E(X \mid \mathbf{B})$ in L^1. Even so, almost sure convergence of $E(X_n \mid \mathbf{B})$ to $E(X \mid \mathbf{B})$ need not follow.

Let $Y_1, Y_2, \ldots$ and $Z_1, Z_2, \ldots$ be two independent sequences of independent random variables such that

$$P(Y_n = 1) = \frac{1}{n}, \qquad P(Y_n = 0) = 1 - \frac{1}{n},$$

$$P(Z_n = n^2) = \frac{1}{n^2}, \qquad P(Z_n = 0) = 1 - \frac{1}{n^2}.$$

Define $X_n = Y_n Z_n$. Because $\Sigma\, P(Z_n \neq 0) = \Sigma\, 1/n^2 < \infty$, by the first Borel-Cantelli Lemma we have $Z_n = 0$ eventually, which implies that $X_n = 0$ eventually. Thus $X_n \to X$ almost surely where we define $X = 0$. Also,

$$E(X_n) = E(Y_n)E(Z_n) = \frac{1}{n} \to 0 = E(X) \qquad \text{as } n \to \infty,$$

which implies that the X_n are uniformly integrable.

Now let $\mathbf{B} = \sigma(Y_1, Y_2, \ldots)$ be the σ-field generated by $Y_1, Y_2, \ldots$. The conditional expectation of X_n is

$$\begin{aligned} E(X_n \mid \mathbf{B}) &= E(Y_n Z_n \mid \mathbf{B}) = Y_n E(Z_n \mid \mathbf{B}) \\ &= Y_n E(Z_n) = Y_n. \end{aligned}$$

However, because $\Sigma\, P(Y_n = 1) = \Sigma\, 1/n = \infty$, by the second Borel-Cantelli Lemma, $Y_n = 1$ infinitely often with probability one. Thus $E(X_n \mid \mathbf{B})$ does not converge almost surely to $0 = E(X \mid \mathbf{B})$.

Another way to look at this counterexample is the following: If X_n is a sequence of random variables such that $X_n \to X$ in probability and $|X_n| \leq Y$ where Y is integrable, then $E(X_n) \to E(X)$ as $n \to \infty$. This is a slight extension of the usual dominated convergence theorem. The conclusion $E(X_n \mid \mathbf{B}) \to E(X \mid \mathbf{B})$ (almost surely) is false in general if $\mathbf{B} \neq \{\phi, \Omega\}$. However, strengthening the hypothesis to $X_n \to X$ almost surely and $|X_n| \leq Y$ does yield the conclusion $E(X_n \mid \mathbf{B}) \to E(X \mid \mathbf{B})$ for any sub-σ-field.

6.13. A situation in which a regular conditional probability distribution does not exist.

Recall the definition of a regular-conditional probability distribution given in the introduction to this chapter. If X is a random variable (or, more generally, a random object that takes values in a complete separable metric space equipped with the Borel σ-field) defined on a probability space $(\Omega, \mathbf{F}, P)$, and $\mathbf{G}$ is a sub-σ-field of $\mathbf{F}$, then there always exists a regular conditional probability for X given $\mathbf{G}$. However, a regular conditional probability distribution need not exist in general. The problem stems from the fact that an uncountable union of sets of probability zero need not have probability zero. Indeed, for any sequence $B_1, B_2, \ldots$ of disjoint sets in $\mathbf{F}$, we can choose versions of the conditional probability distributions so that

$$P\left(\bigcup_{n=1}^{\infty} B_n \mid \mathbf{G}\right)(\omega) = \sum_{n=1}^{\infty} P(B_n \mid \mathbf{G})(\omega)$$

except for a set $N = N(B_1, B_2, \ldots)$ of probability zero. Thus, the set of ω such that the set function $P(\cdot \mid \mathbf{G})(\omega)$ is not countably additive is the union of all such N, where the union extends over all sequences $B_1, B_2, \ldots$ of disjoint sets. This is therefore an uncountable union of null sets.

We outline an example given in Ash (1972, Problem 6.6.4). Let $\Omega = [0, 1]$, let μ be Lebesgue measure, and choose a subset H of $[0, 1]$ such that the inner Lebesgue measure of H is zero:

$$\mu_*(H) = \sup\{\mu(B) : B \text{ is a Borel subset of } H\} = 0$$

and the outer Lebesgue measure of H is one:

$$\mu^*(H) = \inf\{\mu(B) : B \text{ is a Borel set containing } H\} = 1.$$

Clearly such a set H is not Lebesgue measurable. The non-Lebesgue-measurable set given in Example 1.29 will suffice, or, alternatively, see Ash (1972, Problem 6.6.3). Define

$$\mathbf{F} = \{(B_1 \cap H) \cup (B_2 \cap H^c) : B_1 \text{ and } B_2 \text{ are in } \mathbf{B}\}$$

where $\mathbf{B}$ denotes the Borel sets in $[0, 1]$. Note that $\mathbf{F}$ contains $\mathbf{B}$. Next define

$$P[(B_1 \cap H) \cup (B_2 \cap H^c)] = \frac{\mu(B_1) + \mu(B_2)}{2}.$$

Taking $B = B_1 = B_2$ and $B \in \mathbf{B}$ shows that $P(B) = \mu(B)$, so that P agrees with μ on **B**. Also note that P is a well-defined set function in the sense that if

$$(B_1 \cap H) \cup (B_2 \cap H^c) = (B_3 \cap H) \cup (B_4 \cap H^c),$$

then

$$P[(B_1 \cap H) \cup (B_2 \cap H^c)] = P[(B_3 \cap H) \cup (B_4 \cap H^c)].$$

Let $X(\omega) = \omega$ and $\mathbf{G} = \mathbf{B}$. We now argue by contradiction. Suppose $Q(\cdot, \cdot)$ is a regular probability for X given $\mathbf{G}$. If $B \in \mathbf{G}$, then

$$P(H \cap B) = \frac{\mu(B)}{2} = \int_B \left(\frac{1}{2}\right) dP.$$

Thus $P(H \mid \mathbf{G}) = \frac{1}{2}$ almost surely. But $P(H \mid \mathbf{G})(\omega) = Q(\omega, H)$ almost surely, so that $Q(\omega, H) = \frac{1}{2}$ almost surely. Similarly, $Q(\omega, H^c) = \frac{1}{2}$ almost surely. Also, if $B \in \mathbf{G}$ and $G \in \mathbf{G}$, then $P(B \cap G) = \int_G I_B \, dP$, so that $P(B \mid \mathbf{G}) = I_B$ almost surely, which implies that $Q(\omega, B) = I_B(\omega)$ almost surely. This, in turn, implies that $Q(\omega, \{\omega\}) = 1$ almost surely. But, by monotonicity, if $\omega \in H$, then $Q(\omega, \{\omega\}) \le Q(\omega, H) = \frac{1}{2}$ almost surely. Also, if $\omega \in H^c$, then $Q(\omega, \{\omega\}) \le Q(\omega, H^c) = \frac{1}{2}$ almost surely. This is a contradiction.

For a further discussion about the nonexistence of regular conditional probability distributions, see Blackwell and Ryll-Nardzewski (1963).

6.14. Martingales, defined on a common probability space, whose sum is not a martingale.

If $\{X_n, \mathbf{F}_n\}$ and $\{Y_n, \mathbf{F}_n\}$ are martingales, then so is $\{X_n + Y_n, \mathbf{F}_n\}$. This follows immediately from the linearity of conditional expectations; however, it is important that $\{X_n\}$ and $\{Y_n\}$ are both adapted to the same sequence of σ-fields $\{\mathbf{F}_n\}$. This example from Chung (1974, Problem 9.3.6) shows the earlier statement to be false in general.

Let X_1 and Y_1 be independent random variables each taking the values 1 and -1 with probability $\frac{1}{2}$. Define

$$Z = \begin{cases} 1 & \text{if } X_1 + Y_1 = 0 \\ -1 & \text{if } X_1 + Y_1 \neq 0 \end{cases}$$

and set $X_2 = X_1 + Z$ and $Y_2 = Y_1 + Z$. Note that Z is independent of X_1 and also that Z is independent of Y_1. However, Z is not independent of (X_1, Y_1)! The distribution of Z is given by $P(Z = 1) = P(Z = -1) = \frac{1}{2}$.

Hence, $\{X_1, X_2\}$ and $\{Y_1, Y_2\}$ are both two-term martingales (meaning that the defining equality for a martingale in the beginning of this chapter is satisfied for $n = 1, 2$). However, $\{X_1 + Y_1, X_2 + Y_2\}$ is not a two-term martingale. To see this, note that the event $\{X_1 + Y_1 = 0\}$ has probability $\frac{1}{2} > 0$, but

$$P(X_2 + Y_2 = 2 \mid X_1 + Y_1 = 0\} = 1,$$

which implies that $E(X_2 + Y_2 \mid X_1 + Y_1 = 0) = 2$, which is not equal to zero. Hence, $E(X_2 + Y_2 \mid X_1 + Y_1) \neq X_1 + Y_1$ almost surely.

6.15. A convex function of a submartingale that is not a submartingale.

In general, if $\{X_n, \mathbf{F}_n\}$ is a submartingale and ϕ is a convex increasing function from $\mathbf{R}$ to $\mathbf{R}$ with $\phi(X_n)$ integrable for all n, then $\{\phi(X_n), \mathbf{F}_n\}$ is also a submartingale. In addition, if $\{X_n, \mathbf{F}_n\}$ is a martingale and ϕ is assumed to be convex (but not necessarily increasing) only with $\phi(X_n)$ integrable, then $\{\phi(X_n), \mathbf{F}_n\}$ is a submartingale (see Ash 1972, p. 288). Indeed, if $\{X_n, \mathbf{F}_n\}$ is a submartingale but not a martingale, then the requirement that ϕ be increasing cannot be dropped, as is shown by this counterexample.

Let $\{H_n, n \geq 1\}$ be a sequence of independent coin tosses with $p > \frac{1}{2}$ so that $q = 1 - p < p < 1$, and

$$H_n = \begin{cases} 1 & \text{with probability } p \\ -1 & \text{with probability } q = 1 - p. \end{cases}$$

Set $X_0 = 0$ and $X_n = X_{n-1} + H_n$, so that $\{X_n, n \geq 1\}$ is a simple $p - q$ random walk on the integers. If $\mathbf{F}_n = \sigma(X_0, \ldots, X_n)$ and $p > q$, then $\{X_n, \mathbf{F}_n, n \geq 1\}$ is a submartingale (but not a martingale because $p > q$).

Now consider the convex function ϕ from $\mathbf{R}$ to $\mathbf{R}$ defined by $\phi(t) = |t|$. We will show that $Y_n = \phi(X_n) = |X_n|$ is not a submartingale with respect to the σ-fields $\mathbf{F}_n$. Note that $E(|Y_n|) = E(|X_n|) \leq n < \infty$. However,

$$\begin{aligned} E(Y_n \mid \mathbf{F}_{n-1}) &= E[|X_n| \mid \sigma(X_0, \ldots, X_{n-1})] \\ &= \begin{cases} p(X_{n-1} + 1) + q(X_{n-1} - 1) & \text{if } X_{n-1} > 0 \\ 1 & \text{if } X_{n-1} = 0 \\ p(|X_{n-1}| - 1) + q(|X_{n-1}| + 1) & \text{if } X_{n-1} < 0. \end{cases} \end{aligned}$$

Since $p > q$, $p(|X_{n-1}| - 1) + q(|X_{n-1}| + 1) = Y_{n-1} + q - p$ is less than Y_{n-1} almost surely and (since $P\{X_{n-1} < 0\}$ because $p < 1$) then $E(Y_n \mid \mathbf{F}_{n-1})$

is not almost surely greater than or equal to Y_n so that $\{Y_n, \mathbf{F}_n, n \geq 1\}$ is not a submartingale. Of course, when $p = q = \frac{1}{2}$, $\{X_n, \mathbf{F}_n\}$ is a martingale and $\{\phi(X_n), \mathbf{F}_n\}$ is a submartingale even though ϕ is not increasing.

6.16. A convex function of a martingale that is not a submartingale.

As mentioned in Example 6.15, if $\{X_n, \mathbf{F}_n\}$ is a martingale and if ϕ is a convex function from $\mathbf{R}$ to $\mathbf{R}$ with $\phi(X_n)$ integrable for all n, then $\{\phi(X_n), \mathbf{F}_n\}$ is a submartingale. The point of this example is that we cannot ignore the requirement that $\phi(X_n)$ be integrable, especially if we want to apply certain inequalities or convergence theorems to the "submartingale" $\{\phi(X_n), \mathbf{F}_n\}$.

Let $\{Y_n, n \geq 1\}$ be any sequence of independent random variables with zero mean but infinite variances. For example, if the Y_n values are symmetric and identically distributed with density function

$$f(y) = \begin{cases} \dfrac{1}{|y|^3} & \text{if } |y| \geq 1 \\ 0 & \text{otherwise} \end{cases}$$

then $E(|Y_n|) = 2 < \infty$ and $E(Y_n) = 0$, but $E(Y_n^2) = \infty$. The t-distribution with two degrees of freedom also has zero mean but infinite variance. Now set $X_n = Y_1 + \cdots + Y_n$ and $\mathbf{F}_n = \sigma(X_1, \ldots, X_n)$. Then $\{X_n, \mathbf{F}_n, n \geq 1\}$ is a martingale. The function $\phi(t) = t^2$ is convex. However, because

$$E[\phi(X_n)] = E[(Y_1 + \cdots + Y_n)^2] = \infty,$$

it follows that $\{\phi(X_n), \mathbf{F}_n\}$ is not a submartingale.

A similar example could be constructed when $\{X_n, \mathbf{F}_n\}$ is a submartingale and ϕ is an increasing convex function from $\mathbf{R}$ to $\mathbf{R}$. For example, the choice $\phi(t) = t^2 I(t \geq 0)$ would suffice with this counterexample.

6.17. A martingale $X_1, X_2, \ldots$ and an integrable stopping time T such that $E(X_T)$, the expected value of the stopped martingale at time T, is infinite.

Observe that a stopping time T being integrable means that $E(T) < \infty$, which further implies that $T < \infty$ with probability one. However, there is no guarantee that $E(|X_T|) < \infty$, even if T and all the X_n are integrable.

Define a sequence of independent random variables $D_1, D_2, \ldots$ that satisfy

$$D_n = \begin{cases} 4^n & \text{with probability } \frac{1}{2} \\ -4^n & \text{with probability } \frac{1}{2}. \end{cases}$$

Set $X_0 = 0$ and $X_n = X_{n-1} + D_n$. Because $E(|D_n|) < \infty$ and $E(D_n) = 0$, it follows that $\{X_n, n \geq 1\}$ is a zero-mean martingale. Define the stopping time T as

$$T = \inf\{n : X_n < 0\}.$$

Clearly, T has a geometric distribution with $P(T = n) = 2^{-n}$ so that

$$E(T) = \sum_{n=1}^{\infty} n2^{-n} = 2 < \infty.$$

Now we compute the expected absolute value of the stopped sequence:

$$\begin{aligned} E(|X_T|) &= E(-X_T) = -\sum_n E[X_n I(T = n)] \\ &= \frac{4}{2} - \sum_{n=2}^{\infty} 2^{-n}\left[\sum_{k=1}^{n-1} 4^k - 4^n\right] \\ &= 2 - \sum_{n=2}^{\infty} 2^{-n}\left[\frac{-2(4^n) - 4}{3}\right] \\ &= 2 + \sum_{n=2}^{\infty} \frac{2^{n+1} - 2^{-n+2}}{3} = \infty. \end{aligned}$$

Thus $E(|X_T|) = \infty$, even though $E(T) < \infty$.

6.18. Independent random variables $D_1, D_2, \ldots$ with mean zero and an integrable stopping time T for which the expectation of the sum of the first T terms is not zero; that is, $E(D_1 + \cdots + D_T) \neq E(D_1) \cdot E(T)$.

The main point of this example is that Wald's identity is false without the assumption that the sequence of independent random variables is also identically distributed. Indeed, Wald's identity asserts that if $D_1, D_2, \ldots$ is a sequence of independent and identically distributed random variables such that $E(D_1)$ exists and T is an $\{\mathbf{F}_n\}$ stopping time with $E(T) < \infty$, where $\mathbf{F}_n$ contains $\sigma(D_1, \ldots, D_n)$, then $E(D_1 + \cdots + D_T) = E(D_1) \cdot E(T)$. For a proof and some generalizations, see Chow and Teicher (1978, pp. 137–39).

For the counterexample, consider the sequence $\{D_n\}$ and T as defined in Example 6.17. There it was shown that $E(D_n) = 0$ and $E(T) = 2$, even though

$$E(D_1 + \cdots + D_T) = -\infty \neq 0 = E(D_1) \cdot E(T).$$

6.19. A martingale $X_1, X_2, \ldots$ and an integrable stopping time T such that $E(|X_T|)$ is finite, but $E(X_1) \neq E(X_T)$.

The point of this counterexample is that we cannot omit the requirement that

$$\liminf_{n\to\infty} E[|X_n| I(T > n)] = 0$$

in the Optional Sampling Theorem.

Consider a symmetric random walk on the integers; that is, let $Y_1, Y_2, \ldots$ be a sequence of independent random variables that satisfies

$$P(Y_j = 1) = P(Y_j = -1) = \tfrac{1}{2}, \qquad j = 1, 2, \ldots$$

and set $X_0 = 0$, $X_n = X_{n-1} + Y_n$ for $n = 1, 2, \ldots$. Then $\{X_n, n \geq 1\}$ is a martingale. Define the stopping time T to be the first time that $X_n = 1$, precisely

$$T = \inf\{n : X_n = 1\}.$$

It is well known that the random walk $Y_1, Y_2, \ldots$ is recurrent, and so T is finite almost surely (see, for example, Feller 1968, Section XIV.7, p. 360). Therefore,

$$E(X_T) = E(1) = 1 \neq 0 = E(X_1).$$

6.20. A martingale $X_1, X_2, \ldots$ and an integrable stopping time T that satisfy $\liminf E[|X_n| I(T > n)] = 0$, but $E(X_1) \neq E(X_T)$.

The point of this counterexample is that we cannot omit the requirement that $E(|X_T|) < \infty$ in the Optional Sampling Theorem, even if we assume that T has moments of all orders.

Let $Y_1, Y_2, \ldots$ be a sequence of independent random variables that satisfies

$$Y_n = \begin{cases} a_n & \text{with probability } \frac{1}{4} \\ 0 & \text{with probability } \frac{1}{2} \\ -a_n & \text{with probability } \frac{1}{4} \end{cases}$$

where $a_n > 0$. Choose a_n so that $\Sigma\, 2^{-n} a_n = \infty$; for example, $a_n = 2^n$. Define $X_n = Y_1 + \cdots + Y_n$, so that $\{X_n, n \geq 1\}$ is a martingale with respect to the generated σ-fields. Define the stopping time T to be

$$T = \inf\{n : X_n \neq 0\}.$$

Note that $P(T = k) = (\frac{1}{2})^k$, so that T has a geometric distribution that possesses moments of all orders. Because the event $\{T > n\}$ is contained in the event $\{X_n = 0\}$, we have $E[|X_n| I(T > n)] = 0$ and

$$\liminf_{n \to \infty} E[|X_n| I(T > n)] = 0.$$

However, $E(X_1) \neq E(X_T)$. Indeed, $E(X_1) = 0$, whereas X_T is not even integrable because

$$E(|X_T|) = \sum a_k P(T = k) = \sum a_k 2^{-k} = \infty.$$

6.21. A martingale, for which each term is a product of positive independent random variables with mean one, that converges almost surely, but for which the expectations do not converge to the expectation of the limit.

Let $Y_1, Y_2, \ldots$ be a sequence of independent random variables with $P(Y_j = 2) = P(Y_j = 0) = \frac{1}{2}$. Then

$$X_n = \prod_{j=1}^{n} Y_j$$

represents the gambler's fortune at time n in a game of "double or nothing." Note that $E(|X_n|) = E(X_n) = 1$ and

$$E(X_n \mid X_1, \ldots, X_{n-1}) = X_{n-1} E(Y_n) = X_{n-1}$$

almost surely, and therefore $\{X_n\}$ is a martingale.

Let $T = \inf\{n : X_n = 0\}$. Then

$$P(T < \infty) = \sum P(T = k) = \sum_{k=1}^{\infty} (\tfrac{1}{2})^k = 1.$$

Indeed, T has a geometric distribution, and hence X_n is eventually zero with probability one. Thus,

$$\lim_{n\to\infty} X_n = X_\infty = 0 \quad \text{almost surely.}$$

However, in terms of expectations,

$$\lim_{n\to\infty} E(X_n) = 1 \neq 0 = E\left(\lim_{n\to\infty} X_n\right).$$

As an aside, this also says that

$$\prod_{j=1}^{\infty} E(Y_j) \neq E\left(\prod_{j=1}^{\infty} Y_j\right).$$

Thus, we may have a sequence of positive independent integrable random variables whose "expectation of the product" is not equal to the "product of the expectations." Of course, this cannot happen for finite products. In general, if $Y_1, Y_2, \ldots$ is a sequence of positive random variables with mean one, and if $X_\infty = \lim \Pi_{j=1}^n Y_j$ exists almost surely (which is always true if the Y_j are independent by the martingale convergence theorem), then by Fatou's Lemma we always have

$$E\left[\prod_{j=1}^{\infty} Y_j\right] = E\left[\liminf_{n\to\infty} \prod_{j=1}^{n} Y_j\right]$$

$$\leq \liminf_{n\to\infty} \prod_{j=1}^{n} E(Y_j) = 1.$$

This example shows that we may have strict inequality here. For an alternative example, see Billingsley (1979, Problem 35.4).

6.22. A martingale that tends to $-\infty$ with probability one.

This example shows that such a martingale may be taken to be the sum of independent random variables with zero means. Let $X_1, X_2, \ldots$ be a sequence of independent random variables such that $X_1 = 0$, and for

$j = 2, 3, \ldots$ we have

$$X_j = \begin{cases} j^2 & \text{with probability } \dfrac{1}{j^2} \\[2ex] \dfrac{-j^2}{j^2 - 1} & \text{with probability } 1 - \dfrac{1}{j^2}. \end{cases}$$

It is straightforward to verify that $E(|X_j|) < \infty$ and that $E(X_j) = 0$ for all j. Hence the process defined by partial sums $S_n = X_1 + \cdots + X_n$ is a martingale.

To see that $P(S_n \to -\infty) = 1$, note that

$$\sum_{j=1}^{\infty} P(X_j = j^2) = \sum_{j=2}^{\infty} \frac{1}{j^2} < \infty.$$

So, by the first Borel-Cantelli Lemma, $P(X_j = j^2 \text{ infinitely often}) = 0$. This, in turn, implies that

$$P\left[X_j = \frac{-j^2}{j^2 - 1} \text{ eventually}\right] = 1.$$

Therefore, $P(X_j < -1 \text{ eventually}) = 1$, so that with probability one, after a certain point every X_j is less than -1. We conclude that $P(S_n \to -\infty) = 1$.

Finally, since any martingale is trivially a submartingale, we have constructed a submartingale S_n for which $P(S_n \to -\infty) = 1$. Thus, although a martingale may sometimes be thought of as capturing the intuitive notion of a "fair" game, this counterexample shows this interpretation to be misleading. Indeed, if a gambler wins the amount X_n at time n, so that S_n is the total fortune at time n, then the gambler will eventually be in debt for an arbitrarily large amount (unless he or she decides to quit first).

6.23. A martingale that converges in probability but not almost surely.

A well-known theorem from Levy says that if $\{X_n, n \geq 1\}$ is a sequence of independent random variables, then the sum $S_n = X_1 + \cdots + X_n$ converges almost surely if and only if it converges in probability (see, for example, Chow and Teicher 1978, p. 72). At first, one might expect this result to carry over to martingales, but this example shows that it does not. To construct such an example, recall the sequence of independent random variables that converges in probability but not almost surely. If $H_1, H_2, \ldots$ is a sequence of independent coin tosses with $P(H_n = 1) = 1/n$ and $P(H_n = 0) = 1 - 1/n$, then H_n converges to zero in probability but not almost surely. It is just a matter of "spicing up" this example to get a martingale.

Let $\{Z_n, n \geq 1\}$ be a sequence of independent random variables with

$$Z_n = \begin{cases} 1 & \text{with probability } \dfrac{1}{2n} \\ 0 & \text{with probability } 1 - \dfrac{1}{n} \\ -1 & \text{with probability } \dfrac{1}{2n}. \end{cases}$$

To construct a martingale $\{X_n, n \geq 1\}$ that converges in probability to zero, we would like to find a martingale $\{X_n, n \geq 1\}$ such that $P(X_n = 0) = P(Z_n = 0)$. To this end, define $X_1 = Z_1$ and, for $n \geq 2$, define

$$X_n = \begin{cases} Z_n & \text{if } X_{n-1} = 0 \\ nX_{n-1}|Z_n| & \text{if } X_{n-1} \neq 0. \end{cases}$$

Let $\mathbf{F}_n = \sigma(X_1, \ldots, X_n)$, the generated σ-field. Then we compute (for $n \geq 2$)

$$\begin{aligned} E(X_n \mid \mathbf{F}_{n-1}) &= E[Z_n I(X_{n-1} = 0) + nX_{n-1}|Z_n| I(X_{n-1} \neq 0) \mid \mathbf{F}_{n-1}] \\ &= I(X_{n-1} = 0) E(Z_n \mid \mathbf{F}_{n-1}) + nX_{n-1} I(X_{n-1} \neq 0) E(|Z_n| \mid \mathbf{F}_{n-1}) \\ &= I(X_{n-1} = 0) E(Z_n) + nX_{n-1} I(X_{n-1} \neq 0) E(|Z_n|) \\ &= 0 + nX_{n-1} I(X_{n-1} \neq 0)\left(\frac{1}{n}\right) = X_{n-1}. \end{aligned}$$

Note that these equalities are asserted almost surely. Also, $|X_1| \leq 1$ and $E|X_n| \leq nE|X_{n-1}|$ imply that $E(|X_n|) \leq n! < \infty$. Therefore, $\{X_n, \mathbf{F}_n\}$ is a martingale.

Now, by the definition of X_n, $X_n = 0$ if and only if $Z_n = 0$. Hence, $P(X_n = 0) = P(Z_n = 0) = 1 - 1/n \to 1$ as $n \to \infty$, so that X_n converges in probability to zero. So, to show that X_n does not converge almost surely, it suffices to show that X_n does not converge almost surely to zero. To this end, since X_n is integer-valued, it suffices to show that $P(X_n \neq 0$ infinitely often$) = 1$. Because the sequence $\{Z_n, n \geq 1\}$ is independent and $\Sigma P(Z_n \neq 0) = \infty$, the second Borel-Cantelli Lemma gives $P(Z_n \neq 0$ infinitely often$) = 1$. But $P(X_n \neq 0$ infinitely often$) = P(Z_n \neq 0$ infinitely often$) = 1$ and so X_n does not converge almost surely to zero. In fact, $P(X_n \neq 0$

infinitely often) $= 1$ and X_n converges in probability to zero actually imply that X_n diverges almost surely.

6.24. A martingale that converges almost surely even though the supremum of the expected absolute values is infinite.

According to a fundamental (sub)martingale convergence theorem, if X_n is an L^1-bounded (sub)martingale, meaning that $\sup E(|X_n|) < \infty$, then X_n converges almost surely. That the condition that X_n be L^1-bounded is not necessary for almost sure convergence is shown by this example.

Define numbers $\{a_n, n \geq 1\}$ by $a_1 = 2$ and $a_n = 4(a_1 + \cdots + a_{n-1})$ if $n \geq 2$. Let $\{Z_n, n \geq 1\}$ be a sequence of independent random variables defined by

$$Z_n = \begin{cases} a_n & \text{with probability } \dfrac{1}{2n^2} \\ 0 & \text{with probability } 1 - \dfrac{1}{n^2} \\ -a_n & \text{with probability } \dfrac{1}{2n^2}. \end{cases}$$

Note that $E(|Z_n|) = a_n/n^2 < \infty$ and $E(Z_n) = 0$. Now define $X_n = Z_1 + \cdots + Z_n$. Because X_n is a sum of independent random variables with zero means, it follows that X_n is a martingale.

To see that X_n converges almost surely, it is enough to show that $P(Z_n = 0 \text{ eventually}) = 1$. However,

$$\sum_{n=1}^{\infty} P(Z_n \neq 0) = \sum_{n=1}^{\infty} \frac{1}{n^2} < \infty.$$

Thus, by the Borel-Cantelli Lemma, $P(Z_n \neq 0 \text{ infinitely often}) = 0$, and so $P(Z_n = 0 \text{ eventually}) = 1$.

However, we will show that $\sup E(|X_n|) = \infty$. To this end, note that $|X_n| \geq a_n/2$ if and only if $|Z_n| = a_n$. Hence,

$$E(|X_n|) \geq \frac{a_n}{2} P\left(|X_n| \geq \frac{a_n}{2}\right)$$

$$= \frac{a_n}{2} P(|Z_n| = a_n) = \frac{a_n}{2n^2}.$$

But $a_n \geq 2^n$, so that $\sup E(|X_n|) = \infty$.

6.25. Nonuniqueness of last elements for sub- or supermartingales.

Recall that if $\{X_n, \mathbf{F}_n, n = 1, 2, \ldots, \infty\}$ is an s-martingale (that is, either a sub- or a supermartingale), then X_∞ is called a "last element." When the s-martingale is not a martingale, X_∞ need not be unique.

For example, consider the game double or nothing defined by $X_1 = 1$ and conditional on $X_1, \ldots, X_{n-1}$,

$$X_n = \begin{cases} 2X_{n-1} & \text{with probability } \frac{1}{2} \\ 0 & \text{with probability } \frac{1}{2}. \end{cases}$$

That is, a gambler starts out with one dollar and continues doubling the bet until all of his or her money is lost. Let $\mathbf{F}_n = \sigma(X_1, \ldots, X_n)$; since $E(X_n \mid \mathbf{F}_{n-1}) = X_{n-1}$ almost surely, we have that $\{X_n, \mathbf{F}_n, n = 1, 2, \ldots\}$ is a martingale and hence a supermartingale. Let $\mathbf{F}_\infty$ denote the σ-field generated by $\mathbf{F}_1, \mathbf{F}_2, \ldots$. If X_∞ is any nonpositive constant c, then $E(X_\infty \mid \mathbf{F}_n) = c \le X_n$, so that $\{X_n, \mathbf{F}_n, n = 1, 2, \ldots, \infty\}$ has many last elements when regarded as a supermartingale. Note that the martingale $\{X_n, \mathbf{F}_n, n = 1, 2, \ldots\}$ converges almost surely to zero. Since $E(0 \mid \mathbf{F}_n) = 0 \ne X_n$ almost surely, however, the martingale $\{X_n, \mathbf{F}_n, n = 1, 2, \ldots\}$ has no last element (because a martingale with a last element converges almost surely to the last element).

The preceding construction clearly applies to any nonnegative supermartingale or nonpositive submartingale. Also note that an s-martingale with a last element converges almost surely. To see this, if $\{X_n, \mathbf{F}_n, n = 1, 2, \ldots, \infty\}$ is a submartingale, then so is $\{X_n^+, \mathbf{F}_n, n = 1, 2, \ldots, \infty\}$ (where t^+ denotes $\max[t, 0]$) and $\sup E(X_n^+) \le E(X_\infty^+) < \infty$ and so the L^1 submartingale convergence theorem applies. As in this example, however, the limit and the last element may differ almost surely.

6.26. A supermartingale with a last element that is not uniformly integrable.

In general, a martingale with a last element is uniformly integrable and converges almost surely to the last element (see, for example, Theorem 7.6.6 of Ash 1972). The same is not true for sub- or supermartingales.

For the example, consider the martingale $\{X_n, \mathbf{F}_n\}$ as defined in Example 6.25. There it was shown that $X_n \to 0$ almost surely. Since $X_n \ge 0$ if X_∞ is identically zero, then $E(X_\infty \mid \mathbf{F}_n) = 0 \le X_n$, so that X_∞ is a last element when $\{X_n, \mathbf{F}_n\}$ is regarded as a supermartingale. However, the sequence $\{X_n\}$ is not uniformly integrable; if it were, then we would have $E(X_n) \to E(X_\infty)$, but $E(X_n) = 1 \ne 0 = E(X_\infty)$.

6.27. A sequence of independent random variables with zero means and finite variances whose partial sums converge almost surely even though the sum of the variances is infinite.

A fundamental result due to Kolmogorov says that for a sequence of independent random variables with zero means, if the sum of the variances is finite, then the partial sums of the sequence converge with probability one. A proof may be found in Chow and Teicher (1978, p. 114). This counterexample shows that the converse is false as it stands. However, a partial converse says that if the sequence is uniformly bounded almost surely and the sum of the variances is infinite, then the sequence of partial sums diverges almost surely. (Note that Kolmogorov's Zero-One Law implies that if the partial sums did not converge almost surely, then they would have to diverge almost surely.)

For the example, suppose that $X_1, X_2, \ldots$ is a sequence of independent random variables that satisfies

$$X_n = \begin{cases} c_n & \text{with probability } p_n \\ 0 & \text{with probability } 1 - 2p_n \\ -c_n & \text{with probability } p_n \end{cases}$$

for some constants $c_n \geq 0$ and $p_n \in [0, \frac{1}{2}]$, to be determined shortly. Note that $E(X_n) = 0$ and $\text{Var}(X_n) = 2p_n c_n^2$. The series $S_n(\omega) = X_1(\omega) + \cdots + X_n(\omega)$ converges for those ω that satisfy $X_n(\omega) = 0$ eventually. Assuming $\Sigma p_n < \infty$, by the Borel-Cantelli Lemma we find that

$$\sum_{n=1}^{\infty} P(X_n \neq 0) = 2 \sum_{n=1}^{\infty} p_n < \infty$$

implies that $P(X_n \neq 0 \text{ infinitely often}) = 0$, which implies that $P(X_n = 0 \text{ eventually}) = 1$. Thus, S_n converges with probability one if $\Sigma p_n < \infty$.

On the other hand, if p_n and c_n are also chosen so that $\Sigma p_n c_n^2 = \infty$, then $\Sigma \text{Var}(X_n) = \infty$. For example, if $p_n = 1/n^2$ and $c_n = n$, then S_n converges with probability one even though $\Sigma \text{Var}(X_n) = \infty$.

6.28. Independent random variables $Y_1, Y_2, \ldots$ such that $\Sigma P\{|Y_n| > 1\} < \infty$ and $\Sigma \text{Var}\{Y_n I(|Y_n| \leq 1)\} < \infty$, but for which ΣY_n diverges almost surely.

The point of this example is that we cannot remove the requirement that $\Sigma E\{Y_n I(|Y_n| \leq 1)\} < \infty$ in the Three Series Theorem. Simply let $Y_n = \frac{1}{2}$ with probability one. Then $\Sigma P(|Y_n| > 1) = 0$ and $\Sigma \text{Var}\{Y_n I(|Y_n| \leq 1)\} =$

0. However, $S_n = Y_1 + \cdots + Y_n = n/2$ almost surely, so that $S_n \to \infty$ almost surely.

6.29. Independent random variables $Y_1, Y_2, \ldots$ such that $\Sigma E\{Y_n I(|Y_n| \leq 1)\}$ converges and $\Sigma \operatorname{Var}\{Y_n I(|Y_n| < 1)\} < \infty$, but for which ΣY_n diverges almost surely.

This example shows that we cannot remove the requirement in the Three Series Theorem that $\Sigma P(|Y_n| > 1) < \infty$ in order for ΣY_n to converge almost surely, even if the other two series converge. Let $Y_1, Y_2, \ldots$ be independent with

$$Y_n = \begin{cases} n & \text{with probability } \dfrac{1}{2n} \\ 0 & \text{with probability } 1 - \dfrac{1}{n} \\ -n & \text{with probability } \dfrac{1}{2n}. \end{cases}$$

Then $\Sigma E\{Y_n I(|Y_n| \leq 1)\} = 0$ and $\Sigma \operatorname{Var}\{Y_n I(|Y_n| \leq 1)\} = 1$. Note, however, that

$$\sum P(|Y_n| > 1) = \sum_{n=2}^{\infty} \frac{1}{n} = \infty.$$

To see directly that $S_n = Y_1 + \cdots + Y_n$ diverges almost surely, note that by the Borel-Cantelli Lemma, $|Y_n| = n$ infinitely often with probability one, so that S_n diverges almost surely.

6.30. Independent random variables $Y_1, Y_2, \ldots$ such that $\Sigma P(|Y_n| > 1) < \infty$ and $\Sigma E\{Y_n I(|Y_n| \leq 1)\}$ converges, but for which ΣY_n diverges almost surely.

This counterexample shows that we cannot omit the requirement that $\Sigma \operatorname{Var}\{Y_n I(|Y_n| \leq 1)\} < \infty$ in the Three Series Theorem in order for $S_n = \Sigma_{j=1}^{n} Y_j$ to converge almost surely.

For the example, let $Y_1, Y_2, \ldots$ be independent with $P(Y_n = 1) = P(Y_n = -1) = \frac{1}{2}$. Then $\Sigma P(|Y_n| > 1) = 0$ and $\Sigma E\{Y_n I(|Y_n| \leq 1)\} = 0$. However, S_n diverges almost surely because $P(Y_n \to 0) = 0$.

It is of interest to note that care must be taken to use the weak inequality in the series for the variance rather than a strong inequality. If the strong inequality had been used, we would find for this example that

$$\sum \text{Var}\{Y_n I(|Y_n| < 1)\} = 0 < \infty,$$

which would erroneously suggest that all three series converge in the Three Series Theorem. However, using the correct form of this series, we find that

$$\sum \text{Var}\{Y_n I(|Y_n| \leq 1)\} = \infty,$$

which is consistent with $\Sigma\, Y_n$ diverging almost surely.

6.31. A sequence $Y_1, Y_2, \ldots$ of independent random variables, each with finite mean and variance, and positive constants $b_1, b_2, \ldots$ such that $\Sigma\,[\text{Var}(Y_n)]/b_n^2 < \infty$ but $\Sigma_{j=1}^n [Y_j - E(Y_j)]/b_n$ does not converge almost surely to zero.

The point of this example is that we cannot omit the requirement that $b_n \to \infty$ as $n \to \infty$ in Kolmogorov's strong law of large numbers. Let $b_n = 1$ for all n, and let $Y_1, Y_2, \ldots$ be independent Gaussian random variables with mean zero and $\text{Var}(Y_n) = 1/2^n$. Note that

$$\sum_{n=1}^{\infty} \frac{\text{Var}(Y_n)}{b_n^2} = \sum_{n=1}^{\infty} \frac{1}{2^n} = 1.$$

The normalized distribution is

$$\mathbf{L}\left[\frac{\sum_{j=1}^n [Y_j - E(Y_j)]}{b_n}\right] = \mathbf{L}\left[\sum_{j=1}^{n} Y_j\right] = N\left(0, 1 - \frac{1}{2^n}\right),$$

which converges in distribution to a standard Gaussian. Thus the law of the normalized sum

$$\frac{\sum_{j=1}^n [Y_j - E(Y_j)]}{b_n}$$

converges weakly to a standard Gaussian distribution and therefore does not converge to zero almost surely (because convergence almost surely implies convergence in law to the same limit).

6.32. A sequence $Y_1, Y_2, \ldots$ of independent random variables, each with finite mean and variance, and nondecreasing positive constants $b_1, b_2, \ldots$ such that $b_n \to \infty$, satisfying $\Sigma_{j=1}^n [Y_j - E(Y_j)]/b_n \to 0$ almost surely, but $\Sigma \operatorname{Var}(Y_n)/b_n^2 = \infty$.

The point of this example is that the requirement $\Sigma \operatorname{Var}(Y_n)/b_n^2 < \infty$ is not a necessary condition in Kolmogorov's strong law of large numbers.

For example, let $Y_1, Y_2, \ldots$ be independent such that

$$Y_n = \begin{cases} n^2 & \text{with probability } \dfrac{1}{2n^2} \\ 0 & \text{with probability } 1 - \dfrac{1}{n^2} \\ -n^2 & \text{with probability } \dfrac{1}{2n^2}. \end{cases}$$

Then $E(Y_n) = 0$ and $\operatorname{Var}(Y_n) = n^2$. Let $b_n = n$, so that

$$\sum \frac{\operatorname{Var}(Y_n)}{b_n^2} = \sum 1 = \infty.$$

To see that

$$\sum_{j=1}^{n} \frac{Y_j - E(Y_j)}{b_n} = \frac{1}{n} \sum Y_j \to 0 \quad \text{almost surely}$$

note that $\Sigma P(Y_j \neq 0) = \Sigma 1/n^2 < \infty$. So, by the Borel-Cantelli Lemma, $Y_n = 0$ eventually with probability one. This clearly implies that $(\Sigma_{j=1}^n Y_j)/n$ converges to zero with probability one.

CHAPTER SEVEN
PROPERTIES OF STATISTICAL EXPERIMENTS

Introduction

The basis of any statistical analysis is the raw data, which are the observations from a random experiment. The purpose of statistical inference is to obtain information, based on the data, about the object under investigation.

A mathematical description of a statistical experiment is the following: The statistician is given the outcome X of a random experiment, with values in some measurable space $(\mathbf{X}, \mathbf{B})$ usually called the "sample space." Often we will consider the case when $X = (X_1, \ldots, X_n)$ is a vector of real-valued random variables. The probability distribution of X, P_θ, defined on the sample space $(\mathbf{X}, \mathbf{B})$, is known to belong to a certain family of probability distributions $\mathbf{P} = \{P_\theta : \theta \in \Omega\}$. The set Ω is called the "parameter space." In choosing a particular parametrization for the family $\mathbf{P}$, we will assume that the parametrization is identifiable in the sense that if $\theta_1 \in \Omega$, $\theta_2 \in \Omega$, and $\theta_1 \neq \theta_2$, then $P_{\theta_1} \neq P_{\theta_2}$. The abstract triple $\mathbf{E} = \{\mathbf{X}, \mathbf{B}, \{P_\theta, \theta \in \Omega\}\}$ that represents the possible generating mechanisms for the observation X is known as a "statistical experiment." Note that any inference made concerning the family $\mathbf{P}$ is equivalent to inferences concerning the values of $\theta \in \Omega$.

One of the most important problems of statistical inference is point estimation. Suppose $q(\theta)$ is a function defined on Ω. Because θ is unknown, we would like to find a function T of the data X that is close to $q(\theta)$ in some sense. We will deal more specifically with this problem in Chapters 8 and 9. Alternatively, we may be interested in making a yes or no inference about θ. This is the problem of "hypothesis testing," which we will look at more closely in Chapter 10. Comprehensive treatments of point estimation and hypothesis testing can be found in Lehmann (1983) and Lehmann (1959), respectively. Of course, many other problems arise under the heading of statistical inference. A common feature of statistical inference, however, is deciding how to use the data X. In general, the goal is to find a suitable function $T(X)$ to extract the interesting features of the data with the aim of solving the specific problem in mind. Any (measurable) function $T(X)$ is called a "statistic."

Regardless of the actual inference we would like to perform, it is useful to begin by trying to condense the data X by a function $T(X)$ in such a way that $T(X)$ contains all the information about θ in the sample. Herein lies the notion of a sufficient statistic. By definition, a statistic $T(X)$ is called "sufficient" for the family $\mathbf{P} = \{P_\theta : \theta \in \Omega\}$ of distributions of X if, for every $B \in \mathbf{B}$, there is a real random variable, measurable with respect to $\sigma(T)$, the σ-field generated by T, such that

$$g_B = P_\theta\{B \mid \sigma(T)\} \quad \text{a.s. } P_\theta.$$

The function g_B cannot depend on θ, but the exceptional set might. Although this definition seems rather technical, the intuitive notion of "sufficiency" is clear. The idea is that T contains all the relevant information about θ. Indeed, once the value of the sufficient statistic is known, the data can be treated as if generated by a random process that does not depend on the unknown θ. Since this random process does not depend on θ, it is irrelevant and uninformative in inference about θ. (See Lehmann 1983, Section 1.5, for more details.)

The Factorization Theorem, proved in various forms by Fisher, Neyman, Halmos, and Savage, provides a simple criterion for the determination of a sufficient statistic. The family $\mathbf{P} = \{P_\theta : \theta \in \Omega\}$ is said to be "dominated" by a σ-finite measure μ, which is also defined on $(\mathbf{X}, \mathbf{B})$, if each member of $\mathbf{P}$ is absolutely continuous with respect to μ. Thus (by the Radon-Nikodym Theorem), the distribution P_θ has a density with respect to μ given by

$$p_\theta = \frac{dP_\theta}{d\mu}$$

in the sense that, if $B \in \mathbf{B}$, then

$$P_\theta(B) = \int_B p_\theta(x)\, d\mu(x).$$

In such a case, a necessary and sufficient condition for a statistic $T(X)$ to be sufficient for the family $\{P_\theta : \theta \in \Omega\}$ of distributions of X dominated by a σ-finite measure μ is that there exist nonnegative measurable functions g_θ and h (defined on the range of T and X, respectively) such that

$$p_\theta(x) = g_\theta[T(x)]h(x)$$

holds almost everywhere with respect to μ. For a proof, see Lehmann (1959, Section 2.6).

Since we would like to determine a sufficient statistic that provides a maximal reduction in the data, we say that a sufficient statistic T is "minimal sufficient" if, for any other sufficient statistic S, there exists a function f such that $T = f(S)$ almost everywhere with respect to each probability measure in $\mathbf{P}$.

Given any statistic $T(X)$, it will be of interest to look at the distribution of $T(X)$. If X has distribution P_θ, let P_θ^T denote the distribution of $T(X)$. Often the distribution of $T(X)$, P_θ^T, will depend on θ. In particular, if q is real-valued and $T(X)$ is also real-valued, then some simple characteristics of the distribution P_θ^T are its moments. If, for all $\theta \in \Omega$, $E_\theta[T(X)] = q(\theta)$, then T is said to be "unbiased" for $q(\theta)$. The difference $q(\theta) - E_\theta[T(X)]$ is called the "bias" of the statistic $T(X)$ in estimating $q(\theta)$. A statistic T is called "median unbiased" for $q(\theta)$ if, for all $\theta \in \Omega$, there exists a median of the distribution P_θ^T for which $\text{median}_\theta[T(X)] = q(\theta)$.

Sometimes the family of distributions $\{P_\theta^T : \theta \in \Omega\}$ is independent of θ. In this case, the statistic $T(X)$ is said to be "ancillary." When $T(X)$ is real- or vector-valued, $T(X)$ is said to be "first-order ancillary" if $E_\theta[T(X)]$ is independent of θ.

A convenient property that guarantees certain uniqueness properties and hence simplification of the statistical problem is completeness. A statistic T is said to be "complete" if $E_\theta[f(T)] = 0$ for all $\theta \in \Omega$ implies that $f(t) = 0$ almost everywhere with respect to $\mathbf{P}$. Thus, a statistic T is complete if no nonconstant function of T is first-order ancillary.

It is interesting to note that a complete, sufficient statistic is always minimal sufficient. For proofs, see Lehmann and Scheffe (1950) and Bahadur (1958). The converse, however, is false. A weaker property than completeness is bounded completeness. A statistic T is called "boundedly complete" if $E_\theta[f(T)] = 0$ for all $\theta \in \Omega$ and f is bounded imply that $f(t) = 0$ almost everywhere with respect to $\mathbf{P}$. Completeness is really a property of the family of distributions of T, $\{P_\theta^T : \theta \in \Omega\}$. Thus, we say that T is complete if and only if its family of distributions is complete.

Often we will be looking at the case when $X = (X_1, \ldots, X_n)$ is a vector of real-valued independent and identically distributed random variables, each with common distribution F_θ on the real line. The law of X, previously denoted by P_θ, is then simply F_θ^n; that is, the joint distribution of $(X_1, \ldots, X_n)$ is the n-fold product of the marginal distributions. Suppose the family $\{F_\theta : \theta \in \Omega\}$ of distributions of $(X_1, \ldots, X_n)$ is dominated by a σ-finite measure ξ, so that each random variable X_i has a density with respect to ξ given by $f(x_i; \theta)$. The joint density of the product law with respect to ξ^n is then

$$L(X_1, \ldots, X_n; \theta) = \prod_{i=1}^{n} f(X_i; \theta).$$

In statistics, this function L is known as the "likelihood function." In general, when P_θ is dominated by a σ-finite measure μ, the likelihood function is simply the density of P_θ with respect to μ:

$$L(X; \theta) = \frac{dP_\theta}{d\mu}.$$

A family of distributions $\mathbf{P} = \{P_\theta : \theta \in \Omega\}$ is said to be a "k-parameter exponential family" if the distributions P_θ have densities with respect to some dominating measure μ so that the likelihood function has the special form

$$L(X; \theta) = \exp\left[\sum_{i=1}^{k} \eta_i(\theta) T_i(X) - B(\theta)\right] h(X)$$

where η_i and B are real-valued functions defined on Ω. By a reparametrization using the η_i as the parameters, it is conveniently written in the canonical form:

$$L(X; \eta) = \exp\left[\sum_{i=1}^{k} \eta_i T_i(X) - A(\eta)\right] h(X).$$

We will, without loss of generality, assume that neither the T_i nor the η_i values satisfy a linear constraint; otherwise, we would change the representation appropriately. Usually we will consider the case where $X_1, \ldots, X_n$ is a sample of independent and identically distributed random variables, each with the distribution P_θ given earlier. In such a case, the family of distributions of $X = (X_1, \ldots, X_n)$ is still a k-parameter exponential family. An important property of this experiment is that the statistic T defined by

$$T(X) = T(X_1, \ldots, X_n) = \left[\sum_{i=1}^{n} T_1(X_i), \ldots, \sum_{i=1}^{n} T_k(X_i)\right]$$

is sufficient for the family $\mathbf{P}$. Furthermore, if the parameter space $\{(\eta_1(\theta), \ldots, \eta_k(\theta)) : \theta \in \Omega\}$ contains an open rectangle in $\mathbf{R}^k$, the k-parameter exponential family is said to be of "full rank." In such case, $T(X)$ is complete as well as sufficient. Thus, in a sample from a k-parameter family, the data can be reduced to a k-dimensional complete sufficient statistic independent of the sample size. For this reason, statistical experiments when $\mathbf{P}$ is a k-parameter exponential family are much easier to deal with mathematically than more general statistical experiments.

Another common statistical experiment is when the family $\mathbf{P} = \{P_\theta : \theta \in \Omega\}$ is a group family, which is now described. Suppose U is a fixed random object (for example, a real random variable) with probability law P. Let G be a group of transformations so that if $g \in G$, then gU is another random object with probability law denoted by P_g. Then the family $\{P_g : g \in G\}$ is called a "group family." An important special case is a location family; that is, suppose U is a random variable with distribution $F(x)$. Consider the group of transformations defined by $\{g_\theta U = U + \theta : \theta \in \mathbf{R}\}$. The resulting family is $\{F(x - \theta) : \theta \in \mathbf{R}\}$. Alternatively, the group of transformations defined by $\{g_{(\tau,\theta)} U = \tau U + \theta : \tau > 0, \theta \in \mathbf{R}\}$ results in the location-scale family of distributions $\{F[(x - \theta)/\tau] : \tau > 0, \theta \in \mathbf{R}\}$.

7.1. A one-parameter family of distributions for which no single continuous sufficient statistic exists.

We will give two examples. First, suppose $X_1, \ldots, X_n$ is a sample from a Gaussian distribution with known coefficient of variation $v_0 = \sigma/\mu$ so that X_i has the Gaussian distribution with unknown mean μ and variance $\mu^2 v_0^2$. The likelihood function is given by:

$$L(\mathbf{x}; \mu) = \prod_{i=1}^{n} \frac{\exp\left[-\dfrac{(x_i - \mu)^2}{2v_0^2\mu^2}\right]}{v_0\mu(2\pi)^{1/2}}$$

$$= \exp\left[-\sum_{i=1}^{n} \frac{x_i^2}{2v_0^2\mu^2} + \sum_{i=1}^{n} \frac{x_i}{v_0^2\mu} - \frac{n}{2v_0^2} - \frac{n}{2}\log(2\pi v_0^2\mu^2)\right].$$

Now this is the exponential form used in identifying minimal sufficient statistics. Thus we see that

$$S = \left(\sum_{i=1}^{n} X_i, \sum_{i=1}^{n} X_i^2\right)$$

is minimal sufficient for μ; minimality follows because S can be determined from the values that L takes on at several values of μ.

To see that a single continuous function of the data cannot be minimal sufficient, we consider, for example, the case where $n = 2$. To argue by contradiction, first suppose that there is a single continuous real-valued function $T(X_1, X_2)$ that is minimal sufficient. Then by a geometric argument, T provides a one-to-one continuous mapping from $\{(x_1, x_2) : x_1 \leq x_2\}$ to $\mathbf{R}$. But this is impossible, as may be seen by considering the intermediate value theorem: For each α between $T(0, 0)$ and $T(1, 1)$, there must be a value β and a value γ, each between zero and one, such that $T(\beta, \beta) = T(\gamma^2, \gamma) = \alpha$, contradicting the one-to-one property of T.

It is interesting to note that there do exist one-dimensional sufficient statistics in this case, indicating that the restriction to "continuous" functions is very important here. Some related details are given in Example 7.2. Such a mapping cannot be continuous, however. These considerations suggest that care must be exercised when speaking of the "dimensionality" of a minimal sufficient statistic (see also Lehmann 1983, p. 44).

Our second example is the uniform distribution on $(\theta, \theta + 1)$. Given a sample $X_1, \ldots, X_n$, the likelihood function is

$$L(x_1, \ldots, x_n; \theta) = \prod_{i=1}^{n} I_{(\theta, \theta+1)}(x_i) = I_{(x_{(n)}-1, x_{(1)})}(\theta)$$

$$= \begin{cases} 1 & \text{if } x_{(n)} < \theta + 1 \text{ and } x_{(1)} > \theta \\ 0 & \text{otherwise} \end{cases}$$

where $x_{(1)} = \min(x_1, \ldots, x_n)$ and $x_{(n)} = \max(x_1, \ldots, x_n)$. By the Factorization Theorem, $S = (X_{(1)}, X_{(n)})$ is seen to be jointly minimal sufficient. Minimality follows because S can be determined from the values of L at all values of θ. Note that this is a special case of the general situation when the support of the distribution depends on θ (see Kendall and Stuart 1979, p. 29).

7.2. A real-valued sufficient statistic for a sample from a Gaussian distribution with unknown mean and unknown variance.

We usually think of $(\bar{X}, S)$, where $\bar{X}$ denotes the sample mean and $S^2 = [\Sigma (X_i - \bar{X})^2]/(n - 1)$ denotes the sample variance, as being the greatest reduction possible of a sample $X_1, \ldots, X_n$ from a Gaussian with unknown mean μ and unknown variance σ^2 without losing useful information about the parameters. So it may come as a surprise that there is, in fact, a real-valued sufficient statistic for this problem.

Here is one scheme for coding the information about $\bar{X}$ and S within a single real number so that they can be easily reconstructed. Define T in terms of its decimal number representation, which will be assumed to have the form

$$T = \ldots 0t_n 0t_{n-1} 0 \ldots 0t_1 0t_0 . t_{-1} 0t_{-2} 0t_{-3} 0 \ldots$$

which contains extra zeros to avoid problems with rounding. Such a number will code by interleaving the alternate nonzero digits and using t_0 to code the sign of $\bar{X}$ as follows:

$$\bar{X} = \begin{cases} \ldots t_{2n-1}t_{2n-3} \ldots t_3t_1.t_{-1}t_{-3} \ldots & \text{if } t_0 = 1 \\ -\ldots t_{2n-1}t_{2n-3} \ldots t_3t_1.t_{-1}t_{-3} \ldots & \text{if } t_0 = 0 \end{cases}$$

$$S = \ldots t_{2n}t_{2n-2} \ldots t_4t_2.t_{-2}t_{-4} \ldots.$$

For example, $T = 201040601.103010803050\ldots$ is a single real number from which the values $\bar{X} = -16.113\ldots$ and $S = 24.385\ldots$ can be reconstructed. The statistic T defined in this way is indeed one-dimensional and is a sufficient statistic for (μ, σ). Of course, T is not a continuous function of the data, and so the argument given in Example 7.1 is not violated.

7.3. A one-parameter family of distributions for which the order statistics are minimal sufficient.

Given an independent and identically distributed sample $X_1, \ldots, X_n$ of real-valued observations from a family of distributions, the order statistics are always sufficient by an exchangeability argument. Although a greater reduction in the data can often be achieved, this is not always the case.

We will give two examples. Suppose $X_1, \ldots, X_n$ is a sample from the Cauchy distribution with location parameter θ. The likelihood function of the data is given by

$$L(\mathbf{x}; \theta) = \pi^{-n} \prod_{i=1}^{n} [1 + (x_i - \theta)^2]^{-1}.$$

By the Factorization Theorem, if $T = t(X_1, \ldots, X_n)$ is any sufficient statistic (possibly vector-valued), then the ratio

$$\frac{L(\mathbf{x}; \theta)}{L(\mathbf{x}_0; \theta)} = \frac{g(t; \theta)h(\mathbf{x})}{g(t_0; \theta)h(\mathbf{x}_0)}$$

must be independent of θ when $t = t_0$. Intuitively, the sufficient statistic T partitions $\mathbf{R}^n$ into equivalence classes [setting $\mathbf{x}$ and $\mathbf{x}_0$ equivalent if $t(\mathbf{x}) = t(\mathbf{x}_0)$] in such a way that once the value of $T = t$ is known, there is no further information available about θ. In the Cauchy case, the ratio

$$\frac{\prod_{j=1}^{n} [1 + (x_{0j} - \theta)^2]}{\prod_{j=1}^{n} [1 + (x_j - \theta)^2]}$$

must be independent of θ. In general, the sample itself and the order statistics are sufficient, but this is the greatest reduction possible in this case. If the above ratio is to be independent of θ, we must have the ratio of two polynomials in θ of degree $2n$ independent of θ. Thus, all powers of θ must have identical coefficients in the numerator and denominator. It is clear that this is true only when $(x_1, \ldots, x_n)$ is some permutation of $(x_{01}, \ldots, x_{0n})$. Hence the set of order statistics $(X_{(1)}, \ldots, X_{(n)})$ is minimal sufficient, and no further reduction in the data is possible.

The second example is provided by the "change point" model where $u_1, \ldots, u_n$ are fixed known distinct constants and the random variables X_j are chosen from the Gaussian distribution $N(0, 1)$ if $\mu_j < \theta$, but from the distribution $N(\xi, 1)$ if $\mu_j \geq \theta$, for $j = 1, \ldots, n$ and for some unknown parameter θ. Then a sufficient statistic for (ξ, θ) is the entire data set $(X_1, \ldots, X_n)$ and no further reduction is possible (Cox and Hinkley 1974, p. 30). See Lehmann (1983) for additional examples.

7.4. A sufficient statistic (X, C) for a parameter θ such that C does not depend on θ; that is, C is ancillary, yet X alone is not sufficient for θ.

Before giving an example, we note that this phenomenon is actually quite common. The statistic X is sometimes called "conditionally sufficient," since X is used as a sufficient statistic conditional on the value of C (see Cox and Hinkley 1974, p. 32).

Suppose a random variable X is equally likely to come from a Gaussian distribution with mean μ and variance σ_1^2 and from a Gaussian distribution with mean μ and variance σ_2^2. Let the random variable C take on the value one with probability $\frac{1}{2}$, in which case X has the former distribution, and let C take on the value two with probability $\frac{1}{2}$, in which case X has the latter distribution. The likelihood in this case is then:

$$f_{C,X}(c, x) = \frac{\exp\left\{-\dfrac{(x-\mu)^2}{2\sigma_c^2}\right\}}{2(2\pi)^{1/2}\sigma_c}.$$

By the Factorization Theorem, it is clear that (C, X) is sufficient for μ when σ_1^2 and σ_2^2 are known; however, it is not true that X alone is sufficient. If X alone were sufficient, then the ratio $[f_{C,X}(1, x)]/[f_{C,X}(2, x)]$ would be independent of μ for all x. However, if $x = 0$, for example, then this ratio is

$$\frac{\sigma_2}{\sigma_1}\exp\left\{-\mu^2\frac{\sigma_1^{-2} - \sigma_2^{-2}}{2}\right\},$$

which obviously depends on μ when the variances are distinct. However, $P(C = 1) = P(C = 2) = \frac{1}{2}$, independent of the value of μ, so C is indeed ancillary.

7.5. Nonuniqueness of ancillary statistics (Cox and Hinkley 1974, p. 33).

Suppose $X_1, \ldots, X_n$ is a sample from the discrete distribution specified by

X	1	2	3	4
$P(X)$	$\frac{1-\theta}{6}$	$\frac{1+\theta}{6}$	$\frac{2-\theta}{6}$	$\frac{2+\theta}{6}$

where θ is between -1 and 1. Let N_m be the number of times m appears in the sequence $X_1, \ldots, X_n$ for $m = 1, 2, 3$, and 4. Then the joint probability of a particular sequence is given by

$$f_{\mathbf{X}}(\mathbf{x}; \theta) = \frac{(1-\theta)^{N_1}(1+\theta)^{N_2}(2-\theta)^{N_3}(2+\theta)^{N_4}}{6^n}.$$

The statistic $S = (N_1, N_2, N_3, N_4)$ is therefore sufficient. However, there are two possible ancillary statistics C_1 and C_2 given by

$$C_1 = (N_1 + N_2, N_3 + N_4)$$

and

$$C_2 = (N_1 + N_4, N_2 + N_3)$$

because $N_1 + N_2$ is a binomial random variable with success probability $\frac{1}{3}$ in each of n trials and $N_3 + N_4$ is $n - N_1 - N_2$. Similarly, $N_1 + N_4$ is a binomial random variable with success probability $\frac{1}{2}$ in each of n trials and $N_2 + N_3$ is $n - N_1 - N_4$. Thus, C_1 and C_2 are both ancillary.

Although such a problem of nonuniqueness does not seem to arise frequently in practice, the possibility is alarming. The Conditionality Principle, suggested by Fisher, states that inferences about an unknown parameter need only be made conditional on an ancillary statistic (see Kendall and Stuart 1979, p. 232). The arbitrary nature of having to make a choice is theoretically unpleasant. In choosing among the different possibilities, one might use the variance of the conditional information. In this case, it can be shown that C_1 is preferable (Cox and Hinkley 1978, p. 48).

7.6. Nonuniqueness of sufficient statistics.

Sufficient statistics, even minimal sufficient statistics, are far from unique. For example, any nonzero constant multiple of a sufficient statistic is sufficient, as are many other one-to-one transformations of sufficient statistics.

For example, let $X_1, \ldots, X_n$ be a sample from the Gaussian distribution with mean μ and variance σ^2. Then

$$S_1 = \left(\bar{X}, \sum (X_i - \bar{X})^2\right) \qquad \text{where } \bar{X} = \sum \frac{X_i}{n}$$

$$S_2 = \left(\sum X_i, \sum X_i^2\right)$$

and

$$S_3 = \left(37 \sum (X_i + X_i^2), 49 \sum (X_i - X_i^2)\right)$$

are all minimal sufficient statistics for (μ, σ^2). Note, however, that the σ-fields generated by S_1, S_2, and S_3, respectively, are the same.

7.7. A statistic (S, T) that is sufficient for (θ, ϕ) such that S is sufficient for θ when ϕ is fixed and known, but T is not sufficient for ϕ when θ is fixed and known.

Let $X_1, \ldots, X_n$ be a sample from the Gaussian distribution with mean $\theta = \mu$ and variance $\phi = \sigma^2$. The likelihood is given by:

$$L(\mathbf{x}; \mu) = \exp\left\{\frac{\mu S(\mathbf{x})}{\sigma^2} - \frac{T(\mathbf{x})}{2\sigma^2} - \frac{n\mu^2}{2\sigma^2} - \frac{n \log(2\pi\sigma^2)}{2}\right\}$$

where $S(\mathbf{x}) = \Sigma_{i=1}^n x_i$ and $T(\mathbf{x}) = \Sigma_{i=1}^n x_i^2$. By the Factorization Theorem (or by results for identifying sufficient statistics for distributions that form an exponential family), (S, T) is sufficient for (μ, σ^2). Furthermore, by the same reasoning, when σ^2 is fixed and known, S is sufficient for μ. However, it is not true that T is sufficient for σ^2 when μ is fixed and known. There are a couple of ways to see this. The following idea uses the fact that there are points in the sample space $\mathbf{R}^n$ where T cannot point out the difference but the likelihood can. Suppose T is sufficient for σ^2 when μ is known. Then we may write (by the Factorization Theorem)

$$L(\mathbf{x}; \mu, \sigma^2) = g(T(\mathbf{x}); \mu, \sigma^2) h(\mathbf{x}; \mu), \qquad \mathbf{x} \in \mathbf{R}^n.$$

Next, evaluate L at $\mathbf{x} = (a, \ldots, a)$ and at $\mathbf{y} = (-a, \ldots, -a)$. Note that $T(\mathbf{x}) = na^2 = T(\mathbf{y})$. The ratio $L(\mathbf{x}; \mu, \sigma^2)/L(\mathbf{y}; \mu, \sigma^2)$ should then be independent of σ^2. However,

$$\frac{L(\mathbf{x}; \mu, \sigma^2)}{L(\mathbf{y}; \mu, \sigma^2)} = \exp\left(\frac{2na\mu}{\sigma^2}\right),$$

which is a contradiction, since this ratio depends on σ^2. Thus T is not sufficient for σ^2 when μ is known.

7.8. Two statistics that are each ancillary but jointly are not ancillary (Basu 1964).

Suppose $(X_1, Y_1), \ldots, (X_n, Y_n)$ are n independent observations from a bivariate Gaussian distribution with zero means and unit variances and unknown correlation coefficient ρ. Then the distribution of X_i is standard Gaussian, independent of ρ. Hence $T_1 = (X_1, \ldots, X_n)$ is ancillary. Similarly, $T_2 = (Y_1, \ldots, Y_n)$ is also ancillary. Since the joint distribution (T_1, T_2), which is the entire data set, depends on ρ, it follows that (T_1, T_2) is not ancillary.

It is often suggested that inference about an unknown parameter be made conditionally on the value of an ancillary statistic, much in the same way that the sample size is fixed. (See, for example, Cox 1958.) Basu's example here points out that a unique choice of "reference set" is not necessarily possible. For example, in regression, the statistician often treats the X observations as fixed constants, considering the conditional distribution of the Y values given the X values. If, however, the statistician may equally well choose the Y values to be fixed, then there is no unique way to define the sampling error for the estimate of ρ.

Another way to state this phenomenon is that reduction of the data by conditioning on an ancillary statistic cannot always be carried out by a "maximal" ancillary. That is, an ancillary statistic T is said to be "maximal" if there is no ancillary U (not equivalent to T) such that $T = f(U)$. In this example, a unique maximal ancillary statistic does not exist.

7.9. An ancillary statistic that is not location invariant in a location family.

In general, when sampling $X_1, \ldots, X_n$ from a location family with densities $\{f(x - \theta) : \theta \in \mathbf{R}\}$ with respect to Lebesgue measure, any real-valued statistic T that satisfies

$$T(X_1 + c, \ldots, X_n + c) = T(X_1, \ldots, X_n) \qquad \text{for all } c \in \mathbf{R}$$

is ancillary. Any statistic T that satisfies this property is said to be "location invariant." To see why such a T is ancillary, note that if B is a Borel set, then

$$\begin{aligned} P_\theta\{T(X_1, \ldots, X_n) \in B\} &= P_0\{T(X_1 + \theta, \ldots, X_n + \theta) \in B\} \\ &= P_0\{T(X_1, \ldots, X_n) \in B\} \end{aligned}$$

which does not depend on the value of θ.

More generally, when X has a family of distributions $\mathbf{P} = \{P_\theta : \theta \in \Omega\}$ that remains invariant under a group of transformations G such that the induced group on Ω is transitive, then it is easy to see that any invariant statistic T is ancillary, where T is invariant if and only if $T(gx) = T(x)$ for all $g \in G$. That the converse is false is shown here.

This example is from Padmanabhan (1977). Take X_1 and X_2 independent, with the Gaussian distribution having unknown mean $\theta \in \mathbf{R}$ and variance one. Let

$$\delta(X_1, X_2) = \begin{cases} X_1 - X_2 & \text{if } X_1 + X_2 \geq 0 \\ X_2 - X_1 & \text{if } X_1 + X_2 < 0. \end{cases}$$

Then δ is ancillary. To see why, note that $X_1 - X_2$ and $X_1 + X_2$ are independent. Conditional on $X_1 + X_2$, we find that $X_1 - X_2$ and $X_2 - X_1$ are each Gaussian with mean zero and variance two. Thus, for any θ, we will have δ Gaussian with mean zero and variance two.

However, δ is not location invariant. For example,

$$\delta(-1, 1) = -2 \neq 2 = \delta(-2, 0) = \delta(-1 + c, 1 + c) \qquad \text{where } c = -1.$$

7.10. Unbiased estimators need not be unique.

Let $X_1, \ldots, X_n$ be a sample from the uniform distribution on $(0, \theta)$ for some $\theta > 0$. The density function of X_i is given by $f_{X_i}(x) = (1/\theta)I_{(0,\theta)}(x)$. Thus,

$$E(X_i) = \int_0^\theta \left(\frac{x}{\theta}\right) dx = \left.\frac{x^2}{2\theta}\right|_0^\theta = \frac{\theta}{2}.$$

Hence it follows that $2\bar{X} = 2(X_1 + \cdots + X_n)/n$ is unbiased for θ, since

$$E(2\bar{X}) = 2E(X_1) = 2\left(\frac{\theta}{2}\right) = \theta.$$

However, we can also show that $[(n+1)/n]X_{(n)}$, where $X_{(n)}$ denotes the largest order statistic of the sample, is unbiased for θ. First note that the cumulative distribution function of $X_{(n)}$ is given by

$$F_{X_{(n)}}(t) = P(X_{(n)} \le t) = P(\text{all } X_i \le t)$$

$$= \prod_{i=1}^{n} F_{X_i}(t) = \prod_{i=1}^{n} \left(\frac{t}{\theta}\right) = \left(\frac{t}{\theta}\right)^n$$

for $0 < t < \theta$. Thus the density function of $X_{(n)}$ is $(n/\theta)(t/\theta)^{n-1}$, and we may compute

$$E(X_{(n)}) = \int_0^\theta t\left(\frac{n}{\theta}\right)\left(\frac{t}{\theta}\right)^{n-1} dt = n\theta^{-n}\int_0^\theta t^n\, dt$$

$$= \left(\frac{n}{n+1}\right)\theta^{-n}t^{n+1}\Big|_0^\theta = \frac{n\theta}{n+1}.$$

It now follows that $E\{[(n+1)/n]X_{(n)}\} = \theta$.

We have shown by counterexample that unbiased estimators are not unique. In fact, this is generally the case. Another example is when the sample is from a Poisson distribution with mean λ and variance λ. When the first two moments of a distribution from a sample exist, then the sample mean $\bar{X}$ and the sample variance $S^2 = [1/(n-1)]\Sigma(X_i - \bar{X})^2$ are always unbiased for the population mean and variance, respectively. Thus $\bar{X}$ and S^2 are both unbiased estimators of λ in the Poisson case.

A trivial example that applies to estimating the mean of any distributional family with finite mean follows from the fact that not only is the sample average an unbiased estimate of the population mean, but any particular observation from the sample—say, X_1—is also an unbiased estimator of the mean.

7.11. A unique unbiased estimator.

Consider a single Bernoulli trial X, where X takes the value one with probability p and zero with probability $1-p$. Let the estimator T take on the value t_1 if the outcome is one and the value t_0 if the outcome is zero. Then

$$E(T) = t_1 p + t_0(1-p) = p(t_1 - t_0) + t_0.$$

The only way that T can be unbiased for p (for all p) is if $p(t_1 - t_0) + t_0 = p$ always. Equating coefficients of similar powers of p, we solve for $t_1 = 1$

and $t_0 = 0$. Thus,

$$T(X) = \begin{cases} 1 & \text{if } X = 1 \\ 0 & \text{if } X = 0, \end{cases}$$

which we recognize as X itself, is the only unbiased estimator of p.

Generally speaking, several unbiased estimators are usually available for a given estimation problem. This example shows that this need not necessarily be the case, however. In fact, whenever the family of distributions of X is complete, any statistic is the unique unbiased estimator for its expectation (provided it is finite).

7.12. A parameter for which no unbiased estimator exists.

Let $(X_1, \ldots, X_n)$ be the indicators of success in n Bernoulli trials with success probability p, and consider estimating the odds ratio $q(p) = p/(1 - p)$. The possible realizations of $(X_1, \ldots, X_n)$ are the 2^n distinct strings of zeros and ones. Any statistic T takes each string into a real number, t_j, where $j = 1, 2, \ldots, 2^n$ enumerates these strings. Therefore we may compute the expectation directly:

$$E(T) = \sum_{j=1}^{2^n} t_j p^{n_j}(1 - p)^{n-n_j}$$

where n_j is the number of successes obtained in the jth string of zeros and ones. For T to be unbiased, we must have the condition

$$\sum_{j=1}^{2^n} t_j p^{n_j}(1 - p)^{n-n_j} = \frac{p}{1 - p} \qquad \text{for all } p \in (0, 1).$$

This says that a polynomial in p must be equal to the ratio $p/(1 - p)$ for all p in an interval. Clearly this cannot be, and it follows that no unbiased estimator of the odds ratio exists.

7.13. A statistic that is unbiased for a parameter but is not median unbiased for that parameter.

The point of Examples 7.13 and 7.14 is that the properties of median unbiasedness and unbiasedness are distinct.

Let X have the exponential distribution with density function

$$f_X(x) = \lambda e^{-\lambda x} I_{(0,\infty)}(x).$$

Then, using integration by parts, the expectation is found to be

$$E(X) = \int_0^\infty x\lambda e^{-\lambda x}\,dx = \frac{1}{\lambda}$$

so that X is an unbiased estimate of its mean $1/\lambda$. The distribution function of X is $1 - e^{-\lambda t}$ for $t > 0$, so that the median m of X satisfies $1 - e^{-\lambda m} = \frac{1}{2}$, and we solve for the median $m = [\log(2)]/\lambda$. Therefore, X is not median unbiased for $1/\lambda$.

When an estimator has a symmetric distribution and its mean exists, so that its mean and median are identical, it will be unbiased for a parameter if and only if it is median unbiased for that parameter. This need not be the case more generally.

7.14. A statistic that is median unbiased for a parameter but is not unbiased for that parameter.

As in Example 7.13, let X have the exponential distribution with mean $1/\lambda$. Then the median of X is $[\log(2)]/\lambda$. Hence the statistic $T = T(X) = X/\log(2)$ is median unbiased for $1/\lambda$. The expectation is $E(T) = E(X)/\log(2) = 1/[\lambda \log(2)] \neq 1/\lambda$, however, and we see that T is not unbiased for $1/\lambda$.

Also note that if X is an observation from $f(x - \theta)$, where $f(x)$ is the Cauchy density (which is symmetric), then X is median unbiased for θ but is not unbiased for θ.

7.15. A "ridiculous" unique unbiased estimator.

Suppose X has a Poisson distribution with mean $\lambda > 0$. The probability function of X is given by

$$p(x; \lambda) = \frac{\lambda^x e^{-\lambda}}{x!}, \qquad x = 0, 1, 2, \ldots.$$

Consider estimating the parameter $q(\lambda) = e^{-3\lambda}$ based on a sample of size one. Let $T(X) = (-2)^X$. Then the expectation is

$$E(T) = \sum_{x=0}^{\infty} (-2)^x \frac{\lambda^x e^{-\lambda}}{x!} = e^{-\lambda} \sum_{x=0}^{\infty} \frac{(-2\lambda)^x}{x!}$$

$$= e^{-\lambda} e^{-2\lambda} = e^{-3\lambda}.$$

Therefore T is unbiased for $e^{-3\lambda}$. However, T is "ridiculous" in the sense that T may be negative (which happens whenever X is odd) even though $e^{-3\lambda}$ is strictly positive. Clearly, the biased estimator $T^* = \max(T, 0)$ would be preferred over the unbiased estimator T because T^* is sometimes closer to $e^{-3\lambda}$ but never farther from $e^{-3\lambda}$ than T is. In fact, a large value of X suggests a large value for λ, which suggests a small value for $e^{-3\lambda}$. However, if $X = 10$, then $T = 1024$, whereas if $X = 11$, then $T = -2048$. These estimated values are clearly ridiculous.

Note that T is the *only* unbiased estimator for $e^{-3\lambda}$. This may be seen either directly by a power-series argument or by using completeness of the Poisson family of distributions. It should not be automatically assumed that a unique unbiased estimator is necessarily good!

7.16. A statistic that is unbiased for a parameter, but whose square is not unbiased for the square of the parameter.

The point of this counterexample is that the property of unbiasedness is not invariant under transformations. Suppose $X_1, \ldots, X_n$ are independent and identically distributed Poisson random variables with mean λ. Then $T(\mathbf{X}) = \bar{X}$, the sample mean, is clearly unbiased for the population mean λ. Since the second moment of a Poisson random variable is $\lambda^2 + \lambda$, we have

$$E(T^2) = E\left[\frac{(X_1 + \cdots + X_n)^2}{n^2}\right] = \frac{1}{n^2}\left[\sum_{i=1}^{n} E(X_i^2) + \sum_{i<j} E(X_i X_j)\right]$$

$$= \frac{n(\lambda^2 + \lambda) + n(n-1)\lambda^2}{n^2} = \lambda^2 + \frac{\lambda}{n}.$$

From this we see that $E(T^2) = [E(T)]^2$ can happen only when $\lambda = 0$. It follows that T^2 is not unbiased for λ^2.

In fact, we can show in general that if T is unbiased for θ, then T^2 will not be unbiased for θ^2 unless T is constant with probability one. Suppose $E(T) = \theta$. Then $\text{Var}(T) = E(T^2) - [E(T)]^2$, so that $E(T^2) = \text{Var}(T) + \theta^2$. So T^2 is unbiased for θ^2 if and only if $\text{Var}(T) = 0$ or, equivalently, if T is constant with probability one.

In particular, for a sample from a nondegenerate distribution, S and S^2 cannot be unbiased for the standard deviation and for the variance of the population, respectively, where S^2 denotes the sample variance.

7.17. Unbiasedness is not an invariant property.

Suppose T is unbiased for θ and g is a function. Example 7.16 showed that $g(T)$ need not be an unbiased estimator of $g(\theta)$ in the particular case where $g(t) = t^2$. For more general functions, an unbiased estimate of $g(\theta)$ need not even exist.

Consider a Bernoulli trial X with success probability θ, so that $P(X = 1) = \theta$ and $P(X = 0) = 1 - \theta$. Then $E(X) = \theta$, so that X is unbiased for θ. If we use the function $g(t) = t/(1 - t)$, however, we see that $g(\theta) = \theta/(1 - \theta)$ is the odds ratio, for which no unbiased estimate exists, as was shown in Example 7.12.

7.18. A statistic that is not complete.

Consider the family of all distributions with finite mean and variance, and let $X_1, \ldots, X_n$ be a sample from a distribution F in this family. The statistic $T = X_1 - X_2$ is not complete because $E(T) = E(X_1) - E(X_2) = 0$ for any distribution in the family, but $X_1 - X_2$ is not identically zero with probability one.

7.19. A family of distributions that is not complete.

Consider the family $\chi_0^2(\lambda)$ of noncentral chi-squared distributions with zero degrees of freedom and noncentrality parameter λ (Siegel 1979). The density of this distribution does not properly exist with respect to Lebesgue measure because there is a mass of size $e^{-\lambda/2}$ at the origin. The positive part of this distribution has a "density" that is given by

$$f_\lambda(t) = \frac{e^{-(\lambda+t)/2}}{t} \sum_{k=1}^{\infty} \frac{(\lambda t/4)^k}{k!\,(k - 1)!}.$$

If X has the $\chi_0^2(\lambda)$ distribution, let $g(X)$ be a statistic that satisfies $g(0) = 0$. Then the expectation of g may be written as

$$E[g(X)] = \int_0^\infty g(t) f_\lambda(t)\,dt.$$

Now choose $g(t) = e^{-t^{1/4}} \sin(t^{1/4})$. Then we find

$$E[g(X)] = e^{-\lambda/2} \sum_{k=1}^{\infty} \frac{(\lambda/4)^k}{k!\,(k - 1)!} \int_0^\infty t^{k-1} e^{-t^{1/4}} \sin(t^{1/4})\,dt.$$

Note that the interchange of summation and integration is valid here because g is integrable and the sum is integrable, consisting of positive summands, so we may apply a form of the Lebesgue Dominated Convergence Theorem. Now, as in Example 3.15, the integral

$$\int_0^\infty t^{k-1} e^{-t^{1/4}} \sin(t^{1/4})\, dt$$

is identically zero for $k = 1, 2, \ldots$. Thus it follows that $E[g(X)] = 0$ even though $g(X)$ is not identically zero with probability one. Therefore the family $\chi_0^2(\lambda)$ is not complete.

7.20. A discrete minimal sufficient statistic that is not complete (Lehmann and Scheffe 1950).

Suppose X is a discrete random variable with

$$P(X = -1) = p$$

and

$$P(X = n) = (1 - p)^2 p^n, \qquad n = 0, 1, 2, \ldots$$

for some p with $0 < p < 1$. Then, for a sample of size one, X is clearly minimal sufficient for p. However, X is not complete because X has expectation zero always. To see this, we compute

$$E(X) = (-1)p + \sum_{j=1}^\infty j(1 - p)^2 p^j = -p + p(1 - p)^2 \sum_{j=1}^\infty j p^{j-1}$$

$$= -p + p(1 - p)^2 \frac{d}{dp}(1 - p)^{-1} = -p + p(1 - p)^2 (1 - p)^{-2} = 0.$$

Thus, taking $g(X) = X$, we see that X is not complete.

7.21. A continuous minimal sufficient statistic that is not complete (Kendall and Stuart 1967, p. 202).

Let X have the uniform distribution on $(\theta, \theta + 1)$. For a sample of size one, X is minimal sufficient. However, let $h(x)$ be any bounded periodic function of period one that is not almost everywhere zero but satisfies

$$\int_0^1 h(x)\,dx = 0.$$

For example, the function $h(x) = \sin(2\pi x)$ will do. Then the expectation of $h(X)$ is

$$E[h(X)] = \int_\theta^{\theta+1} h(x)\,dx = \int_0^1 h(x)\,dx = 0.$$

But $h(X)$ is not zero with probability one, and so this shows that X is not complete.

7.22. A statistic that is boundedly complete but not complete.

Although completeness guarantees bounded completeness, the converse is false, as this counterexample shows. As in Example 7.20, suppose X has the discrete probability function given by

$$P(X = -1) = p$$
$$P(X = n) = (1 - p)^2 p^n, \qquad n = 0, 1, 2, \ldots$$

for some p in $(0, 1)$. It was shown that X is not complete; however, X is boundedly complete. To see this, let $h(X)$ be a statistic. Then $E[h(X)] = 0$ implies that

$$E[h(X)] = ph(-1) + (1 - p)^2 \sum_{n=0}^{\infty} h(n)p^n = 0$$

which is equivalent to

$$\sum_{n=0}^{\infty} h(n)p^n = -\frac{ph(-1)}{(1 - p)^2} = -h(-1)[p + 2p^2 + 3p^3 + \cdots].$$

Boundedness of h implies that the series on the left-hand side converges for $0 < p < 1$. Because two convergent power series can be equal in an interval only if the corresponding coefficients are equal, it follows that $h(n) = nh(-1)$. But if h is bounded, it must then follow that $h(-1) = 0$ and hence that h is identically zero. Thus, X is indeed boundedly complete, although it is not complete.

7.23. A minimal sufficient statistic that is not boundedly complete.

Although completeness and sufficiency together imply minimal sufficiency, the converse is false. Suppose $X_1, \ldots, X_n$ is a sample from a Gaussian distribution with mean μ and variance σ_X^2, and $Y_1, \ldots, Y_m$ is a sample from a Gaussian distribution with mean μ and variance σ_Y^2. Also assume that the samples $X_1, \ldots, X_n$ and $Y_1, \ldots, Y_m$ are independent. Then $T = (\bar{X}, \bar{Y}, S_X^2, S_Y^2)$, consisting of the sample averages and sample variances, is minimal sufficient, as may be seen by examining the exponential form of the likelihood. However, T is not boundedly complete (and hence is not complete) because the statistic

$$T' = \begin{cases} \bar{X} - \bar{Y} & \text{if } |\bar{X} - \bar{Y}| < 1 \\ 0 & \text{otherwise} \end{cases}$$

has expectation zero, since the distribution of $\bar{X} - \bar{Y}$ is symmetric about zero. Lack of bounded completeness follows because T' is a bounded function of T but is not identically zero.

7.24. A family of univariate distributions indexed by a parameter (θ, ϕ) such that the family is complete with respect to θ when ϕ is known but is not even boundedly complete with respect to ϕ when θ is known.

Let X have a Gaussian distribution with mean μ and variance σ^2. Then the likelihood of X is

$$f(x; \mu, \sigma^2) = \exp\left[-\frac{x^2}{2\sigma^2} + \frac{\mu x}{\sigma^2} - \frac{\mu^2}{2\sigma^2} - \frac{1}{2}\log(2\pi\sigma^2)\right].$$

Because this density function is expressed in its exponential form, we find that X is complete when σ^2 is fixed. On the other hand, when $\mu = \mu_0$ is fixed and known, define

$$T(X) = \begin{cases} X - \mu_0 & \text{if } |X - \mu_0| < 1 \\ 0 & \text{otherwise.} \end{cases}$$

Then T is bounded with zero expectation because $X - \mu_0$ is symmetrically distributed about zero. Thus it follows that X is not even boundedly complete in this case when μ is fixed and known.

7.25. In an exponential family, a natural minimal sufficient statistic that is not complete (Cox and Hinkley 1974, p. 31).

As noted in the introduction to this chapter, if $X_1, \ldots, X_n$ is a sample from a distribution that is a member of a k-parameter exponential family with minimal sufficient statistic T, and if the dimension of the parameter space is k, then T is complete. The requirement that the dimension of the parameter space be equal to k is necessary, as this example shows.

As in Example 7.1, suppose $X_1, \ldots, X_n$ is a sample from a Gaussian distribution with mean μ and variance $\mu^2 v_0^2$ for some known v_0, the coefficient of variation. We saw that $T = (\Sigma X_i, \Sigma X_i^2)$ is minimal sufficient for μ. We will show that this minimal sufficient statistic T is not complete, however. Consider the statistic S defined by

$$S = \frac{v_0^2}{1 + v_0^2} \sum_{i=1}^{n} X_i^2 - \left[\sum_{i=1}^{n} X_i \right]^2 .$$

Since $E(X_i^2) = \text{Var}(X_i) + [E(X_i)]^2 = \mu^2 v_0^2 + \mu^2$, some algebra will show that S has expectation zero for all μ. Thus T is not complete. This could happen because, whereas there is only one unknown parameter, the family of distributions comes from a two-parameter exponential family.

CHAPTER EIGHT
CONSTRUCTION OF ESTIMATORS

Introduction

Standard methods for finding estimators proceed as follows. The data observed are a sample from a parent population indexed by a parameter $\theta \in \Omega$, where θ is often real- or vector-valued. We would like to consider estimators T of θ or some function $q(\theta)$. The question of how "good" the estimators are will be dealt with in Chapter 9. However, the following constructions of estimators seem "reasonable" on intuitive grounds.

Suppose $X_1, \ldots, X_n$ is a sample from a family of distributions indexed by real parameters $\theta_1, \ldots, \theta_k$. The "method of moments" prescribes the simple idea of equating population moments to sample moments. Precisely, define the jth sample moment m_j as

$$m_j = \sum_{i=1}^{n} \frac{X_i^j}{n}, \qquad j = 1, 2, \ldots.$$

In applying the method of moments to estimate the parameter $q(\theta_1, \ldots, \theta_k)$, we must be able to express q as a function of the first r, say, population moments. In such a case, we have

$$q(\theta_1, \ldots, \theta_k) = g[E(X_i), E(X_i^2), \ldots, E(X_i^r)].$$

The method of moments then takes the estimator $T(\mathbf{X})$ of $q(\boldsymbol{\theta})$ given by

$$T(\mathbf{X}) = g(m_1, m_2, \ldots, m_r).$$

The esteemed method of "maximum likelihood," first proposed by C. F. Gauss in 1821 (see Bickel and Doksum 1977, p. 99), is constructed as follows. Suppose $X_1, \ldots, X_n$ is a sample from a family of distributions $\mathbf{P} = \{P_\theta : \theta \in \Omega\}$ dominated by a σ-finite measure μ, and $L(\mathbf{x}; \theta)$ is the likelihood function (that is, the density function of $\mathbf{X}$ evaluated at $\mathbf{x}$ viewed as a function of the parameter θ). (See the introduction to Chapter 7 for a definition of the likelihood function.) The method of maximum likelihood proposes an estimate T of θ that is "most likely" to have produced the data. In other words, if T maximizes the likelihood function over all values of θ so that

$$L[\mathbf{X}; T(\mathbf{X})] = \sup\{L(\mathbf{X}; \theta) : \theta \in \Omega\}$$

then T is a maximum likelihood estimator of θ. A maximum likelihood estimator of a function $q(\theta)$ is defined to be $q(T)$. The T is a maximum likelihood estimator of θ. Under suitable regularity conditions [Ω is an open set of $\mathbf{R}^k$, the maximum likelihood estimator $T(\mathbf{X})$ exists in Ω, and $L(\mathbf{x}; \boldsymbol{\theta})$ is differentiable in $\boldsymbol{\theta} = (\theta_1, \ldots, \theta_k)$ for fixed $\mathbf{x}$], it is necessary that the maximum likelihood estimator satisfy the "likelihood equations"

$$\frac{\partial \log[L(\mathbf{x}; \boldsymbol{\theta})]}{\partial \theta_j} = 0 \qquad \text{for } j = 1, \ldots, k.$$

Note that the maximum likelihood method adheres to the Likelihood Principle, which says that only the likelihood need be used in inference.

8.1. A biased method of moments estimator.

Examine $X_1, \ldots, X_n$ from a Gaussian distribution with mean μ and variance σ^2. Consider estimating σ^2. Since $\sigma^2 = E(X_i^2) - [E(X_i)]^2$, the method of moments prescribes the estimator

$$T(\mathbf{X}) = \sum_{i=1}^{n} \frac{X_i^2}{n} - \left[\sum_{i=1}^{n} \frac{X_i}{n}\right]^2.$$

However, the expected value of this estimator is

$$E[T(\mathbf{X})] = \frac{1}{n}\sum_{i=1}^{n} E(X_i^2) - [E(\bar{X}^2)]$$

$$= \sigma^2 + \mu^2 - \left[\frac{\sigma^2}{n} + \mu^2\right]$$

$$= \left(\frac{n-1}{n}\right)\sigma^2 \neq \sigma^2.$$

Thus T is not an unbiased estimator of σ^2. A close look shows that this example does not depend on the fact that the family considered is Gaussian because, in general, T is a method of moments estimator for the variance of the parent population. The adjusted estimator

$$S^2 = \frac{n}{n-1}T = \frac{1}{n-1}\sum_{i=1}^{n}(X_i - \bar{X})^2$$

is then an unbiased estimator of the variance of the population.

8.2. Nonuniqueness of method of moments estimators.

Consider the problem of estimating the intensity, λ, of a Poisson distribution. The first two moments are

$$E(X_i) = \lambda \quad \text{and} \quad E(X_i^2) = \lambda + \lambda^2.$$

If we use the first moment to construct a method of moments estimator T_1, we are led to the estimator $T_1(\mathbf{X}) = \bar{X}$, the sample average. On the other hand, if we use the second moment to construct a method of moments estimator T_2, then we must solve

$$\frac{1}{n}\sum_{i=1}^{n} X_i^2 = T_2(\mathbf{X}) + [T_2(\mathbf{X})]^2.$$

Solving this equation (by taking the positive root), we find that

$$T_2(\mathbf{X}) = -\frac{1}{2} + \left[\frac{1}{4} + \frac{1}{n}\sum_{i=1}^{n} X_i^2\right]^{1/2}.$$

It is clear that the two estimators T_1 and T_2 are different, despite the fact that the first two central moments of a Poisson distribution, $E(X)$ and $\mathrm{Var}(X)$, are the same. Thus the method of moments technique does not guarantee unique estimators.

Note that this is, unfortunately, a general property of the method of moments. A second example is when the parent population follows the noncentral chi-squared distribution with zero degrees of freedom, $\chi_0^2(\lambda)$, with noncentrality parameter λ (see Example 7.19). The moments are calculated in Siegel (1979):

$$E(X_i) = \lambda \quad \text{and} \quad E(X_i^2) = \lambda^2 + 4\lambda.$$

Using each of the first two moments again, we get different estimators.

8.3. A method of moments estimator that may produce a value that could not be the true parameter (Bickel and Doksum 1977, p. 93).

Consider the problem of estimating the size of a population. Suppose the population has θ members labeled from 1 to θ. The population is sampled with replacement. Thus, the sample mean is trivially calculated to be

$$\mu = E(\bar{X}) = \frac{\theta + 1}{2}.$$

Because $\theta = 2\mu - 1$, the method of moments prescribes $2\bar{X} - 1$ as an estimator of θ. This estimator, of course, behaves "badly" if $X_{(n)}$, the maximum order statistic of the sample, is greater than $2\bar{X} - 1$, since we know that θ must be at least $X_{(n)}$. Indeed, $\max(X_{(n)}, 2\bar{X} - 1)$ is a "better" estimator than just $2\bar{X} - 1$ because it is never farther from, but is sometimes closer to, θ.

8.4. A method of moments estimator that is not a function of complete sufficient statistics.

We will give two examples. First, suppose X_i is uniform on $(\theta_1 - \theta_2, \theta_1 + \theta_2)$ with density function

$$f(x; \theta_1, \theta_2) = \begin{cases} \dfrac{1}{2\theta_2} & \text{if } \theta_1 - \theta_2 < x < \theta_1 + \theta_2 \\ 0 & \text{otherwise} \end{cases}$$

where $\theta_2 > 0$. Method of moments estimators for θ_1 and θ_2 based on the first two moments are easily calculated to be

$$\bar{X} \quad \text{and} \quad \left[3 \sum_{i=1}^{n} \frac{(X_i - \bar{X})^2}{n}\right]^{1/2}.$$

However, the complete and sufficient statistics for this family of distributions are $(X_{(1)}, X_{(n)})$ (see Patel, Kapadia, and Owen 1976, p. 147).

As a second example, suppose X has a noncentral chi-squared distribution with zero degrees of freedom and noncentrality parameter λ. Then, by Example 8.2, X is itself a method of moments estimator for λ. On the other hand, by Example 7.19, the statistic X is not complete.

8.5. Nonexistence of a method of moments estimator.

Although the method of moments is sometimes justified as a quick, easily computed prescription for constructing estimators, they may not even exist, as the following two examples show.

First, let $X_1, \ldots, X_n$ be a sample from the Cauchy distribution with location parameter θ. From Example 3.2, we know that no moments of the Cauchy distribution exist. Clearly, the method of moments does not help in finding an estimator for the center of the distribution.

A second way that the method of moments may not yield a well-defined estimator is exemplified by the following. Suppose $X_1, \ldots, X_n$ is a sample from the uniform distribution on $(-\theta, \theta)$ for some $\theta > 0$. The first moment of X_i is identically zero, so that no method of moments estimator exists using the first population moment, since the expectation of the sample mean does not depend on θ. Of course, method of moments estimators may be constructed for this distribution by using higher moments.

8.6. An uninformative likelihood (Rao 1973, p. 355).

Suppose there are N balls labeled $1, 2, \ldots, N$ with unknown distinct values $\theta_1, \ldots, \theta_N$ written on them. Draw n balls at random without replacement (where $n < N$) and observe (a_i, b_i) for $i = 1, 2, \ldots, n$, where a_i is the number of the ith ball drawn and b_i is the value θ_{a_i} written on the ball labeled a_i. Thus the sample S consists of the outcomes $S = \{(a_1, b_1), \ldots, (a_n, b_n)\}$. Now consider estimating the parameter $\theta = \theta_1 + \cdots + \theta_N$. If inference is to be based on the likelihood of the data (by using the method of maximum likelihood, for instance), the likelihood function must be examined.

Given an observation S, the likelihood of the unknown parameters $\theta_1, \ldots, \theta_N$ is zero if the parameters are not consistent with the observations and is constant with value $\binom{N}{n}^{-1}$ otherwise, irrespective of the parameter values of unobserved balls. Thus, the likelihood principle does not by itself give preference to any set of parametric values for the unobserved balls. Maximum likelihood estimators do exist, however, and one could be chosen, but not based on the likelihood of the sample.

It should be pointed out that although the likelihood itself is uninformative on unknown values, it is wrong to say that the sample does not contain any information about θ. For example, the estimator $N(b_1 + \cdots + b_n)/n$ is unbiased for θ, which indicates that the sample has information about θ. In any case, it is clear that the likelihood approach does have its limitations. One can expect to run into trouble in a situation like this one in which there are more parameters than data values.

8.7. A biased maximum likelihood estimator.

Let $X_1, \ldots, X_n$ be a sample from the uniform distribution on $(0, \theta]$ for some $\theta > 0$. The likelihood function is given by

$$L(\mathbf{x}; \theta) = \prod_{i=1}^{n} \frac{1}{\theta} I_{(0,\theta]}(x_i) = \theta^{-n} I_{(0,\theta]}(x_{(n)})$$

where $x_{(n)} = \text{maximum}(x_1, \ldots, x_n)$. Maximizing $L(\mathbf{x}; \theta)$ over all possible $\theta > 0$ gives $X_{(n)}$ as the unique maximum likelihood estimator of θ. However, from Example 7.10 we find that

$$E[X_{(n)}] = \frac{n}{n+1}\theta.$$

Thus maximum likelihood estimators, in general, need not be unbiased.

Even for the Gaussian distribution the maximum likelihood estimators need not be unbiased. Let $X_1, \ldots, X_n$ be a random sample from the Gaussian distribution with mean μ and variance σ^2. Then it can be calculated that the maximum likelihood estimators are

$$\frac{1}{n}\sum_{i=1}^{n} X_i \qquad \text{for } \mu$$

$$\frac{1}{n}\sum_{i=1}^{n} (X_i - \bar{X})^2 \qquad \text{for } \sigma^2.$$

Note that whereas the estimate of the mean μ is unbiased, the estimate of the variance σ^2 is not the unbiased estimator S^2 mentioned in Example 8.1 but is the biased estimator $[(n - 1)/n]S^2$ instead.

8.8. Two data sets that each provide the same value of a maximum likelihood estimate but, when combined into one data set, provide a different value.

If two data sets provide the same values for all components of the maximum likelihood estimate of a vector of parameters, then combining the two data sets into one will result in the same maximum likelihood estimate. This may be seen by writing the likelihood function as a product of the likelihoods of the two original independent samples and observing that if each term is maximized by the same choice of parameters, then their product will also be maximized there.

If we fix attention on just one parameter, however, the preceding argument does not hold. Consider sampling from the Gaussian distribution with mean μ and variance σ^2, and focus attention on the maximum likelihood estimator $\Sigma(X_i - \bar{X})^2/n$ for σ^2. Consider the two data sets (9, 10, 11) and (29, 30, 31). For the first data set, we estimate σ^2 as

$$\frac{(9 - 10)^2 + (10 - 10)^2 + (11 - 10)^2}{3} = \frac{2}{3}.$$

For the second data set, we find the same estimate of σ^2:

$$\frac{(29 - 30)^2 + (30 - 30)^2 + (31 - 30)^2}{3} = \frac{2}{3}.$$

However, for the combined data set (9, 10, 11, 29, 30, 31), we find a different estimator for σ^2:

$$\frac{(9-20)^2 + (10-20)^2 + (11-20)^2 + (29-20)^2 + (30-20)^2 + (31-20)^2}{6}$$

$$= 100\frac{2}{3} \neq \frac{2}{3}.$$

8.9. Nonuniqueness of solutions to the likelihood equation.

Suppose $X_1, \ldots, X_n$ is a sample from the Cauchy distribution with location parameter θ and density function given by

$$f(x;\theta) = \frac{1}{\pi[1 + (x - \theta)^2]} \qquad \text{for } -\infty < x < \infty.$$

The log of the likelihood function is given by

$$\log[L(\mathbf{x};\theta)] = -n\log(\pi) - \sum_{i=1}^{n} \log[1 + (x_i - \theta)^2].$$

Taking the partial derivative with respect to θ, we find

$$\frac{\partial[\log(L)]}{\partial\theta} = \sum_{i=1}^{n} \frac{2(x_i - \theta)}{1 + (x_i - \theta)^2}.$$

Unfortunately, although maximum likelihood estimators are often easy to compute, it is inconvenient to find the solution(s) to this particular likelihood equation. For a given sample, the values of θ that correspond to relative maxima of $\log(L)$ must be found. If more than one occurs, the maximum likelihood estimate corresponds to the value that yields the absolute maximum of $\log(L)$.

Barnett (1966) has found that multiple solutions of the likelihood equation do indeed occur in this Cauchy location case. This is not surprising, since (by putting terms over a common denominator) solving the likelihood equation is equivalent to finding the roots of a polynomial in θ of degree $2n - 1$. Depending on the sample itself, the maximum likelihood estimator may not even be unique. More details may be found in Barnett (1966) and in Lehmann (1983, Section 6.3, p. 423).

8.10. Nonuniqueness of the maximum likelihood estimator.

This example shows that there may be literally an infinite number of distinct maximum likelihood estimators for a parameter. Let the sample $X_1, \ldots, X_n$ come from a uniform distribution on $(\theta - \frac{1}{2}, \theta + \frac{1}{2})$ with density given by

$$f(x; \theta) = I_{(\theta-1/2,\theta+1/2)}(x)$$

for some real number θ. The likelihood function of the data is then given by

$$L(\mathbf{x}; \theta) = \prod_{i=1}^{n} I_{(\theta-1/2,\theta+1/2)}(x_i) = I_{(x_{(n)}-1/2,x_{(1)}+1/2)}(\theta)$$

where $x_{(1)}$ and $x_{(n)}$ are, respectively, the minimum and the maximum values in the sample.

Note that the likelihood function takes on its maximum value of one whenever θ falls between $x_{(n)} - 0.5$ and $x_{(1)} + 0.5$. It follows that any estimator T that satisfies $x_{(n)} - 0.5 < T < x_{(1)} + 0.5$ will be a maximum likelihood estimator of θ. A whole range of possibilities for T then exists, with the cardinality of the continuum, and therefore an uncountably infinite number of maximum likelihood estimators exist.

We see that the method of maximum likelihood does not necessarily prescribe a unique estimator (or even a finite number of values) to estimate a parameter. In this connection, also see Example 8.6.

8.11. Nonexistence of a maximum likelihood estimator.

We will give two examples. First, let X be a sample of size one from the Gaussian distribution with mean μ and variance σ^2. Consider the bivariate parameter $\theta = (\mu, \sigma^2)$ within the parameter space $\{(\mu, \sigma^2): -\infty <$

$\mu < \infty, \sigma^2 > 0\}$. The log likelihood function is given by

$$\log[L(x; \mu, \sigma^2)] = -\frac{1}{2}\log(2\pi\sigma^2) - \frac{(x-\mu)^2}{\sigma^2}.$$

No maximum likelihood estimate of (μ, σ^2) exists because the log likelihood function becomes unbounded when $x = \mu$ and σ^2 tends to zero. Of course, it would be ridiculous to conclude that σ^2 is small after observing only one observation.

One way to avoid problems like this is to do as Kempthorne (1966) recommended and require that a properly defined likelihood function be bounded.

Our second example (from Bickel and Doksum 1977, Problem 3.3.12) is originally due to Kiefer and Wolfowitz. Suppose $X_1, \ldots, X_n$ is a sample from a mixture of Gaussian distributions, with density given by

$$f(x; \mu, \sigma^2) = \frac{9}{10}g\left(\frac{x-\mu}{\sigma}\right) + \frac{1}{10}g(x-\mu)$$

where $g(t)$ is the density of the standard Gaussian distribution with mean zero and variance one; then $g(t) = (2\pi)^{-1/2}\exp(-t^2/2)$. Consider estimating the parameters (μ, σ^2) within the parameter space $\Omega = \{(\mu, \sigma^2): -\infty < \mu < \infty, \sigma^2 > 0\}$. We will show that the maximum likelihood estimates do not exist because the likelihood function is unbounded. The likelihood function is

$$L(\mathbf{x}; \mu, \sigma^2) = \prod_{i=1}^{n}\left[\frac{9\exp[-(x_i-\mu)^2/(2\sigma^2)]}{10\sigma(2\pi)^{1/2}} + \frac{\exp[-(x_i-\mu)^2/2]}{10(2\pi)^{1/2}}\right].$$

Next, we would like to expand this product. In general, the expansion of $\Pi_{i=1}^{n}(a_i + b_i)$ has 2^n terms, each consisting of a product of a_i or b_i for each i. Some of the terms look like this:

$$\prod_{i=1}^{n}(a_i + b_i) = a_1a_2\cdots a_n + b_1a_2\cdots a_n + b_1b_2a_3\cdots a_n + \cdots$$
$$+ b_1b_2\cdots b_{n-1}a_n + b_1b_2\cdots b_n.$$

Thus we may write

$$L(\mathbf{x}; \mu, \sigma^2) = \sum_{j=1}^{2^n}\left[\prod_{i\in A_j}\left[\frac{9\exp[-(x_i-\mu)^2/(2\sigma^2)]}{10\sigma(2\pi)^{1/2}}\right]\right.$$
$$\left.\cdot\prod_{i\in A_j^c}\left[\frac{\exp[-(x_i-\mu)^2/2]}{10(2\pi)^{1/2}}\right]\right]$$

where the products are taken over all i in A_j and over all i in A_j^c, the complement of A_j, respectively. The A_j denote the 2^n distinct subsets of the integers $\{1, \ldots, n\}$.

Now we show that the likelihood L is unbounded. First of all, note that every term in the product expansion is nonnegative. Thus, if we can show that any one term is unbounded within the parameter space Ω, we are done. In particular, look at the $j = 2^n - 1$ term:

$$\frac{9\exp[-(x_1 - \mu)^2/(2\sigma^2)]}{10\sigma(2\pi)^{1/2}} \prod_{i=2}^{n} \frac{\exp[-(x_i - \mu)^2/2]}{10(2\pi)^{1/2}}.$$

Let $\tilde{\mu} = x_1$. The product on the right side from $i = 2$ to n evaluated at $\tilde{\mu} = x_1$ is some positive finite number. However, the first factor evaluated at $\tilde{\mu}$ becomes

$$\frac{9\exp[-(x_1 - \tilde{\mu})^2/(2\sigma^2)]}{10\sigma(2\pi)^{1/2}} = \frac{9}{10(2\pi)^{1/2}} \cdot \frac{1}{\sigma}.$$

Clearly, this becomes unbounded as σ approaches zero, and hence the likelihood is unbounded by setting $\tilde{\mu} = x_1$ and letting σ approach zero. Thus, for no $(\tilde{\mu}, \tilde{\sigma}^2)$ in Ω does the likelihood attain a maximum. Furthermore, we could have set $\tilde{\mu} = x_i$ for any i and the same argument would have shown that the likelihood becomes unbounded as σ tends to zero.

8.12. A maximum likelihood estimator that is discontinuous.

Let the random variable X have the noncentral chi-squared distribution with zero degrees of freedom and noncentrality parameter $\lambda \geq 0$. Consider finding a maximum likelihood estimator of λ with a sample of size one. The distribution of X has a discrete mass of $e^{-\lambda/2}$ at zero and a continuous part on $(0, \infty)$ with a density given by

$$f_\lambda(t) = \frac{1}{t} e^{-(\lambda + t)/2} \sum_{k=1}^{\infty} \frac{(\lambda t/4)^k}{k!(k-1)!}.$$

This mixture of discrete and continuous variables has a density with respect to the dominating measure consisting of Lebesgue measure on $(0, \infty)$ plus a point mass at zero. The likelihood function is then given by

$$L(t; \lambda) = \begin{cases} f_\lambda(t) & \text{if } t > 0 \\ e^{-\lambda/2} & \text{if } t = 0. \end{cases}$$

First, suppose the observation is $X = 0$. Then the likelihood, $e^{-\lambda/2}$, is maximized when $\lambda = \lambda^*(0) = 0$. If, on the other hand, the observation is $X = t > 0$, then the maximum likelihood estimator will be given by the unique solution to the equation

$$\frac{\partial f_\lambda(t)}{\partial \lambda} = \sum_{k=1}^{\infty} \frac{(\lambda/2)^{k-1}(t/2)^k}{(k-1)!(k-1)!}\left[1 - \frac{\lambda}{2k}\right] = 0.$$

Thus, for any value of $t > 0$, the solution $\lambda^*(t)$ must be greater than two. Otherwise, all the terms in the preceding summation would be positive and their sum would not be zero. Therefore we have

$$\lim_{t \to 0^+} \lambda^*(t) \geq 2, \quad \text{but} \quad \lambda^*(0) = 0.$$

Thus the maximum likelihood estimator $\lambda^*(t)$ for λ is discontinuous at $t = 0$. Furthermore, no values of λ between zero and two can possibly be estimated using the method of maximum likelihood.

8.13. A maximum likelihood estimator that is not a function of a given sufficient statistic (Moore, 1971).

Although it is often claimed that a maximum likelihood estimator must be a function of any sufficient statistic, this proposition is false in general. The "proof," as often stated, is easily derived. Suppose $T = t(X_1, \ldots, X_n)$ is sufficient for a parameter θ, where both T and θ may possibly be vector-valued. Then the factorization criterion for sufficiency says that we may write the likelihood of the sample as

$$L(\mathbf{x}; \theta) = g(t(\mathbf{x}); \theta)h(\mathbf{x}).$$

A maximum likelihood estimate θ^* maximizes L over all possible θ for fixed $\mathbf{x} = (x_1, \ldots, x_n)$. Since $\mathbf{x}$ is fixed, so is $h(\mathbf{x})$, so that maximizing L over θ is equivalent to maximizing g over θ. Thus θ^* will depend on $\mathbf{x}$ only through t. So what is the flaw? The problem with this argument is that the set S of values of θ that maximize the likelihood L depends on the sample only through t. On the other hand, if this set S does not consist of one unique value, a particular maximum likelihood estimate must be chosen from S. Because this selection may be done in a manner independent of the sample, the "proof" is wrong.

This incorrect "proof" may even be found in some of the best introductory statistics textbooks. The correctly stated theorem is as follows (for example, Moore 1971). If T is sufficient for θ and a unique maximum likelihood estimate θ^* of θ exists, then θ^* must be a function of T. Moreover, if several maximum likelihood estimators exist, a maximum likelihood estimate may be chosen to be a function of the sufficient statistic T.

To give a concrete example, suppose $X_1, \ldots, X_n$ is a sample from the uniform distribution on $(\theta - \frac{1}{2}, \theta + \frac{1}{2})$. As in Example 7.10, we find that $(X_{(1)}, X_{(n)})$ is a jointly sufficient statistic for θ. Furthermore, from Example 8.10, any θ^* that satisfies $x_{(n)} - 0.5 < \theta^* < x_{(1)} + 0.5$ is a maximum likelihood estimator of θ. For instance, define θ^* by

$$\theta^*(\mathbf{X}) = (X_{(n)} - 0.5) + [\cos(X_1)]^2[X_{(1)} - X_{(n)} + 1].$$

Then θ^* as defined is a maximum likelihood estimator because it satisfies the preceding inequality. However, θ^* does not depend on the sample only through the jointly sufficient statistic $(X_{(1)}, X_{(n)})$. Note that $(X_{(1)} + X_{(n)})/2$ is also a maximum likelihood estimator that does depend on the sample only through $(X_{(1)}, X_{(n)})$, in accord with the true theorem as previously stated.

8.14. A maximum likelihood estimator that does not satisfy the likelihood equation.

Suppose $X_1, \ldots, X_n$ is a sample from the uniform distribution on $(0, \theta]$ for some $\theta > 0$. The likelihood of the sample is given by

$$L(\mathbf{x}; \theta) = \prod_{i=1}^{n} \left(\frac{1}{\theta}\right) I_{(0,\theta]}(x_i) = \left(\frac{1}{\theta^n}\right) I_{[x_{(n)},\infty)}(\theta).$$

Thus the likelihood is a strictly monotone decreasing function of θ beyond $\theta = x_{(n)}$, where it attains its unique maximum. The likelihood is, in fact, discontinuous at this point, and hence its derivative cannot be zero at $x_{(n)}$. Thus the maximum likelihood estimator is not a solution of the likelihood equation.

8.15. A likelihood equation that yields a spurious solution for a maximum likelihood estimator.

As in Example 8.14, if $X_1, \ldots, X_n$ is a sample from the uniform distribution on $(0, \theta]$, then $X_{(n)}$ is the maximum likelihood estimator of θ. On the other hand, the derivative of the likelihood is identically zero for any value of θ such that $0 < \theta < X_{(n)}$. Of course, these solutions all correspond to a minimum value of zero for the likelihood function and hence do not correspond to maximum likelihood estimators.

8.16. A likelihood equation with a unique root, but no maximum likelihood estimator exists.

Consider the location family with density $f(x - \theta)$, where θ is the unknown (real) parameter and f is given by

$$f(x) = \begin{cases} \dfrac{e^{-|x|}}{4(\pi|x|)^{1/2}} & \text{if } x < 0 \\ \dfrac{1}{2\pi[x(1 - x)]^{1/2}} & \text{if } 0 < x < 1 \\ \dfrac{e^{-(x-1)}}{4[\pi(x - 1)]^{1/2}} & \text{if } x > 1 \\ 0 & \text{if } x = 0 \text{ or } 1. \end{cases}$$

A plot of this density function is shown in Figure 8.16.1. With a sample of size one, the likelihood function will take the same basic form as f itself. From the figure, we see that there is just one place where the likelihood

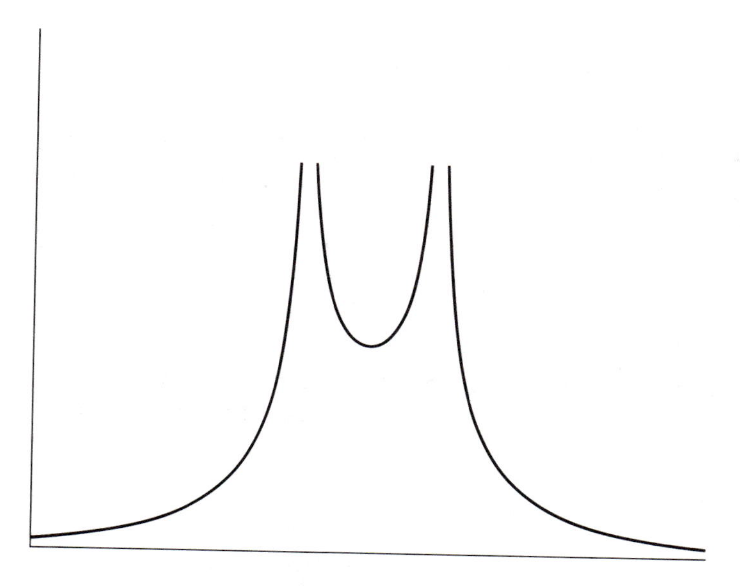

FIGURE 8.16.1

equation $\partial f(X - \theta)/\partial\theta = 0$ is satisfied and that this is a local minimum of the likelihood rather than a maximum. No maximum likelihood estimator exists here because the maximum value of the likelihood, ∞, is not attained by the function.

8.17. A maximum likelihood estimator that either does not satisfy the likelihood equation or is not unique (Cox and Hinkley 1974, p. 285).

Let $X_1, \ldots, X_n$ be a sample from the double exponential (Laplace) distribution with density given by

$$f(t; \theta) = \tfrac{1}{2}e^{-|t-\theta|}.$$

The log of the likelihood function is given by

$$\log[L(\mathbf{x}; \theta)] = -n\log(2) - \sum_{i=1}^{n} |x_i - \theta|.$$

Note that maximizing the likelihood function with respect to θ is equivalent to minimizing the sum of absolute deviations from θ. A peculiar feature of the log likelihood is that, although it is continuous everywhere, it is differentiable except at the sample points $x_1, \ldots, x_n$.

Because the order of summation does not affect the sum, we may use the order statistics:

$$\log[L(\mathbf{x}; \theta)] = -n\log(2) - \sum_{i=1}^{n} |x_{(i)} - \theta|.$$

Thus, maximizing the likelihood with respect to θ is equivalent to minimizing $g(\theta)$, the sum of the absolute deviations of the order statistics from θ, given by

$$g(\theta) = \sum_{i=1}^{n} |x_{(i)} - \theta|.$$

If $n = 2k + 1$ is odd, then the minimum will occur at $\theta^* = x_{(k+1)} = \text{median}(x_1, \ldots, x_n)$. To see this, suppose θ is between $x_{(j)}$ and $x_{(j+1)}$. Then

$$g(\theta) = \sum_{i=1}^{j-1} (\theta - x_{(i)}) + x_{(j+1)} - x_{(j)} + \sum_{i=j+2}^{n} (x_{(i)} - \theta).$$

Increasing θ by some small amount ε will then increase the left-hand sum by $(j-1)\varepsilon$ and decrease the right-hand sum by $(n-j-1)\varepsilon$. Thus the total sum $g(\theta)$ will decrease if and only if $n - j - 1 > j - 1$—that is, if and only if $n > 2j$. Increasing j to decrease the sum will work until $j = k$, since $n = 2k + 1 > 2k$. But if θ is between $x_{(k)}$ and $x_{(k+1)}$, since $n > 2k$, we can increase θ by ε until $\theta = x_{(k+1)}$ and then $g(\theta)$ decreases. For $j \geq k + 1$, we will have $n > 2j$, and so $g(\theta)$ increases in θ when $\theta > x_{(k+1)}$. Thus, $\theta^* = x_{(k+1)}$ minimizes $g(\theta)$. Therefore we have a unique maximum likelihood estimator $\theta^* = x_{(k+1)}$ that does not satisfy the likelihood equation because the likelihood does not have a derivative at $\theta^* = x_{(k+1)}$ (or at any other data value).

A similar argument shows that if $n = 2k$ is even, then $g(\theta)$ is minimized for any θ in the interval $(x_{(k)}, x_{(k+1)})$ and thus [provided $x_{(k)} \neq x_{(k+1)}$] an infinite number of maximum likelihood estimators exist. All these non-unique estimators satisfy the likelihood equation because the likelihood function is constant in this region and therefore has derivative zero. Should it happen that $x_{(k)} = x_{(k+1)}$, an event of probability zero, the maximum likelihood estimate will be unique but will not satisfy the likelihood equation, as in the case of odd sample size.

8.18. A maximum likelihood estimator of the population mean that is not the sample mean.

Often the maximum likelihood estimator of the population mean is $\bar{X}$, the sample mean, as happens in samples from binomial, exponential, Poisson, and Gaussian populations, to name a few. In fact, the method of moments (using the first moment) always prescribes estimating the population mean by the sample mean. However, this is not always the case with the method of maximum likelihood.

As in Example 8.13, let $X_1, \ldots, X_n$ be a sample from the uniform distribution on $(\theta - \frac{1}{2}, \theta + \frac{1}{2})$. The population mean is θ. However, a maximum likelihood estimator of θ was calculated to be $(X_{(1)} + X_{(n)})/2 \neq \bar{X}$. Note that because of the nonuniqueness of the maximum likelihood estimator in this case any value between $X_{(n)} - 0.5$ and $X_{(1)} + 0.5$ is a maximum likelihood estimator. Whereas $\bar{X}$ may fall within this range, $\bar{X}$ may well not be a maximum likelihood estimator. For example, with the data set 5.1, 5.3, 5.2, 5.9, we have the range of maximum likelihood estimators from $5.9 - 0.5 = 5.4$ to $5.1 + 0.5 = 5.6$. However, the sample mean is $\bar{X} = 5.375$, which does not fall within (5.4, 5.6) and is therefore not a maximum likelihood estimator.

An alternative was given in Example 8.17, where the sample is taken from the double exponential distribution with mean θ. Instead of $\bar{X}$ being the maximum likelihood estimator of θ, the sample median (determined uniquely for odd sample sizes) is the maximum likelihood estimator.

CHAPTER NINE
OPTIMALITY OF ESTIMATORS

Introduction

Given a sample from a family of distributions $\{\mathbf{P}_\theta : \theta \in \Omega\}$ indexed by a parameter θ, we would like to find a statistic whose value will be "close" to $q(\theta)$. Chapter 8 was devoted to the construction of estimators. Now we would like to measure the performance of competing estimators. For example, the "mean squared error" of a real-valued estimator T of a real parameter $q(\theta)$ based on a sample $\mathbf{X} = (X_1, \ldots, X_n)$ is defined to be

$$\text{MSE}(T; \theta) = E_\theta\{[T(\mathbf{X}) - q(\theta)]^2\}.$$

Naturally, we would like an estimator T to have a small value for the mean squared error. Conveniently, the mean squared error of an estimator T can be decomposed into two components:

$$\text{MSE}(T; \theta) = \text{Var}_\theta(T) + [\text{Bias}(T; \theta)]^2$$

where the "bias" is defined as $\text{Bias}(T; \theta) = q(\theta) - E_\theta(T)$. An estimator is said to be "unbiased" if its bias is always zero. In general, we can define a loss function $L(T; \theta)$ with the properties that $L(T; \theta)$ is nonnegative and $L(q(\theta); \theta) = 0$. Then $L(t; \theta)$ is the loss incurred if $q(\theta)$ is estimated by $T = t$ when θ is the true parameter. Clearly, the loss depends on the sample and on θ as well as on the estimator. For a given loss function L, it is then natural to define the risk R of an estimator as its expected loss:

$$R(T; \theta) = E_\theta[L(T; \theta)].$$

In general, an estimator is said to be "risk unbiased" if $E_\theta[L(T; \theta)] \leq E_\theta[L(T; \theta')]$ for all $\theta' \neq \theta$.

Unfortunately, given two estimators T_1 and T_2, their respective risks depend on the unknown parameter θ. Often T_1 has smaller risk for some values of θ, but T_2 has smaller risk for other values of θ. If T_1 has no greater risk than T_2 for all values of θ in the parameter space but it does have a smaller risk for some value of θ, then T_2 is said to be "inadmissible" because T_1 would be preferred as an estimator of θ.

In general, there will not be an estimator with uniformly smallest risk for all possible values of θ. The simple reason is that the class of possible estimators is too big. If T_0 is the constant estimator $T_0 = q(\theta_0)$, then $R(T_0; \theta)$ will be zero when $\theta = \theta_0$. Hence, an estimator with uniformly smallest risk must have zero risk when $\theta = \theta_0$. But we can define such estimators for any value of $q(\theta)$ so that an estimator with uniformly smallest risk must have risk equal to zero for all θ. This can happen only in trivial situations where $q(\theta)$ is known with certainty from any sample.

Therefore, it is natural to restrict attention to a smaller class of estimators. An elegant theory is obtained by restricting the class of estimators T to those that are unbiased. Using a squared error loss function, we are led to finding an estimator with the uniformly smallest risk or mean squared error for all θ or, equivalently, the unbiased estimator with uniformly minimum variance for all θ. Such an estimator, if it exists, is called "uniformly minimum variance unbiased." The Lehmann-Scheffe Theorem used to construct uniformly minimum variance unbiased estimates is stated as follows: Suppose $X_1, \ldots, X_n$ is a sample from a family of distributions $\{\mathbf{P}_\theta : \theta \in \Omega\}$. If T is a complete sufficient statistic and S is an unbiased estimator of $q(\theta)$, then $T^* = E(S|T)$ is a uniformly minimum variance unbiased estimator. Furthermore, if $\operatorname{Var}(T^*)$ is finite for some θ, then T^* is the unique uniformly minimum variance unbiased estimator.

An alternative method for finding estimators with good risk properties is to appeal to the principle of invariance. The situation is the following: We observe a random object X with range $\mathbf{X}$ and distribution from $\mathbf{P} = \{P_\theta : \theta \in \Omega\}$. Let G be a group of one-to-one transformations from $\mathbf{X}$ onto itself such that if $g \in G$, then the distribution $P_{\theta'}$ of $X' = gX$ is in $\mathbf{P}$. We then say that G leaves the model invariant. Furthermore, if X has distribution P_θ and gX has distribution $P_{\theta'}$, then $\theta' = \bar{g}\theta$ is a bijection (one-to-one and onto) from Ω to Ω. These transformations $\bar{g}$ then also form a group of transformations from Ω to Ω, denoted by $\bar{G}$. Assume that, in estimating a function $q(\theta)$, for any $\bar{g} \in \bar{G}$, $q(\theta_1) = q(\theta_2)$ implies that $q(\bar{g}\theta_1) = q(\bar{g}\theta_2)$. The common value of $q(\bar{g}\theta)$ for values of θ to which q assigns the same value will be denoted by

$$g^*q(\theta) = q(\bar{g}\theta).$$

Then, if $\mathbf{K} = \{h(\theta) : \theta \in \Omega\}$, the transformations g^* form a group G^* of one-to-one transformations from $\mathbf{K}$ onto $\mathbf{K}$.

We may think of a transformation as a change in the coordinate system, which should not affect the estimation procedure. The estimated value d of $q(\theta)$ when expressed in new coordinates is g^*d. Since the loss incurred in estimating $q(\theta)$ should not depend on the choice of coordinate system, it should satisfy

$$L(g^*d, \bar{g}\theta) = L(d, \theta).$$

When all the above holds, we say that the problem of estimating $q(\theta)$ "remains invariant" under the group G. An estimator δ is said to be "equivariant" under G if $\delta(gX) = g^*\delta(X)$ for all $g \in G$. Within the class of equivariant estimators, an estimator with uniformly minimum risk is called a "minimum risk equivariant estimator." The success in determining such an estimator is in the fact that, when a problem remains invariant under a group of transformations G and $\bar{G}$ is transitive, any equivariant estimator will have constant risk. In such a case, we are in a position to minimize this constant. Further details may be found in Lehmann (1983, Chapter 3).

The "Bayesian" method for choosing an estimator reduces the problem to optimizing a single measure of risk that is an average over the parameter space. This is often much more tractable than attempting to find an estimator that is uniformly optimal for all values in the parameter space. Let Λ denote a given probability distribution (called the "prior distribution") on the parameter space Ω, so that θ is now a random object with distribution Λ. Let the observed data $X_1, \ldots, X_n$ (conditional on the value of θ) be a sample from the family of distributions $\{\mathbf{P}_\theta : \theta \in \Omega\}$. This is much as before, except that θ is not a fixed value in Ω but is instead previously chosen according to Λ. Moreover, there is now a joint probability structure on the sample space $\mathbf{X} \times \Omega$. Because θ is random, for any particular estimator T the risk $R(T; \theta)$ is a real random variable, and we may define the "Bayes risk" of T with respect to the prior distribution Λ to be the average risk:

$$\text{Bayes risk of } T = E_\Lambda[R(T; \theta)].$$

We may now define a "Bayes estimator" with respect to the prior distribution Λ to be any estimator T_Λ that minimizes this average risk so that

$$E_\Lambda[R(T_\Lambda; \theta)] = \inf_T E_\Lambda[R(T; \theta)].$$

The minimax approach tries to control the worst-case behavior of an estimator. A "minimax" estimator is defined as any estimator T^* whose maximum risk is the lowest possible—that is, such that

$$\sup_\theta R(T^*; \theta) = \inf_T \sup_\theta R(T; \theta).$$

A very general lower bound for the variance of an estimator is given by the "Cramer-Rao Inequality," sometimes called the "Information Inequality." Suppose the observation X represents the outcome of the statistical experiment $(\mathbf{X}, \mathbf{B}, \{P_{\boldsymbol{\theta}} : \boldsymbol{\theta} \in \Omega\})$. Here Ω is a subset of $\mathbf{R}^k$ and the probability measures $P_{\boldsymbol{\theta}}$ are dominated by a σ-finite measure λ on $\mathbf{B}$ with $p_{\boldsymbol{\theta}}(x)$ denoting the density of $P_{\boldsymbol{\theta}}$ wrt λ. Then, under certain "regularity assumptions" (see Section 1.7 of Ibragimov and Has'minskii 1981 or Section 2.7 of Lehmann 1983), if $T(X)$ is any real-valued statistic, $\text{Var}_{\boldsymbol{\theta}}(T) \geq \alpha'[I(\boldsymbol{\theta})]'\alpha$, where α' is the row matrix with ith element

$$\alpha_i = \frac{\partial}{\partial \theta_i} E_{\boldsymbol{\theta}}[T(X)]$$

and $I(\boldsymbol{\theta})$ is the $k \times k$ matrix (called the "Information Matrix") with (i, j) entry

$$I_{ij}(\boldsymbol{\theta}) = E\left[\left[\frac{\partial}{\partial \theta_i} \log(p_{\boldsymbol{\theta}}(X))\right]\left[\frac{\partial}{\partial \theta_j} \log(p_{\boldsymbol{\theta}}(X))\right]\right].$$

An alternative method for measuring the performance of estimators is to look at their large sample or asymptotic properties. Consider a sequence $T_1, T_2, \ldots$ of estimators that are "similarly constructed" in the sense that T_n is computed for a sample of size n according to a procedure that is basically the same for different sample sizes. (This is not a precise definition, but the idea is often clear.) For example, T_n could be the maximum likelihood estimator of $q(\theta)$ based on a sample of size n. When this is understood, we may refer to the estimator T_n rather than the sequence of estimators $T_1, T_2, \ldots$.

By definition, T_n is "consistent" for $q(\theta)$ if T_n converges in probability to $q(\theta)$—that is, if for all $\varepsilon > 0$ and all θ, we have

$$P_\theta(|T_n - q(\theta)| \geq \varepsilon) \to 0 \qquad \text{as } n \to \infty.$$

A stronger notion of consistency is "mean squared error consistency," which holds if T_n converges to $q(\theta)$ in quadratic mean—that is, if for every $q(\theta)$, we have

$$E_\theta[(T_n - q(\theta))^2] \to 0 \qquad \text{as } n \to \infty.$$

We will often be interested in approximating the distribution of T_n. If T_n is an estimator of $q(\theta)$, then there may exist a sequence σ_n of positive normalizing constants such that $[T_n - q(\theta)]/\sigma_n$ has a nondegenerate limiting distribution. If this limiting distribution has zero mean, then T_n is said to be "asymptotically unbiased." Otherwise, T_n is said to be "asymptotically biased." We often have convergence in distribution to a Gaussian, so that

$$\mathbf{L}\{n^{1/2}[T_n - q(\theta)]\} \to N(0, \tau^2(\theta)).$$

In this case, $\tau^2(\theta)$ is referred to as the "asymptotic variance" of T_n.

Let T_n be based on a sample $X_1, \ldots, X_n$ from a family $\{P_\theta : \theta \in \Omega\}$ of distributions that satisfy the regularity conditions for the Information Inequality, where Ω is a subset of the real numbers. If T_n is unbiased for $q(\theta)$, then

$$\mathrm{Var}_\theta(T_n) \geq \frac{[q'(\theta)]^2}{nI(\theta)}$$

where $I(\theta)$ is the information in a single observation as defined earlier. Suppose furthermore that $n^{1/2}[T_n - q(\theta)]$ tends in distribution to $N(0, \tau^2(\theta))$. If

$$\tau^2(\theta) = \frac{[q'(\theta)]^2}{I(\theta)}$$

then T_n is said to be "asymptotically efficient" or "Fisher efficient." More modern approaches (such as "locally asymptotically minimax") have been developed by LeCam and Hajek. They avoid the restriction to asymptotically normal estimators and also avoid the problem of "superefficiency." The interested reader can find an introductory account in Lehmann (1983, Section 6.8).

9.1. A uniformly minimum variance unbiased estimator whose variance does not attain the Cramer-Rao lower bound (Bickel and Doksum 1977, Problem 4.3.3).

Let X have a Poisson distribution with mean $\lambda > 0$, so that its discrete probability function is given by:

$$p(x) = e^{-\lambda}\frac{\lambda^x}{x!}, \qquad x = 0, 1, 2, \ldots.$$

Consider estimating the parameter $q(\lambda) = e^{-\lambda}$. Let T be defined by

$$T(X) = \begin{cases} 1 & \text{if } X = 0 \\ 0 & \text{otherwise.} \end{cases}$$

Then $E(T) = e^{-\lambda}$, and so T is unbiased for λ. Now note that X is always sufficient in a sample of size one. Moreover, X is complete because, if a function g satisfies $E[g(X)] = 0$ for all $\lambda > 0$, then

$$e^{-\lambda} \sum_{i=0}^{\infty} g(i) \frac{\lambda^i}{i!} = 0 \qquad \text{for all } \lambda > 0.$$

A power series that converges and is identically zero in an interval must have all coefficients zero, however, forcing $g(i) = 0$ for all $i = 0, 1, 2, \ldots,$ and thus showing that X is complete. By the Lehmann-Scheffe Theorem, it follows that $T = E(T \mid X)$ is a uniformly minimum variance unbiased estimator of λ. Furthermore, since $\text{Var}(T) = e^{-\lambda}[1 - e^{-\lambda}] < \infty$ for all $\lambda > 0$, it follows that T is the unique uniformly minimum variance unbiased estimator of λ.

We will now show that T does not attain the Cramer-Rao lower bound. The Fisher information number for the family of Poisson distributions indexed by the mean λ is given by:

$$I(\lambda) = E\left[\left[\frac{\partial \log[p(X; \lambda)]}{\partial \lambda}\right]^2\right] = E\left\{\left(-1 + \frac{X}{\lambda}\right)^2\right\}$$

$$= \frac{1}{\lambda^2} E(X^2) - \frac{2}{\lambda} E(X) + 1 = \frac{1}{\lambda}.$$

Then the Cramer-Rao lower bound for the variance of unbiased estimators of $q(\lambda) = e^{-\lambda}$ is given by:

$$\frac{[q'(\lambda)]^2}{I(\lambda)} = \lambda e^{-2\lambda}.$$

On the other hand, $\text{Var}(T) = e^{-\lambda}[1 - e^{-\lambda}] > \lambda e^{-2\lambda}$, which follows from an application of the Taylor Series expansion with remainder. Thus the variance of the uniformly minimum variance unbiased estimator is larger than the Cramer-Rao lower bound for all $\lambda > 0$.

9.2. A "poor" uniformly minimum variance unbiased estimator.

Suppose X has a Poisson distribution with mean $\lambda > 0$. Consider estimating the parameter $q(\lambda) = e^{-3\lambda}$. In Example 7.15, we showed that the estimator $T(X) = (-2)^X$ is the unique unbiased estimator for $q(\lambda)$. Because it is the only unbiased estimator, it must have uniformly minimum variance among all unbiased estimators. (The Lehmann-Scheffe Theorem provides an alternative method for showing uniformly minimum variance.)

However, T is a very "poor" estimator in the sense that it is inadmissible. This is because T may be arbitrarily large and negative even though $q(\lambda)$ must be strictly positive. The estimator $T^* = \max(T, 0)$ would be preferred over T because T^* is sometimes closer to $e^{-3\lambda}$ but is never farther from $e^{-3\lambda}$ than T is. Hence, using a squared error loss function, it follows that T^* has uniformly smaller mean squared error than T, showing that the uniformly minimum variance unbiased estimator is inadmissible.

9.3. A parameter for which no uniformly minimum variance unbiased estimator exists.

Let $(X_1, \ldots, X_n)$ be the indicators of n Bernoulli trials, each with success probability p. Consider finding a uniformly minimum variance unbiased estimator of the odds ratio $q(p) = p/(1 - p)$. By Example 7.12, no unbiased estimator of $q(p)$ exists, and hence no uniformly minimum variance unbiased estimator exists.

9.4. T is the uniformly minimum variance unbiased estimator for θ, but $q(T)$ is not the uniformly minimum variance unbiased estimator for $q(\theta)$.

As in Example 9.1, let X have a Poisson distribution with mean $\lambda > 0$. Since X is an unbiased estimator of λ as well as a complete sufficient statistic, X is a uniformly minimum variance unbiased estimator of λ by the Lehmann-Scheffe Theorem. Now consider estimating $q(\lambda) = \exp(-\lambda)$. Then the estimator T defined by

$$T(X) = \begin{cases} 1 & \text{if } X = 0 \\ 0 & \text{otherwise} \end{cases}$$

is the unique uniformly minimum variance unbiased estimator of $q(\lambda)$, as was shown in Example 9.1.

However, $q(X) = \exp(-X) \neq T(X)$, showing that uniformly minimum variance unbiased estimators are not invariant under transformations. That is, the uniformly minimum variance unbiased estimator of a transformation of a parameter need not be the transformation of the uniformly minimum variance unbiased estimator of the parameter itself.

This actually follows from the fact that unbiasedness is not an invariant property of estimators, as was shown in Example 7.17.

9.5. A uniformly minimum variance unbiased estimator exists for θ but not for a function of θ.

This example is even a stronger case of the fact that uniformly minimum variance unbiased estimators are not invariant in general. In fact, T may be a uniformly minimum variance unbiased estimator of θ, but $g(T)$ may not be a uniformly minimum variance unbiased estimator of $g(\theta)$ because there may be no uniformly minimum variance unbiased estimator of $g(\theta)$.

Let X be the indicator of a Bernoulli trial with success probability p, so that $P(X = 1) = p$ and $P(X = 0) = 1 - p$. Then $E(X) = p$, so that X is unbiased as well as sufficient for p. To show that X is complete, suppose that $E[h(X)] = 0$. Then we would have

$$\begin{aligned} E[h(X)] &= ph(1) + (1 - p)h(0) \\ &= p[h(1) - h(0)] + h(0) = 0 \qquad \text{for all } p \end{aligned}$$

which forces $h(0) = h(1) = 0$, so that $h(X)$ is identically zero and X is complete. By the Lehmann-Scheffe Theorem, it follows that X is a uniformly minimum variance unbiased estimator of p. Moreover, since $\operatorname{Var}(X) = p(1 - p) < \infty$, it follows that X is the unique uniformly minimum variance unbiased estimator of p.

Now consider the transformation $g(p) = p/(1 - p)$, the odds ratio. As was shown in Example 7.12, no unbiased estimator of $g(p)$ exists, and therefore no uniformly minimum variance unbiased estimator can exist.

9.6. Nonexistence of a uniformly minimum variance unbiased estimator for a discrete distribution even though an infinite class of unbiased estimators exists (Rao 1973, Problem 5.11).

Suppose X has the discrete probability function defined by:

$$\begin{aligned} P(X = -1) &= p \\ P(X = n) &= (1 - p)^2 p^n, \qquad n = 0, 1, 2, \ldots \end{aligned}$$

for some p in $(0, 1)$. Then the estimator T defined by

$$T(X) = \begin{cases} 1 & \text{if } X = -1 \\ 0 & \text{otherwise} \end{cases}$$

is unbiased for p. In fact, we can describe all unbiased estimators. If

$T(X) = t_X$ is an estimator that takes on the value t_j when $X = j$, then we may compute

$$E(T) = pt_{-1} + \sum_{n=0}^{\infty} (1 - p)^2 p^n t_n$$

$$= pt_{-1} + \sum_{n=0}^{\infty} p^n t_n - 2 \sum_{n=0}^{\infty} p^{n+1} t_n + \sum_{n=0}^{\infty} p^{n+2} t_n$$

$$= t_0 + \sum_{n=0}^{\infty} p^{n+1}(t_{n+1} - 2t_n + t_{n-1}).$$

The condition that T be unbiased for p requires that

$$E(T) = p = t_0 + (t_1 - 2t_0 + t_{-1})p + (t_2 - 2t_1 + t_0)p^2 + \cdots$$

for all $p \in (0, 1)$. Consider t_{-1} as fixed, and equate terms with identical powers of p to find that $t_k = k(1 - t_{-1})$. Thus any t_{-1} defines an unbiased estimator of p. Computing the variance of such an estimate proceeds as follows:

$$E(T^2) = pt_{-1}^2 + \sum_{n=0}^{\infty} (1 - p)^2 p^n n^2 (1 - t_{-1})^2$$

$$= pt_{-1}^2 + (1 - p)^2 (1 - t_{-1})^2 \sum_{n=0}^{\infty} p^n n^2$$

$$= pt_{-1}^2 + (1 - t_{-1})^2 \frac{p^2 + p}{1 - p}$$

and so the variance of T is

$$\text{Var}(T) = pt_{-1}^2 + (1 - t_{-1})^2 \frac{p^2 + p}{1 - p} - p^2.$$

Now fix $p = p_0$. Minimizing $\text{Var}(T)$ over all possible unbiased estimates may be done by differentiating the above variance with respect to t_{-1}, setting it equal to zero, and then verifying that the solution is indeed a minimum. The result of this procedure is that

$$t_{-1} = \frac{p_0 + 1}{2}$$

must hold in order for the variance of T to be a minimum when $p = p_0$. Because different unbiased estimators T, defined by their values of t_{-1}, minimize the variance at different values of p, there can be no single uniformly minimum variance unbiased estimator of p.

9.7. Nonexistence of a uniformly minimum variance unbiased estimator for a continuous distribution even though an infinite class of unbiased estimators exists (Mood, Graybill, and Boes 1974, p. 330).

Suppose $X_1, \ldots, X_n$ is a random sample from the uniform distribution on $(\theta, \theta + 1)$ for some real parameter θ. Since $E(X_i) = \theta + \frac{1}{2}$, it follows that $\bar{X} - \frac{1}{2}$ is unbiased for θ. By symmetry, if $X_{(1)} = \min(X_1, \ldots, X_n)$ and $X_{(n)} = \max(X_1, \ldots, X_n)$, then $[(X_{(1)} + X_{(n)})/2] - \frac{1}{2}$ is also unbiased for θ. Even though several unbiased estimators exist (any of the infinite convex combinations of these two would be unbiased), we will now show that no uniformly minimum variance unbiased estimator exists.

For some p in $[0, 1)$, consider the estimator $g(X_1 - p) + p$, where the function $g(t)$ is defined as the greatest integer less than t. Then

$$E[g(X_1 - p) + p] = \int_{\theta}^{\theta+1} [g(t - p) + p]\, dt = p + \int_{\theta-p}^{\theta+1-p} g(t)\, dt.$$

Next, observe that given any θ and any p, there exists some integer $N = N(\theta, p)$ with the property that $\theta - p < N \leq \theta + 1 - p$. Therefore,

$$\begin{aligned} E[g(X_1 - p) + p] &= p + \int_{\theta-p}^{N} (N - 1)\, dt + \int_{N}^{\theta+1-p} N\, dt \\ &= p + (N - 1)(N - \theta + p) + N(\theta + 1 - p - N) \\ &= \theta \end{aligned}$$

so that $g(X_1 - p) + p$ is unbiased for θ for any p in $[0, 1)$. Now suppose $\theta + 1 - p = k$ is some integer. Then $g(x_1 - p) = k - 1$ for all x_1 that satisfy $\theta - p < x_1 - p \leq \theta + 1 - p$. So

$$g(X_1 - p) + p = k - 1 + p = \theta + 1 - p - 1 + p = \theta$$

for all x_1 in $(\theta, \theta + 1]$. In other words, $g(X_1 - p) + p$ estimates θ with no error and hence has zero variance if $\theta + 1 - p$ is an integer. But, for any fixed $\theta = \theta_0$, we can find a $p_0 \in [0, 1)$ such that this happens. Since some unbiased estimator has zero variance for any parameter θ, the uniformly minimum variance unbiased estimator would have to have zero variance for all θ. But this implies that the uniformly minimum variance unbiased estimator would be constant with probability one and hence could not be unbiased. This contradiction shows that no uniformly minimum variance unbiased estimator of θ can exist in this situation.

9.8. An inefficient uniformly minimum variance unbiased estimator.

Under suitable conditions, when we estimate $q(\theta)$ from a k-dimensional exponential family, minimum variance unbiased estimators (when they exist) are asymptotically efficient. For a precise statement, see Portnoy (1977), who also uses a counterexample to show the result to be false without such assumptions.

Let μ be a measure supported on the set $\{-1, 0, \pi\}$ with $\mu(\{-1\}) = \frac{1}{2}$, $\mu(\{\pi\}) = 1/(2\pi)$, and $\mu(\{0\}) = (\pi - 1)/(2\pi)$. Consider the exponential family of distributions P_θ with densities

$$\frac{dP_\theta}{d\mu} = \exp[\theta x - c(\theta)], \qquad x \in \mathbf{R}, \quad \theta \in \mathbf{R}$$

where $c(\theta)$ is the appropriate integrating constant. Note that

$$\exp[c(\theta)] = \frac{e^{-\theta}}{2} + \frac{\pi - 1}{2\pi} + \frac{e^{\pi\theta}}{2\pi}.$$

Based on a sample $X_1, \ldots, X_n$ from P_θ, we would like to estimate $q(\theta) = P_\theta(\{0\}) = \mu(\{0\})/\exp[c(\theta)]$. By Theorem 5.6 of Lehmann (1983), $\bar{X}_n$ is complete and sufficient. Let $p_n(z)$ denote the proportion of observations in the sample, which equal z.

Next we show that $p_n(0)$ is a function of $\bar{X}_n$. To see this, note that $\bar{X}_n = \pi p_n(\pi) - p_n(-1)$. For rational a and b, consider the equation $a + \pi b = \bar{X}_n$. Clearly, $a = -p_n(-1)$ and $b = p_n(\pi)$ are solutions. To see that they are the only solutions, if (a', b') were another rational solution, we would have $(b' - b)\pi + (a' - a) = 0$, or $\pi = -(a' - a)/(b' - b)$, a contradiction because π is irrational. Hence, from $\bar{X}_n$ we know $p_n(-1)$ and $p_n(\pi)$, and thus we know $p_n(0) = 1 - p_n(\pi) - p_n(-1)$.

Thus, $p_n(0)$ is a function of $\bar{X}_n$, but $E_\theta[p_n(0)] = q(\theta)$, which implies that $p_n(0)$ is a uniformly minimum variance unbiased estimator for $q(\theta)$. Furthermore, $np_n(0)$ is a binomial random variable based on n trials with success probability $q(\theta)$. Thus, $n^{1/2}[p_n(0) - q(\theta)]$ converges in distribution to a Gaussian with mean zero and variance $[q(\theta)][1 - q(\theta)]$. On the other hand, according to Lehmann (1983, Example 6.2.5), if $\tilde{\theta}_n$ is the maximum likelihood estimator for θ, then

$$\mathbf{L}[n^{1/2}(\tilde{\theta}_n - \theta)] \to N(0, [c''(\theta)]^{-1})$$

and so

$$\mathbf{L}\{n^{1/2}[q(\tilde{\theta}_n) - q(\theta)]\} \to N\left(0, \frac{[q'(\theta)]^2}{c''(\theta)}\right).$$

Indeed, $I(\theta) = c''(\theta)$ is the Fisher information number for this parametric family. The asymptotic variances $[q(\theta)][1 - q(\theta)]$ and $[q'(\theta)]^2/c''(\theta)$ are different. For example, when $\theta = 1$, they are 0.0353 and 0.0744, respectively. Thus the uniformly minimum variance unbiased estimator is inefficient, whereas the maximum likelihood estimator is efficient.

9.9. A Pitman estimator that has smaller mean squared error loss than the uniformly minimum variance unbiased estimator (Mood, Graybill, and Boes 1974, p. 338).

In general, whenever a uniformly minimum variance unbiased estimator exists and is location-equivariant for a location parameter under squared error loss, it is also the Pitman estimator of location. This does not extend to scale families, however, as is shown by this example. The simple reason is that squared error is not a scale-invariant loss function.

Let $X_1, \ldots, X_n$ be a random sample from the exponential distribution with mean $\lambda > 0$ and variance λ^2 with density function:

$$f(x) = \frac{1}{\lambda} e^{-x/\lambda} I_{(0,\infty)}(x).$$

The likelihood, in its one-parameter exponential form, is given by

$$L(\mathbf{x}; \lambda) = \exp\left[-\frac{1}{\lambda}\sum_{i=1}^{n} x_i - n\log(\lambda)\right]\prod_{i=1}^{\infty} I_{(0,\infty)}(x_i)$$

showing that the natural sufficient statistic $X_1 + \cdots + X_n$ is also complete (see the introduction to Chapter 7). Then $\bar{X}$ is an unbiased estimator of λ that is also a function of a complete, sufficient statistic. Therefore, by the Lehmann-Scheffe Theorem, $\bar{X}$ is the unique uniformly minimum variance unbiased estimator of λ because it has finite variance λ^2/n.

We will now consider the Pitman estimator (i.e., the minimum risk equivariant estimator) of the scale parameter λ. An estimator $T = t(\mathbf{X})$ is said to be scale invariant if $t(c\mathbf{X}) = ct(\mathbf{X})$ for all $c > 0$. Then the Pitman estimator is defined to be the estimator with the uniformly smallest risk with respect to the loss function $(t - \lambda)^2/\lambda^2$ within the class of scale-invariant estimators. Note that λ is a scale parameter if the distribution of X/λ does not depend on λ, as happens here.

The Pitman estimator for scale can be calculated by the formula (see, for example, Mood, Graybill, and Boes 1974, p. 337):

$$t(\mathbf{X}) = \frac{\int_0^\infty (1/\lambda^2) L(\mathbf{X}; \lambda)\, d\lambda}{\int_0^\infty (1/\lambda^3) L(\mathbf{X}; \lambda)\, d\lambda}.$$

In this case, we find that the Pitman estimator of λ is

$$t(\mathbf{X}) = \frac{\int_0^\infty (1/\lambda^{n+2}) \exp\left[-\sum_{i=1}^n X_i/\lambda\right] d\lambda}{\int_0^\infty (1/\lambda^{n+3}) \exp\left[-\sum_{i=1}^n X_i/\lambda\right] d\lambda}$$

$$= \sum_{i=1}^n X_i \frac{\int_0^\infty t^n e^{-t}\, dt}{\int_0^\infty t^{n+1} e^{-t}\, dt} = \frac{n}{n+1}\bar{X}.$$

Because the uniformly minimum variance unbiased estimator $\bar{X}$ is also a scale estimator, we know that the risk of the Pitman estimator $[n/(n + 1)]\bar{X}$ is uniformly smaller than the risk of $\bar{X}$ for all λ with respect to the loss function $L(t; \lambda) = (t - \lambda)^2/\lambda^2$. But the risk here is simply $1/\lambda^2$ times the mean squared error, so that the mean squared error of the Pitman estimator is also uniformly smaller than that of $\bar{X}$. This implies that the uniformly minimum variance unbiased estimator is inadmissible.

Note that the maximum likelihood estimator is given by the solution of the likelihood equation, or

$$\frac{\partial}{\partial \lambda}\left\{-\sum_{i=1}^n \left[\frac{X_i}{\lambda}\right] - n \log(\lambda)\right\} = 0.$$

Thus the maximum likelihood estimator is also the uniformly minimum variance estimator $\bar{X}$, which shows that the maximum likelihood estimator of λ is also inadmissible here.

9.10. A Pitman estimator of location that is neither admissible nor minimax.

Although the best equivariant estimators are frequently minimax (even sometimes when they are inadmissible), the following counterexample from Blackwell and Girshick (1954) shows that the minimum risk equivariant estimator may not even be minimax. The reader may consult Kiefer (1957) for general conditions for a minimum risk equivariant estimator to be minimax.

Suppose $F(x)$ is the cumulative distribution function that places mass $1/[k(k+1)]$ on the values $1, 2, \ldots$, and suppose X is an observation from the location family $F(x-\theta)$ with real unknown θ. Let the loss function be

$$L(\theta, d) = \begin{cases} d - \theta & \text{if } d > \theta \\ 0 & \text{if } d \leq \theta. \end{cases}$$

The problem remains invariant under the group of location changes, and the only equivariant estimators of θ are of the form $X - c$. Since $\mathrm{Var}_\theta(X)$ is infinite for all θ, any equivariant estimator has infinite risk.

To exhibit an estimator with bounded risk function, let $\delta_M(X) = X - M|X|$ with $M > 1$. Using the hint given in Lehmann (1983, Problem 4.13, p. 316), it is possible to show that the risk of this estimator is bounded.

9.11. A minimum risk equivariant estimator that is neither minimax nor admissible.

Various theorems state suitable assumptions for which a minimum risk equivariant estimator is minimax. In particular, for a location family under squared error loss, it is minimax if there exists some estimator with finite risk. This result is from Girshick and Savage (1951). Although a minimum risk equivariant estimator frequently is minimax, this is false in general. The following example is taken from Lehmann (1983, pp. 285–86).

Let (X_1, X_2) and (Y_1, Y_2) be independent, each with bivariate Gaussian distribution with mean zero and covariance matrices S and ΔS, respectively, where $\Delta > 0$. Let the loss be one when $|d - \Delta|/\Delta > \frac{1}{2}$ and zero otherwise. The problem of estimating Δ remains invariant under the group of non-singular transformations of the form

$$X_1' = a_1 X_1 + a_2 X_2 \qquad Y_1' = c(a_1 Y_1 + a_2 Y_2)$$

$$X_2' = b_1 X_1 + b_2 X_2 \qquad Y_2' = c(b_1 Y_1 + b_2 Y_2).$$

By an argument analogous to that of Example 10.30, the only equivariant estimator of Δ is $\delta(X, Y) = 0$, a constant estimator. However, the risk of Y_2^2/X_2^2 is constant and less than the risk of the zero estimator, which is one. Hence, the minimum risk equivariant estimator is inadmissible and not minimax.

9.12. A problem that remains invariant under two different groups, each leading to different minimum risk equivariant estimators.

An example showing this possibility is given explicitly in Lehmann (1983, Example 2.3, pp. 170–172). It is analogous to Example 10.31.

9.13. A minimum risk equivariant estimator that is not risk unbiased.

In general, when a problem of estimating $\tau(\theta)$ remains invariant under a group G, if $\bar{G}$ is transitive and G^* is commutative, then a minimum risk estimator is risk unbiased. This may fail if either condition is dropped. The following examples are taken from Lehmann (1983, pp. 172–73). Let X be an observation from the Gaussian distribution with mean θ and variance σ^2, so that $\Omega = \{(\theta, \sigma) : \theta \in \mathbf{R}, \sigma > 0\}$. Let the loss be $L[d; (\theta, \sigma)] = (d - \theta)^2/\sigma^2$.

First, consider the group G of location changes, $gx = x + c$. Then X is minimum risk equivariant for θ under G (which follows from Lehmann 1983, Example 3.8). However, X is not risk unbiased because, for example,

$$1 = E_{(0,1)}L[X; (0, 1)] > E_{(0,1)}[X; (1, 5)] = \tfrac{2}{25}.$$

In this case $\bar{G}$, the group of transformations $\bar{g}(\theta, \sigma) = (\theta + c, \sigma)$ is not transitive.

It is interesting to note that if the loss is replaced by $(d - \theta)^2$, then X is still the minimum risk equivariant estimator for θ and is risk unbiased. Thus we may have a minimum risk equivariant estimator that is risk unbiased even though the induced group $\bar{G}$ is not transitive.

Note also that this problem [with loss $(d - \theta)^2/\sigma^2$] also remains invariant under the larger group H such that $hx = ax + c$ with $a > 0$. Since X is equivariant under H and minimum risk equivariant under G, it follows that

X must also be minimum risk equivariant under H. As before, X is not risk unbiased even though $\bar{H}$ is transitive. The problem here is that H^*, with $h^*d = ad + c$, is not commutative.

9.14. The completeness condition cannot be omitted in the Lehmann-Scheffe Theorem.

Let $X_1, \ldots, X_n$ (with $n > 1$) be a sample from the Gaussian distribution with unknown mean μ and variance one. The estimator X_1 is unbiased for μ. Also, the sample itself $(X_1, \ldots, X_n)$ is always sufficient. However, the estimator defined by

$$E[X_1 \mid (X_1, \ldots, X_n)] = X_1$$

is not a uniformly minimum variance unbiased estimator of μ. Note that $\text{Var}(X_1) = 1$, but $\bar{X}$, which is unbiased for μ, has smaller variance $1/n < 1$. The reason the estimator constructed in the manner of the Lehmann-Scheffe Theorem is not uniformly minimum variance unbiased here is that the sample $(X_1, \ldots, X_n)$ is not complete (as was shown in general in Example 7.18).

9.15. Bounded completeness cannot replace completeness as a condition in the Lehmann-Scheffe Theorem.

As in Example 9.6, let X be discrete with distribution

$$P(X = -1) = p$$
$$P(X = n) = (1 - p)^2 p^n \qquad \text{for } n = 0, 1, 2, \ldots$$

for some p in $(0, 1)$. X is clearly sufficient. Define

$$T(X) = \begin{cases} 1 & \text{if } X = -1 \\ 0 & \text{otherwise.} \end{cases}$$

Then $E(T \mid X) = T$, but T is not a uniformly minimum variance unbiased estimator of p, as was shown in Example 9.6. In this case, T is boundedly complete but not complete, as was shown in Example 7.22. Thus we see that the strong property of being complete cannot be relaxed when uniformly minimum variance unbiased estimators are constructed by the Lehmann-Scheffe Theorem.

9.16. Regularity conditions cannot be omitted in the Information Inequality.

Suppose X is uniform on $(0, \theta)$ for some $\theta > 0$. Note that $\log[p(x; \theta)]$ is differentiable for all x in $(0, \theta)$ so that the log likelihood is differentiable with probability one for any $\theta > 0$. Hence, we can define moments of the partial derivative of the log likelihood with respect to θ. Let us deliberately neglect to verify the regularity conditions, and instead see what happens if we proceed to compute the information. We find that

$$I(\theta) = \operatorname{Var}\left[\frac{\partial \log p(X; \theta)}{\partial \theta}\right] = \operatorname{Var}\left(\frac{-1}{\theta}\right) = 0$$

because the derivative $-1/\theta$ is a constant. For a case like this, the Cramer-Rao lower bound $1/[I(\theta)]$ would be infinite.

Consider the unbiased estimator $T(X) = 2X$, however. It has variance

$$\operatorname{Var}(T) = E(T^2) - [E(T)]^2 = \int_0^\theta \frac{(2t)^2}{\theta}\,dt - \theta^2 = \frac{\theta^2}{3}.$$

So T has finite variance, which shows that the Information Inequality is not applicable here.

The reason is that the required regularity conditions do not hold. For example, the operations of differentiation and integration with respect to θ do not hold because

$$\frac{-1}{\theta} = E\left\{\frac{\partial}{\partial\theta}\log[p(X; \theta)]\right\} = \int_0^\theta \frac{\partial p(x; \theta)}{\partial \theta}\,dx$$

$$\neq \frac{\partial}{\partial\theta}\int_0^\theta p(x; \theta)\,dx = \frac{\partial}{\partial\theta}1 = 0.$$

Furthermore, if we define $I(\theta)$ as

$$I(\theta) = E\left[\left[\frac{\partial \log[p(X; \theta)]}{\partial \theta}\right]^2\right]$$

then we find that $I(\theta) = 1/\theta^2$. Even so,

$$\operatorname{Var}(T) = \frac{\theta^2}{3} < \frac{1}{I(\theta)} = \theta^2$$

for $\theta < 3^{1/2}$. Using either definition, we see that the Information Inequality cannot be applied even in such a simple case as a uniform distribution when the appropriate moments and derivatives do exist.

9.17. A randomized minimax estimator.

As a consequence of the Rao-Blackwell Theorem, given any randomized estimator of some parameter $q(\theta)$, there exists a nonrandomized estimator with a uniformly smaller risk function if the loss function is strictly convex and at least as small a risk function if the loss function is convex (see, for example, Corollary 1.6.2 on p. 51 of Lehmann 1983). It follows that when the loss function $L(\theta, d)$ is convex in d, only nonrandomized estimators need be considered in the search for a minimax estimator. This example from Lehmann (1983, p. 252) shows that this is not true without the assumption of convexity of the loss function.

Consider the estimation of the success probability p based on the number X of successes in n independent Bernoulli trials. Suppose the loss function is given by

$$L(p, d) = \begin{cases} 0 & \text{if } |d - p| \leq \alpha \\ 1 & \text{otherwise.} \end{cases}$$

If $\alpha < 1/[2(n + 1)]$, then since any nonrandomized estimator can take on at most $n + 1$ possible values, there will exist some value of p for which the risk is one. Thus, any nonrandomized estimator has a maximum risk equal to one.

On the other hand, consider the randomized estimator U, which, independent of the data, is a uniform random number in $(0, 1)$. Its risk function is

$$\begin{aligned} R(p, U) &= 1 - P(|U - p| \leq \alpha) \\ &= \begin{cases} 1 - (p + \alpha) & \text{if } p < \alpha \\ 1 - 2\alpha & \text{if } \alpha \leq p \leq 1 - \alpha \\ p - \alpha & \text{if } p > 1 - \alpha. \end{cases} \end{aligned}$$

Thus the supremum of the risk function is $1 - \alpha$ and is less than one.

The loss function of this example may seem artificial, particularly because the loss function depends on the sample size. However, for the estimation of p when the loss function is $|d - p|^r$, with $0 < r < 1$, no nonrandomized estimator is minimax. For details, see Hodges and Lehmann (1950).

9.18. An inadmissible method of moments estimator.

As in Example 8.3, suppose a population has θ members labeled from one to θ. The population is sampled with replacement. Then a method of moments estimator of θ is $2\bar{X} - 1$. Because we know that θ must be at least $X_{(n)}$, however, it follows that $\max(X_{(n)}, 2\bar{X} - 1)$ has a smaller mean squared error than does $2\bar{X} - 1$, since it is never farther from θ and is sometimes closer. Thus the method of moments estimator is inadmissible.

9.19. An inadmissible maximum likelihood estimator and the James-Stein Rule (Stein 1956).

This example offers a remarkable case of the inadmissibility of the maximum likelihood estimator with squared error loss function. Let $X_1, \ldots, X_n$, with $n \geq 3$, be independent random variables, and suppose X_i has a Gaussian distribution with mean θ_i and known variance one. We would like to estimate the vector of means $\boldsymbol{\theta} = (\theta_1, \ldots, \theta_n)$. The maximum likelihood estimator can be simply calculated as the intuitive estimator, which is

$$\boldsymbol{\theta}_M^* = \mathbf{X} = (X_1, \ldots, X_n).$$

Because the loss function L is squared error loss given by

$$L(\boldsymbol{\theta}^*; \boldsymbol{\theta}) = \sum_{i=1}^{n} (\theta_i^* - \theta_i)^2,$$

the risk of the maximum likelihood estimator is constant and is given by

$$R(\boldsymbol{\theta}_M^*; \boldsymbol{\theta}) = E\left[\sum_{i=1}^{n} (X_i - \theta_i)^2\right] = \sum_{i=1}^{n} \operatorname{Var}(X_i) = n.$$

James and Stein (1961) introduced the estimator

$$\boldsymbol{\theta}_{JS}^* = \left[1 - \frac{n-2}{S}\right]\boldsymbol{\theta}_M^*$$

where $S = \Sigma_{i=1}^{n} X_i^2$; that is, instead of estimating θ_i by X_i, the proposed estimator uses $[1 - (n-2)/S]X_i$. The James-Stein estimator has a risk given by (see Efron and Morris 1973a, or Cox and Hinkley 1974, p. 445):

$$R(\boldsymbol{\theta}_{JS}^*; \boldsymbol{\theta}) \leq n - \frac{n-2}{n - 2 + \sum_{i=1}^{n} \theta_i^2} < n = R(\boldsymbol{\theta}_M^*; \boldsymbol{\theta}).$$

This says that the mean squared error of the James-Stein estimator is uniformly smaller for all possible parameter values than the maximum likelihood estimator. In other words, the maximum likelihood estimator is inadmissible. Geometrically, the James-Stein estimator is shrinking the vector $\boldsymbol{\theta}_M^* = \mathbf{X}$ in $\mathbf{R}^n$ closer to the origin in such a way as to achieve a reduction in squared error loss. It is surprising, however, that the estimate of each θ_i depends not only on X_i but also on all the other X_j whose distributions appear to be unrelated to θ_i, since the shrinkage factor of the estimate of θ_i is a function of all the X_j values.

This paradox is in part resolved by the fact that the seemingly unrelated parameters are considered jointly in the loss function. Also, the James-Stein estimator does worse in each component separately, even though it does better overall. Because of this, the James-Stein estimator has raised some serious criticism, stemming from the mistrust in the use of the squared error loss function and the generally good performance of the maximum likelihood estimators in applied problems. Efron and Morris (1973b) comment:

> The James-Stein estimator seems to do the impossible ... the result is an improvement over the maximum likelihood estimator no matter what the values of $\theta_1, \theta_2, \ldots, \theta_k$. The reaction of the statistical community to this *tour de force* has been generally hostile, the usual suggestion being that this is some sort of mathematical trick devoid of genuine statistical merit. Thus we have the "speed of light" rhetorical question, "do you mean that if I want to estimate tea consumption in Taiwan I will do better to estimate simultaneously the speed of light and the weight of hogs in Montana?"

Barnard's comment on the James-Stein estimator is (as quoted by Efron and Morris 1973b):

> The difficulty here is to know what problems are to be combined together—why should not all our estimation problems be lumped together into one grand melee?

The use of the James-Stein estimator in practical situations is discussed by Cox and Hinkley (1974, p. 450) and by Efron and Morris (1973b). Generalizations to more complex situations are given. It seems that whether or not the James-Stein estimator should be used depends on whether one is prepared to accept an overall reduction in the risk of a composite loss function at the expense of worsening individual components. There are also indications that the largest improvements are achieved when similar, rather than widely disparate, problems are combined.

9.20. An observation from a one-parameter exponential family such that the natural sufficient statistic is inadmissible for its expectation.

A sufficient condition for the admissibility (under squared error loss) of an estimator of the form $aT(X) + b$ of $E_\theta(T)$, when T is the natural sufficient statistic for a one-parameter exponential family, was reached by Karlin

(1958). It is perhaps surprising that T is not necessarily admissible for $E(T)$. The example is taken from Lehmann (1983, Problem 4.3.14).

Let X have density

$$\beta(\theta)e^{\theta x}\frac{e^{-|x|}}{2} \qquad \text{where } |\theta| < 1$$

and $\beta(\theta) = 1 - \theta^2$. To see that X is not admissible for estimating $E_\theta(X)$, we will show that it is dominated by $\delta = 0$ in the case of squared error loss. We compute

$$E_\theta(X) = -\frac{\beta'(\theta)}{\beta(\theta)} = \frac{2\theta}{1 - \theta^2}$$

$$\text{Var}_\theta(X) = \frac{d}{d\theta}E_\theta(X) = 2\frac{1 + \theta^2}{(1 - \theta^2)^2}.$$

Hence, $E_\theta\{[X - E_\theta(X)]^2\} = 2(1 + \theta^2)/(1 - \theta^2)^2$ and

$$E_\theta\{[aX - E_\theta(X)]^2\} = 2a^2\frac{1 + \theta^2}{(1 - \theta^2)^2} + 4(1 - a)^2\frac{\theta^2}{(1 - \theta^2)^2}.$$

When $a = 0$, this reduces to $4\theta^2/(1 - \theta^2)^2$. Since $2(1 + \theta^2) > 4\theta^2$ for all $\theta \in (-1, 1)$, it follows that X is dominated by the constant estimator zero.

In fact, it can be shown that $aX + b$ is admissible for estimating $E_\theta(X)$ with squared error loss if and only if $0 \le a \le \frac{1}{2}$.

9.21. A situation in which the mean squared error of the maximum likelihood estimator is always worse than using a constant estimator (Lehmann 1983, Problem 6.2.5, p. 473).

Let X have a Bernoulli distribution with success probability p so that $P(X = 1) = p$ and $P(X = 0) = 1 - p$. Restrict the parameter p to be in the parameter space $\Omega = \{p : \frac{1}{3} \le p \le \frac{2}{3}\}$. The maximum likelihood estimate is then given by

$$p_M(X) = \begin{cases} \frac{1}{3} & \text{if } X = 0 \\ \frac{2}{3} & \text{if } X = 1. \end{cases}$$

The mean squared error of the maximum likelihood estimate is

$$E\{[p_M(X) - p]^2\} = p\left[\left(\frac{2}{3}\right) - p\right]^2 + (1-p)\left[\left(\frac{1}{3}\right) - p\right]^2$$

$$= \frac{3p^2 - 3p + 1}{9}.$$

Now consider the constant estimator $\frac{1}{2}$. Its mean squared error is

$$E\left\{\left[\left(\frac{1}{2}\right) - p\right]^2\right\} = \frac{4p^2 - 4p + 1}{4}.$$

The difference in mean squared errors is

$$\text{MSE}[p_M(X)] - \text{MSE}\left[\frac{1}{2}\right] = \left[\frac{3p^2 - 3p + 1}{9}\right] - \left[\frac{4p^2 - 4p + 1}{4}\right]$$

$$= \left(\frac{1}{36}\right) - \left(\frac{2}{3}\right)\left(p - \frac{1}{2}\right)^2$$

which takes its smallest value of $\frac{1}{108}$ when $p = \frac{1}{3}$ or $\frac{2}{3}$. Therefore the mean squared error of the constant estimator $\frac{1}{2}$ is uniformly smaller than that of the maximum likelihood estimator in this situation.

9.22. A situation where the maximum likelihood estimator and the uniformly minimum variance unbiased estimator differ and both are inadmissible.

Suppose $X_1, \ldots, X_n$ is a sample from the uniform distribution on $(0, \theta]$ for some positive parameter θ. The likelihood is given by

$$L(\mathbf{x}; \theta) = \prod_{i=1}^{n} \left(\frac{1}{\theta}\right) I_{(0,\theta]}(x_i) = \begin{cases} \theta^{-n} & \text{if } \theta \geq x_{(n)} \\ 0 & \text{otherwise.} \end{cases}$$

Thus we see that $X_{(n)}$ is sufficient for θ, and the maximum likelihood estimator is $X_{(n)}$, the largest-order statistic. From Example 7.10 we know that $X_{(n)}$ has a density function given by

$$f(t) = \begin{cases} \dfrac{nt^{n-1}}{\theta^n} & \text{if } 0 < t \leq \theta \\ 0 & \text{otherwise} \end{cases}$$

and $E(X_{(n)}) = n\theta/(n + 1)$. We claim that $T(\mathbf{X}) = (n + 1)X_{(n)}/n$ is the unique uniformly minimum variance unbiased estimator of θ. Clearly, T is unbiased and depends on the sufficient statistic $X_{(n)}$. The result will follow from the Lehmann-Scheffe Theorem if we can show that $X_{(n)}$ is complete. To this end, suppose $E[g(X_{(n)})] = 0$ for all $\theta > 0$; that is,

$$\int_0^\theta g(t)t^{n-1}\,dt = 0 \qquad \text{for all } \theta > 0.$$

Differentiating with respect to θ (as in Royden 1968, p. 103), we see that $t^{n-1}g(t) = 0$ with probability one. Thus, $g(X_{(n)})$ is zero with probability one, and $X_{(n)}$ is therefore complete and T is uniformly minimum variance unbiased.

To find the variance of T, note that

$$E[(X_{(n)})^2] = \int_0^\theta n\theta^{-n}t^{n+1}\,dt = \frac{n\theta^2}{n + 2}$$

so that

$$\begin{aligned}\mathrm{Var}(X_{(n)}) &= E[(X_{(n)})^2] - [E(X_{(n)})]^2 \\ &= \frac{n\theta^2}{n + 2} - \left[\frac{n\theta}{n + 1}\right]^2 = \frac{n\theta^2}{(n + 1)^2(n + 2)}.\end{aligned}$$

Thus the mean squared error is

$$\mathrm{MSE}(T; \theta) = \mathrm{Var}(T) = \left[\frac{(n + 1)^2}{n^2}\right]\mathrm{Var}(X_{(n)}) = \frac{\theta^2}{n(n + 2)}.$$

The net result of all this is that T is the unique uniformly minimum variance unbiased estimator of θ.

Next we calculate the mean squared error of the maximum likelihood estimator, $X_{(n)}$:

$$\mathrm{MSE}(X_{(n)}; \theta) = \mathrm{Var}(X_{(n)}) + [\mathrm{Bias}(X_{(n)}; \theta)]^2 = \frac{2\theta^2}{(n + 2)(n + 1)}.$$

However, we will now show that the estimator $T^*(\mathbf{X}) = [(n + 2)/(n + 1)]X_{(n)}$ beats both the uniformly minimum variance unbiased estimator and the maximum likelihood estimator in having the uniformly smallest risk with respect to a squared error loss function. We calculate

$$\text{MSE}(T^*; \theta) = \text{Var}(T^*) + [\text{Bias}(T^*; \theta)]^2$$

$$= \frac{(n+2)^2}{(n+1)^2} \text{Var}(X_{(n)}) + \left[\frac{n+2}{n+1} E(X_{(n)}) - \theta\right]^2.$$

Substituting for the variance and the expectation and simplifying, we obtain

$$\text{MSE}(T^*; \theta) = \frac{\theta^2}{(n+1)^2}.$$

Since $(n + 1)^2 > n(n + 2)$, we have $\text{MSE}(T^*; \theta) < \text{MSE}(T; \theta)$ for all $\theta > 0$. Also, since $2(n + 1)^2/[(n + 2)(n + 1)] = 2/(n + 2) < 1$, we have $\text{MSE}(T^*; \theta) < \text{MSE}(X_{(n)}; \theta)$ for all $\theta > 0$. We have thus shown by finding a "better" estimator that both the maximum likelihood estimator and the uniformly minimum variance unbiased estimator are inadmissible.

Although it seems that we have magically constructed a "better" estimator by "pulling it out of a hat," we could have constructed T^* by directly considering all estimators of the form $cX_{(n)}$ and then choosing the value of c that minimizes the mean squared error of $cX_{(n)}$.

9.23. A uniformly minimum variance unbiased estimator that renders the maximum likelihood estimator inadmissible.

Let X_i be independent Gaussian $N(\theta_i, 1)$ for $i = 1, \ldots, p$, and let $q(\theta_1, \ldots, \theta_p) = \Sigma\, \theta_i^2/p$. Then $(X_1, \ldots, X_p)$ is a complete and sufficient statistic. Because $T = \Sigma\,(X_i^2/p) - p$ is unbiased for $q(\theta_1, \ldots, \theta_p)$, it follows that T is uniformly minimum variance unbiased. The maximum likelihood estimator, $\Sigma\, X_i^2/p$, has the same variance as T but is biased and so is inadmissible with respect to squared error loss.

9.24. An admissible uniformly minimum variance unbiased estimator that is not minimax.

Consider the problem of estimating p with squared error loss based on the number X of successes in n independent Bernoulli trials, each with success probability p. Since X has a binomial distribution that can be expressed as

$$P(X = s) = \binom{n}{s}(1 - p)^n \exp\left[s \log\left(\frac{p}{1-p}\right)\right]$$

we see that the family of distributions of X is a complete one-parameter exponential family. Because X/n is unbiased for p, it follows that the "natural" estimator, X/n, is uniformly minimum variance unbiased. By a simple application of Karlin's Theorem on sufficient conditions for admissibility in one-parameter exponential families (see, for example, Example 4.3.6 of Lehmann 1983), it follows that X/n is also admissible for p.

However, X/n is not minimax. Because the risk is

$$R\left(\frac{X}{n}, p\right) = E_p\left[\left(\frac{X}{n}\right) - p\right]^2 = \frac{p(1-p)}{n}$$

and

$$\sup_p \frac{p(1-p)}{n} = \frac{1}{4n},$$

we see that X/n has a maximum risk of $1/(4n)$ when $p = \frac{1}{2}$. However, the estimator

$$\delta^* = \frac{X}{n}\frac{n^{1/2}}{1+n^{1/2}} + \frac{1}{2(1+n^{1/2})}$$

actually has a constant risk of

$$R(\delta^*; p) = \frac{1}{4(1+n^{1/2})^2} < \frac{1}{4n}.$$

To see this, we compute

$$\begin{aligned}
E_p(\delta^* - p)^2 &= E_p\left[\frac{X}{n}\frac{n^{1/2}}{1+n^{1/2}} - \frac{pn^{1/2}}{1+n^{1/2}} + \frac{pn^{1/2}}{1+n^{1/2}}\right. \\
&\qquad \left. + \frac{1}{2(1+n^{1/2})} - p\right]^2 \\
&= \left[\frac{n^{1/2}}{1+n^{1/2}}\right]^2 E_p\left[\left(\frac{X}{n}\right) - p\right]^2 \\
&\qquad + \left[\frac{pn^{1/2}}{1+n^{1/2}} + \frac{1}{2(1+n^{1/2})} - p\right]^2 \\
&= \left[\frac{n^{1/2}}{1+n^{1/2}}\right]^2 \frac{p(1-p)}{n} + \left[\frac{1-2p}{2(1+n^{1/2})}\right]^2 \\
&= \frac{1}{4(1+n^{1/2})^2}.
\end{aligned}$$

Therefore, the estimator δ^* has a smaller maximum risk than X/n, and so X/n cannot be minimax. In fact, δ^* is actually minimax because it has constant risk and is Bayes with respect to a beta prior distribution. For a detailed proof of this result and a comparison of δ^* and X/n, see Lehmann (1983, Example 4.2.1).

9.25. An admissible estimator that is a pointwise and L^2 limit of inadmissible estimators.

Suppose $X_1, \ldots, X_n$ is a sample from a Gaussian population with unknown mean θ and variance one, and consider the problem of estimating θ with squared error loss function. The risk function of the estimator

$$\delta_c(X_1, \ldots, X_n) = \bar{X}_n + c$$

is given by

$$R(\delta_c, \theta) = \mathrm{Var}_\theta(\bar{X}_n) + [\mathrm{Bias}(\delta_c, \theta)]^2$$

$$= \frac{1}{n} + c^2 \geq \frac{1}{n} = R(\delta_0, \theta)$$

with equality if and only if $c = 0$. Thus, if $c \neq 0$, δ_c is inadmissible because δ_0 is "better."

Note, however, that $\lim_{c\to 0} \delta_c = \delta_0 = \bar{X}_n$, and also

$$\lim_{c\to 0} E[(\delta_c - \delta_0)^2] = \lim_{c\to 0} c^2 = 0$$

and so δ_c converges to $\delta_0 = \bar{X}$ both pointwise and in L^2. However, $\bar{X}_n$ is admissible for this one-dimensional problem. Two proofs of this well-known result are given in Lehmann (1983, pp. 265–67).

9.26. Nonexistence of a Bayes estimator.

Let X have a Gaussian distribution $N(\theta, 1)$ and let the prior for θ be $N(0, 1)$. Let the loss function be

$$L(d; \theta) = \begin{cases} 0 & \text{if } d \geq \theta \\ 1 & \text{if } d < \theta. \end{cases}$$

Consider the constant estimator $d_n = n$. Its risk function is

$$R(d_n; \theta) = I_{(\theta > n)}.$$

Thus, the average risk of d_n with respect to the prior distribution is $P(\theta > n)$ where $\mathbf{L}(\theta) = N(0, 1) = G$. Hence,

$$0 \le \inf_{\delta} \int R(\delta; \theta)\, dG(\theta) \le \inf_{n} \int R(d_n; \theta)\, dG(\theta)$$

$$= \inf_{n} P(\theta > n) = 0.$$

So, in order for any estimator δ to be a Bayes estimator, it must have average risk equal to zero. This cannot happen (unless we allow $\delta = \infty$).

9.27. A pointwise and L^2 limit of unique Bayes estimators that is not Bayes.

Once more we consider the problem of estimating the mean from a sample $X_1, \ldots, X_n$ from a Gaussian population with known variance one. Suppose the loss function is squared error. According to Lehmann (1983, Example 4.1.3, p. 243), the estimators $a\bar{X}_n$ are Bayes estimators if $0 < a < 1$, each with respect to a Gaussian prior with mean zero and some variance, and each being the unique (almost surely) Bayes estimator for that prior. It is clear that

$$\lim_{a \to 1} a\bar{X}_n = \bar{X}_n$$

and

$$\lim_{a \to 1} E[(a\bar{X}_n - \bar{X}_n)]^2 = E[(\bar{X}_n)]^2 \lim_{a \to 1} (a - 1)^2 = 0$$

so that $\bar{X}_n$ is the pointwise and L^2 limit of unique Bayes estimators.

However, $\bar{X}_n$ is not itself a Bayes estimator. To see this, from Theorem 4.1.2 of Lehmann (1983), no unbiased estimator $\delta(\mathbf{X})$ for a parameter $q(\theta)$ can be Bayes unless $E\{[\delta(\mathbf{X}) - q(\theta)]^2\} = 0$ where the expectation is taken with respect to the joint distribution of $\mathbf{X}$ and θ. Here, $\bar{X}_n$ is unbiased for θ and $E[(\bar{X}_n - \theta)^2] = 1/n$, so that $\bar{X}_n$ is not a Bayes estimator. Note, however, that $\bar{X}_n$ is a Bayes estimator with respect to an improper prior measure—namely, Lebesgue measure on the real line.

9.28. An admissible estimator that is not Bayes.

If the parameter space is finite, then any admissible estimator must be Bayes (see Ferguson 1967, Theorem 1, p. 86). This is not quite true for an infinite parameter space.

Take the parameter space to be $\Omega = \mathbf{R}$, and consider estimating $\theta \in \Omega$ given a sample $X_1, \ldots, X_n$ from the Gaussian distribution $N(\theta, 1)$. If loss is squared error, then the sample mean $\bar{X}$ is admissible for θ (see Lehmann 1983, Section 4.3). Because $\bar{X}$ is unbiased for θ with nonzero variance, however, it cannot be Bayes with respect to any prior distribution (see Lehmann 1983, Theorem 1.2, p. 244).

9.29. A Bayes estimator that has constant risk, so that it is minimax, but it is inadmissible.

A Bayes estimator is admissible if it is unique Bayes; hence, the estimator given here is not unique Bayes.

The counterexample is adapted from Ferguson (1967, p. 61). Consider an observation X from the Gaussian distribution $N(\theta, 1)$. Suppose it is known that θ is in $\Omega = \{0, 1\}$. Let the loss function $L(T; \theta)$ be one except when $\theta = T = 1$, in which case $L(1; 1) = 0$. For the prior distribution that assigns probability one to $\theta = 0$, $\delta(X) = 0$ is a Bayes estimator. Furthermore, δ has a constant risk equal to one. However, the estimator $\gamma(X) = 1$ has a risk function that is never larger, and is sometimes smaller, than one. Hence, δ is inadmissible.

9.30. An inadmissible estimator that is a pointwise and L^2 limit of unique Bayes admissible estimators.

Suppose $X_1, \ldots, X_n$ are independent and identically distributed Gaussian random variables with unknown mean θ and known variance σ^2. We will be estimating $q(\theta) = \theta$ with squared error loss. To construct an admissible estimator of θ, it will suffice to construct a unique Bayes estimator with respect to some prior distribution. To this end, suppose we consider the prior distribution for θ to be Gaussian with mean μ and variance τ^2. The posterior distribution for θ is proportional to the likelihood times the prior; that is, $f(\theta \mid \mathbf{X})$ is proportional to

$$\exp\left[-\frac{1}{2\sigma^2}\sum_{i=1}^{n}(X_i - \theta)^2\right]\exp\left[-\frac{1}{2\tau^2}(\theta - \mu)^2\right]$$

which (with respect to θ) is proportional to

$$\exp\left[-\frac{1}{2}\left[\left(\frac{n}{\sigma^2}\right) + \left(\frac{1}{\tau^2}\right)\right]\left[\theta^2 - 2\theta\frac{(n\bar{X}/\sigma^2) + (\mu/\tau^2)}{(n/\sigma^2) + (1/\tau^2)}\right]\right].$$

Thus the posterior distribution for θ given X is seen to be Gaussian with mean

$$E(\theta \mid X) = \frac{(n\bar{X}/\sigma^2) + (\mu/\tau^2)}{(n/\sigma^2) + (1/\tau^2)}$$

and variance

$$\operatorname{Var}(\theta \mid X) = \frac{1}{(n/\sigma^2) + (1/\tau^2)}.$$

Because the loss is squared error, the Bayes estimator of θ is $E(\theta \mid X)$, which can be written in the form $a\bar{X} + b$ where $a = (n/\sigma^2)/[(n/\sigma^2) + (1/\tau^2)]$ and $b = (\mu/\tau^2)/[(n/\sigma^2) + (1/\tau)^2]$. Furthermore, since the Bayes risk is easily checked to be $\operatorname{Var}(\theta \mid X)$, which is finite, the parameter space is $\Omega = \mathbf{R}$, and the function $P_\theta(X \in A)$ is continuous in θ, it follows that the Bayes estimators are unique (by Corollary 4.1.2 of Lehmann 1983).

Also, since μ and τ^2 are allowed to vary over $\mathbf{R}$ and $\mathbf{R}^+$, respectively, this proves the admissibility of the estimator $a\bar{X} + b$ for any a in (0, 1) and any real number b.

Now consider the sequence of estimators defined by

$$\delta_j(X_1, \ldots, X_n) = \left[1 - \left(\frac{1}{j}\right)\right]\bar{X} + 1, \qquad j = 2, 3, \ldots.$$

From the preceding paragraph, δ_j is admissible for all $j = 2, 3, \ldots$. Clearly, $\delta_j(X_1, \ldots, X_n) \to \bar{X} + 1$ as $j \to \infty$, both pointwise and in L^2. The estimator $\bar{X} + 1$ is inadmissible, however. To see this, the risk of $\bar{X} + 1$ is

$$R(\bar{X} + 1; \theta) = \frac{\sigma^2}{n} + 1 > \frac{\sigma^2}{n} = R(\bar{X}; \theta)$$

so that $\bar{X} + 1$ is dominated by $\bar{X}$. Thus we have shown that a limit of unique Bayes estimators need not be admissible.

9.31. A minimax uniformly minimum variance unbiased estimator that is inadmissible.

We will give two examples. First, let X_i be independent with mean θ_i and variance one for $i = 1, 2, \ldots, n$ with $n \geq 3$. Consider estimating $\boldsymbol{\theta} = (\theta_1, \ldots, \theta_n)$ with squared error loss. Then the natural estimator $\mathbf{X} = (X_1, \ldots, X_n)$ is the maximum likelihood estimator and is also minimax and uniformly minimum variance unbiased. However, $\mathbf{X}$ is not admissible because the James-Stein Rule (see Example 9.19) has uniformly smaller mean squared error.

As a second example, let $X_1, \ldots, X_n$ be independent and identically distributed from a Gaussian distribution with unknown mean θ and variance known to be one. Suppose we know that $\theta > 0$, so that the parameter space is $\Omega = \{\theta : \theta > 0\}$. Then the naive estimator $\bar{X}$ is uniformly minimum variance unbiased and minimax (see Lehmann 1983, Example 4.3.3, p. 267). However, $\bar{X}$ is dominated by the simple improvement of requiring that the estimated value be nonnegative. Defining $\delta = \max(\bar{X}, 0)$, we see that δ dominates $\bar{X}$ with respect to squared error loss, and therefore $\bar{X}$ is inadmissible.

9.32. A constant risk minimax estimator that is inadmissible.

In general, a unique Bayes estimator (with respect to any prior) with constant risk is both minimax and admissible. If a constant risk minimax estimator is not a unique Bayes estimator, however, then it may be inadmissible.

Consider the second part of Example 9.31. In estimating the unknown mean θ of a Gaussian distribution $N(\theta, 1)$, where it is assumed that $\Omega = \{\theta : \theta > 0\}$, from a sample $X_1, \ldots, X_n$, it was shown that the estimator $\bar{X}_n$ is minimax and inadmissible for squared error loss. Furthermore,

$$R(\bar{X}_n, \theta) = \frac{1}{n}$$

does not depend on θ.

9.33. A statistic that is asymptotically unbiased but not unbiased for any finite sample size.

Let $X_1, \ldots, X_n$ be a sample from the Gaussian distribution with mean μ and variance σ^2. Define the estimator T_n of μ by:

$$T_n(\mathbf{X}) = \bar{X}_n + \frac{c}{n}$$

for some $c \neq 0$. Then $E(T_n) = \mu + (c/n)$ and $\mathrm{Var}(T_n) = \sigma^2/n$. Note that $\mathrm{Bias}(T_n; \mu) = -c/n \neq 0$ so that T_n is not unbiased for any finite sample size n. However, $n^{1/2}(T_n - \mu) = n^{1/2}(\bar{X}_n - \mu) + c/n^{1/2}$. Since $\mathbf{L}[n^{1/2}(\bar{X}_n - \mu)] \rightarrow$

$N(0, \sigma^2)$ and $c/n^{1/2} \to 0$, by Slutsky's Theorem it follows that

$$\mathbf{L}[n^{1/2}(T_n - \mu)] \to N(0, \sigma^2).$$

Thus T_n (or, strictly speaking, the sequence of estimators $T_1, T_2, \ldots$) is asymptotically unbiased.

9.34. A consistent estimator that is not unbiased for any finite sample size.

Suppose $X_1, \ldots, X_n$ is a sample from the uniform distribution on $(0, \theta]$ for some positive θ. Consider the maximum likelihood estimator T_n given by

$$T_n = X_{(n)}$$

as in Example 9.22. Then T_n has expectation $n\theta/(n + 1)$ and is therefore biased for any finite sample size n. We will show, however, that T_n is mean squared error consistent.

From Example 9.22, the mean squared error of T_n is:

$$E[(T_n - \theta)^2] = \text{MSE}(T_n; \theta) = \frac{2\theta^2}{(n + 1)(n + 2)}$$

which tends to zero as n tends to infinity. Hence the sequence of estimators $T_1, T_2, \ldots$ is mean squared error consistent. Moreover, this implies that T_n is simply consistent; that is, the sequence of random variables $T_1, T_2, \ldots$ converges to θ in probability.

9.35. A consistent estimator that is asymptotically biased.

If $X_1, \ldots, X_n$ is a sample from the Gaussian distribution with mean μ and variance σ^2, let the estimator $S_n(\mathbf{X})$ be defined by

$$S_n(\mathbf{X}) = \bar{X} + \frac{c}{n^{1/2}}$$

for some $c \neq 0$. We compute the mean squared error:

$$\text{MSE}(S_n; \mu) = E[(S_n - \mu)^2] = \text{Var}(S_n) + [\text{Bias}(S_n; \mu)]^2$$

$$= \frac{\sigma^2}{n} + \left[\frac{c}{n^{1/2}}\right]^2.$$

Because this tends to zero as $n \to \infty$, both types of consistency now follow. However, $n^{1/2}[S_n(\mathbf{X}) - \mu] = n^{1/2}(\bar{X}_n - \mu) + c$, which tends in distribution to $N(c, \sigma^2)$, so that S_n is an asymptotically biased sequence of estimators for μ.

9.36. A unique maximum likelihood estimator that is not asymptotically Gaussian.

Let $X_1, \ldots, X_n$ be a random sample from the uniform distribution on $(0, \theta)$. Then the maximum likelihood estimator exists, is unique, and is given by the largest order statistic $X_{(n)}$ of the sample. To find the asymptotic distribution of $X_{(n)}$, observe that

$$P[n(\theta - X_{(n)}) \le t] = P\left[X_{(n)} \ge \theta - \frac{t}{n}\right]$$

$$= 1 - \left[\frac{\theta - (t/n)}{\theta}\right]^n \to 1 - e^{-t/\theta} \qquad \text{as } n \to \infty.$$

Therefore, in this case, the unique maximum likelihood estimator has an asymptotically exponential distribution instead of an asymptotically Gaussian distribution.

9.37. A unique maximum likelihood estimator whose rate of convergence is not the square root of n.

Consider a sample $X_1, \ldots, X_n$ from the uniform distribution on $(0, \theta)$. In Example 9.36 we saw that the unique maximum likelihood estimator $X_{(n)}$, suitably normalized as

$$n[\theta - X_{(n)}],$$

converges in distribution to the exponential distribution with parameter θ. This implies that

$$n^{1/2}[\theta - X_{(n)}]$$

converges in distribution to zero. Thus the rate of convergence of this unique maximum likelihood estimator is order $1/n$, not order $1/n^{1/2}$.

9.38. An example where S^2 and $[(n-1)/n]S^2$ have different asymptotic distributions (from Serfling, 1980, p. 195).

Let $X_1, \ldots, X_n$ be a sample from the binomial distribution $B(1, \frac{1}{2})$. Define the usual estimate of variance and the usual second sample moment about the average:

$$S^2 = \frac{1}{n-1}\sum_{i=1}^{n}(X_i - \bar{X}_n)^2, \qquad m_2 = \frac{n-1}{n}S^2 = \frac{1}{n}\sum_{i=1}^{n}(X_i - \bar{X}_n)^2.$$

By the Central Limit Theorem, $2n^{1/2}(\bar{X}_n - \frac{1}{2})$ converges in distribution to a standard Gaussian distribution, and therefore $4n(\bar{X}_n - \frac{1}{2})^2$ converges in distribution to χ_1^2, a chi-squared distribution with one degree of freedom. From this, we may compute

$$n\left(S^2 - \frac{1}{4}\right) = -\frac{1}{4}\left\{\frac{n}{n-1}\left[4n\left(\bar{X}_n - \frac{1}{2}\right)^2\right] - \frac{n}{n-1}\right\} \xrightarrow{d} -\frac{1}{4}(\chi_1^2 - 1)$$

and

$$n\left(m_2 - \frac{1}{4}\right) = n\left(\frac{n-1}{n}S^2 - \frac{1}{4}\right) = -n\left(\bar{X}_n - \frac{1}{2}\right)^2 \xrightarrow{d} -\frac{1}{4}\chi_1^2.$$

These two distributions are different! From this we see that m_2 is asymptotically biased, whereas S^2 is asymptotically unbiased.

9.39. A maximum likelihood estimator that is asymptotically biased.

Again consider a sample $X_1, \ldots, X_n$ from a uniform distribution on $(0, \theta)$. The unique maximum likelihood estimator for θ is $X_{(n)}$. From Example 9.36, $n(\theta - X_{(n)})$ converges in distribution to the exponential distribution with mean θ. Because $\theta > 0$, $X_{(n)}$ is asymptotically biased for θ.

9.40. A Fisher efficient sequence of estimators whose bias and variance tend to infinity.

Let $X_1, \ldots, X_n$ be independent and identically distributed from the Gaussian distribution with mean θ and variance one. An estimator δ_n of θ will be efficient if $n^{1/2}(\delta_n - \theta)$ tends in distribution to the standard Gaussian.

Define the modified estimator

$$\delta_n = \begin{cases} \bar{X} & \text{with probability } 1 - \dfrac{1}{n} \\ n^2 & \text{with probability } \dfrac{1}{n} \end{cases}$$

where the randomization is accomplished by an independent toss of a biased coin. Then the expectation of this estimator is

$$E(\delta_n) = \theta\left[1 - \frac{1}{n}\right] + n$$

so that the bias $E(\delta_n - \theta)$ tends to infinity. The variance of δ_n is

$$\begin{aligned} \operatorname{Var}(\delta_n) &= E(\delta_n^2) - [E(\delta_n)]^2 \\ &= \left[1 - \frac{1}{n}\right]E(\bar{X}^2) + \left(\frac{1}{n}\right)n^4 - \left\{\theta\left[1 - \frac{1}{n}\right] + n\right\}^2 \\ &= n^3 + (\text{lower-order terms}) \end{aligned}$$

and so the variance also tends to infinity.

However, δ_n is nonetheless efficient; that is,

$$n^{1/2}(\delta_n - \theta) \xrightarrow{\text{d}} N(0, 1)$$

because convergence in distribution is unaffected by these small-probability perturbations of $\bar{X}$.

In general, if T_n is efficient for $g(\theta)$ and satisfies $\mathbf{L}\{n^{1/2}[T_n - g(\theta)]\} \rightarrow N(0, \sigma^2(\theta))$, then

$$T_n' = \begin{cases} T_n & \text{with probability } 1 - \dfrac{1}{n} \\ n^2 & \text{with probability } \dfrac{1}{n} \end{cases}$$

is also Fisher efficient but has bias and variance that tend to infinity.

9.41. An inconsistent sequence of estimators that are unbiased for any finite sample size.

Let $X_1, \ldots, X_n$ be a sample from the Laplace distribution with density function

$$f(x) = \tfrac{1}{2}e^{-|x-\theta|}.$$

Consider estimating the mean θ by the sequence of estimators $T_n(\mathbf{X}) = X_1$ for $n = 1, 2, \ldots$. Because X_1 is unbiased for θ, it follows that T_n is an unbiased estimator and is therefore asymptotically unbiased. Since the variance of X_1 is 2, however, we have

$$E[(T_n - \theta)^2] = \mathrm{Var}(X_1) = 2 \neq 0$$

and so the variance of T_n does not tend to zero as $n \to \infty$, and T_1 is not a consistent estimator of θ.

The above example is inconsequential to statistical practice, for we think of a sequence of estimators based on a sample of size n to be "similarly generated." Even so, we can find examples like the preceding one. Define the midrange

$$T_n(\mathbf{X}) = \frac{X_{(1)} + X_{(n)}}{2},$$

the average of the extreme order statistics. Because the Laplace distribution is symmetric about θ, T_n is unbiased for θ. The sequence of estimators T_1, $T_2, \ldots$ is not consistent, however. In addition, one can show that

$$E[(T_n - \theta)^2] = \mathrm{Var}(T_n) = \frac{\pi^2}{12}$$

is always a constant and does not tend to zero (Kendall and Stuart 1977, p. 365).

9.42. Nonuniqueness of consistent estimators (Patel, Kapadia, and Owen 1976, p. 142).

Suppose T_n is a consistent estimator of θ. Then for any real numbers a and b (with b a noninteger), the estimator

$$T_n' = \frac{n-a}{n-b}T_n$$

is also consistent by Slutsky's Theorem. Also, T_n is mean squared error consistent if and only if both the variance of T_n and its squared bias tend to zero as $n \to \infty$. If this is the case, then

$$\mathrm{Var}(T_n') = \frac{(n-a)^2}{(n-b)^2}\mathrm{Var}(T_n) \to 0 \qquad \text{as } n \to \infty$$

and

$$[\mathrm{Bias}(T_n'; \theta)]^2 = \left[\frac{n-a}{n-b}E(T_n) - \theta\right]^2 \to 0 \qquad \text{as } n \to \infty.$$

Hence T_n' is also mean squared error consistent for θ.

9.43. A useless consistent estimator (Rao 1973, p. 344).

The notion of consistency is an asymptotic property of a sequence of estimators so that one must be prudent in using it as a criterion of estimation in practical situations when only a finite sample of data is available. For instance, if T_n is a consistent estimator, then so is T_n', defined by

$$T_n' = \begin{cases} 0 & \text{if } n \le 10^{20} \\ T_n & \text{if } n > 10^{20}. \end{cases}$$

Clearly, T_n' is useless in any practical situation because 10^{20} is the order of magnitude of the estimated number of seconds in the lifetime of the universe (Abell 1980, p. 333).

9.44. An inconsistent maximum likelihood estimator (Bickel and Doksum 1977, Problem 4.4.14).

Suppose X_{ij} (with $i = 1, 2, \ldots, p$ and $j = 1, 2, \ldots, k$ and $k > 1$) are independent Gaussian random variables with unknown means μ_i and unknown variance σ^2. The likelihood of the sample is given by:

$$L(\mathbf{x}; \boldsymbol{\mu}, \sigma^2) = \prod_{j=1}^{k}\prod_{i=1}^{p}(2\pi\sigma^2)^{-1/2}\exp\left[-\frac{1}{2}\frac{(x_{ij}-\mu_i)^2}{\sigma^2}\right].$$

Taking logs and partial derivatives with respect to the unknown parameters, we get the likelihood equations:

$$\frac{\partial}{\partial \mu_i} \log L(\mathbf{x}; \boldsymbol{\mu}, \sigma^2) = \sum_{j=1}^{k} \frac{x_{ij} - \mu_i}{\sigma^2} = 0, \qquad i = 1, \ldots, p$$

and

$$\frac{\partial}{\partial (\sigma^2)} \log L(\mathbf{x}; \boldsymbol{\mu}, \sigma^2) = \sum_{j=1}^{k} \sum_{i=1}^{p} \left[\frac{-1}{2\sigma^2} + \frac{1}{2} \frac{(x_{ij} - \mu_i)^2}{\sigma^4} \right] = 0.$$

Solving these equations, we find the estimators

$$\tilde{\mu}_i = \frac{1}{k} \sum_{j=1}^{k} X_{ij} \quad \text{and} \quad \tilde{\sigma}^2 = \frac{1}{kp} \sum_{i=1}^{p} \sum_{j=1}^{k} (X_{ij} - \tilde{\mu}_i)^2.$$

We can apply Theorem 3.3.2 of Bickel and Doksum (1977) to see that these solutions, as expected, are indeed the unique maximum likelihood estimators.

Now suppose k is fixed but p tends to infinity. We will show that $\tilde{\sigma}^2$ is not consistent because, in fact, we have

$$\tilde{\sigma}^2 \xrightarrow{P} \frac{k-1}{k} \sigma^2.$$

Noting that $\Sigma_{j=1}^{k} (X_{ij} - \tilde{\mu}_i)^2/\sigma^2$ has a chi-squared distribution with $k - 1$ degrees of freedom (Bickel and Doksum 1977, p. 21) and hence has expectation $k - 1$, we have:

$$E(\tilde{\sigma}^2) = \frac{1}{kp} \sum_{i=1}^{p} \sum_{j=1}^{k} E[(X_{ij} - \tilde{\mu}_i)^2] = \frac{k-1}{k} \sigma^2.$$

Because we can write $\tilde{\sigma}^2$ as the average of the p independent random variables $Y_j = (1/k) \Sigma_{j=1}^{k} (X_{ij} - \tilde{\mu}_i)^2$, and checking that $E(Y_j^2)$ exists for $j = 1, 2, \ldots, p$, we find that the weak law of large numbers applies, so that

$$\frac{Y_1 + \cdots + Y_p}{p} \xrightarrow{P} E(Y_i)$$

or equivalently,

$$\tilde{\sigma}^2 \xrightarrow{P} \frac{k-1}{k} \sigma^2.$$

Thus, $\tilde{\sigma}^2$ is not consistent.

This example was originally by Neyman and Scott (1948). The problem here exemplifies the statement that maximum likelihood estimators may not behave well when the number of incidental parameters is of the order of the total sample size. It should be pointed out that consistent estimators of σ^2 do exist; for example, $k\tilde{\sigma}^2/(k - 1)$ is a consistent estimator of σ^2.

9.45. A family of distinct distributions with density functions $f(x; \theta)$ that are continuous in θ for all x, yet the maximum likelihood estimator for a sample of size n converges almost surely to one, regardless of the true value of θ.

This example, from Ferguson (1982), shows a smooth family of distributions for which we expect the method of maximum likelihood to succeed, but the estimator is inconsistent.

Let $X_1, \ldots, X_n$ be a sample from the distribution with density

$$f(x; \theta) = \frac{1 - \theta}{\delta(\theta)}\left[1 - \frac{|X - \theta|}{\delta(\theta)}\right] I_A(x) + \frac{\theta}{2} I_{[-1,1]}(x)$$

where $A = [\theta - \delta(\theta), \theta + \delta(\theta)]$ and $\delta(\theta)$ is a continuous decreasing function of θ with $\delta(0) = 1$ and $0 < \delta(\theta) \leq 1 - \theta$ for $0 < \theta < 1$. Note that $\theta = 1$ corresponds to the uniform distribution on $[-1, 1]$. The densities $f(x; \theta)$ are continuous in θ for all x. Because a continuous function defined on $[0, 1]$ achieves its maximum, a maximum likelihood estimate of θ exists. We will now show that any maximum likelihood estimator converges to one almost surely if we define $\delta(\theta)$ appropriately.

Let $\tilde{\theta} = \tilde{\theta}(X_1, \ldots, X_n)$ denote any maximum likelihood estimator of θ. Denote the log likelihood function by

$$l_n(\theta) = \sum_{i=1}^{n} \log[f(X_i; \theta)].$$

If $\theta < 1$, then $f(x; \theta) \leq [(1 - \theta)/\delta(\theta)] + \theta/2 < [1/\delta(\theta)] + \frac{1}{2}$. Hence, for any $\alpha < 1$,

$$\max_{0 \leq \theta \leq \alpha} \frac{l_n(\theta)}{n} \leq \log\left\{\left[\frac{1}{\delta(\theta)}\right] + \frac{1}{2}\right\} < \infty.$$

If we choose $\delta(\theta)$ so that

$$\max_{0 \leq \theta \leq 1} \frac{l_n(\theta)}{n} \to \infty \quad \text{almost surely}$$

then δ_n will eventually be greater than α for any $\alpha < 1$ and the result will be shown. To show that this is possible, let $M_n = \max(X_1, \ldots, X_n)$. Note that $M_n \to 1$ almost surely, and

$$\max_{0 \le \theta \le 1} \frac{l_n(\theta)}{n} \ge \frac{l_n(M_n)}{n} \ge \frac{n-1}{n} \log\left(\frac{M_n}{2}\right) + \frac{1}{n} \log\left[\frac{1 - M_n}{\delta(M_n)}\right].$$

Thus,

$$\lim_{n\to\infty} \max_{0 \le \theta \le 1} \frac{l_n(\theta)}{n} \ge \log\left(\frac{1}{2}\right) + \lim_{n\to\infty} \log\left[\frac{1 - M_n}{\delta(M_n)}\right] \quad \text{almost surely.}$$

We need to choose $\delta(\theta)$ so that this last limit is almost surely infinite. For all θ, $M_n \to 1$ almost surely, with the slowest rate corresponding to $\theta = 1$ because this distribution has the smallest mass in a sufficiently small neighborhood of one. Hence, it will suffice to choose $\delta(\theta) \to 0$ as $\theta \to 1$ fast enough so that this limit is almost surely infinite when $\theta = 0$.

Now, if $0 < \varepsilon < 1$, then

$$\begin{aligned} \sum_n P_{(\theta=0)}[n^{1/4}|1 - M_n| > \varepsilon] &= \sum_n P_{(\theta=0)}[M_n < 1 - \varepsilon n^{-1/4}] \\ &= \sum_n \left[1 - \varepsilon^2 \frac{n^{-1/2}}{2}\right]^n \\ &\le \sum_n \exp\left[-\varepsilon^2 \frac{n^{1/2}}{2}\right] < \infty. \end{aligned}$$

Thus by the Borel-Cantelli Lemma, $n^{1/4}[1 - M_n] \to 0$ almost surely. Taking $\delta(\theta) = (1 - \theta)\exp[-(1 - \theta)^{-4}] + 1$, we get a function that is continuous decreasing with $\delta(\theta) = 1, 0 < \delta(\theta) < 1 - \theta$ for all $\theta \in (0, 1)$, and such that

$$\frac{1}{n} \log\left[\frac{1 - M_n}{\delta(M_n)}\right] = \frac{1}{n(1 - M_n)^4} - \frac{1}{n} \to \infty \quad \text{almost surely.}$$

The net result is that any maximum likelihood estimator must tend to one almost surely.

Note that the failure of the maximum likelihood estimator to be consistent is not due to the difficulty of estimating θ. Indeed, the method of moments will provide estimators of θ that are consistent.

Ferguson shows how this example may be modified so that Cramer's assumptions for the existence of a consistent and efficient sequence of roots of the likelihood equation are satisfied, and so that any maximum likelihood estimator satisfies the likelihood equation, but still the maximum likelihood estimator tends to one almost surely for all θ.

For additional related examples, see Kraft and LeCam (1956), Bahadur (1958), and Berkson (1980).

9.46. An inconsistent maximum likelihood estimator for a function of the location parameter of a Gaussian distribution.

Suppose $X_1, \ldots, X_n$ is a sample from $N(\theta, 1)$, the Gaussian distribution with mean θ and variance one. Let

$$g(\theta) = \begin{cases} -\theta & \text{if } \theta \text{ is irrational} \\ \theta & \text{if } \theta \text{ is rational.} \end{cases}$$

Since $\bar{X}_n$ is the maximum likelihood estimator for θ, $g(\bar{X}_n)$ is the maximum likelihood estimator for $g(\theta)$. Note that $\bar{X}_n$ is $N(\theta, 1/n)$ and so is almost surely irrational. Thus, $g(\bar{X}_n) = -\bar{X}_n$ almost surely. By the strong law of large numbers, $g(\bar{X}_n) \to -\theta$ almost surely. Hence, if $\theta \neq 0$ is a rational number, we will have

$$g(\bar{X}_n) \to -\theta \neq \theta = g(\theta) \quad \text{almost surely.}$$

9.47. A superefficient estimator.

It is conceivable to have a sequence of estimators T_n of θ with asymptotic variance of order $1/n$ so that

$$n \operatorname{Var}_\theta(T_n) \to [\sigma(\theta)]^2 > 0 \qquad \text{as } n \to \infty$$

and with T_n asymptotically unbiased so that

$$n^{1/2}[E_\theta(T_n) - \theta] \to 0 \qquad \text{as } n \to \infty$$

but $[\sigma(\theta)]^2 \leq 1/[I(\theta)]$ with strict inequality for some θ. That is, the asymptotic variance does no worse than the Cramer-Rao lower bound and sometimes beats it! Of course, this does not contradict the Information Inequality, since that pertains to only finite samples and requires finite-sample unbiasedness. Nevertheless, this situation is somewhat anomalous. In such a case, the estimator T_n is called "superefficient" if T_n is asymptotically Gaussian with mean θ and variance never greater than, and sometimes smaller than, the Cramer-Rao lower bound. The first example of superefficiency, as given below, is from J. Hodges (see LeCam 1953).

To construct such a sequence of estimators, suppose $X_1, X_2, \ldots, X_n$ is a sample from the Gaussian distribution with mean μ and variance one. The estimator $\bar{X}_n$, the sample mean, is easily checked to be the maximum likelihood estimator of μ and also the uniformly minimum variance unbiased estimator of μ with variance $1/n$, as

$$I(\mu) = E\left[\frac{\partial}{\partial\mu}\log\left\{(2\pi)^{-1/2}\exp\left[-\frac{1}{2}(X_1 - \mu)^2\right]\right\}\right]^2$$

$$= E[(X_1 - \mu)^2] = 1.$$

So, in fact, $\bar{X}_n$ has a variance that achieves the Cramer-Rao lower bound. Define a new estimator:

$$T_n(\mathbf{X}) = \begin{cases} \bar{X}_n & \text{if } |\bar{X}_n| \geq n^{-1/4} \\ b\bar{X}_n & \text{otherwise} \end{cases}$$

for some real b with $|b| < 1$. Since $\bar{X}_n$ has the Gaussian distribution with mean μ and variance $1/n$, the asymptotic distributions of $\bar{X}_n$ and T_n are the same if $\mu \neq 0$. If $\mu = 0$, however, then T_n is asymptotically Gaussian with mean zero and variance b^2/n, an improvement in variance over $\bar{X}_n$ because $|b| < 1$. The value of the parameter $\mu = 0$ is called a "point of superefficiency."

LeCam (1953) has shown that the set of points of superefficiency is countable under certain conditions. Superefficiency is not important in statistical practice because, for any fixed sample size, the reduction in mean squared error at points of superefficiency is counterbalanced by an increase in mean squared error at points near superefficient points.

9.48. An inefficient method of moments estimator (Bickel and Doksum 1977, Problem 4.4.2).

Suppose $X_1, \ldots, X_n$ is a sample from a beta distribution with density given by:

$$f(x; \theta) = \theta(\theta + 1)x^{\theta-1}(1 - x)I_{(0,1)}(x)$$

for some positive θ. To find a method of moments estimator of θ, note that

$$E(X) = \theta(\theta + 1)\int_0^1 x^{\theta}(1 - x)\,dx = \frac{\theta}{\theta + 2}.$$

Then, setting this equal to the first sample moment, $\bar{X}_n$, and solving for $T_n(\mathbf{X})$, we find $\bar{X}_n = T_n(\mathbf{X})/[T_n(\mathbf{X}) + 2]$ and so

$$T_n(\mathbf{X}) = \frac{2\bar{X}_n}{1 - \bar{X}_n}$$

is a method of moments estimator of θ. Furthermore, we may apply Theorem 4.4.2 of Bickel and Doksum (1977) to find that the asymptotic distribution of this estimator is given by

$$n^{1/2}\frac{T_n - \theta}{\sigma_n(\theta)} \xrightarrow{\text{d}} G$$

where G has the standard Gaussian distribution and $\sigma_n^2 = \theta(\theta + 2)^2/[2(\theta + 3)]$. We will show by calculating the Cramer-Rao lower bound that T_n is inefficient, however. To this end, note that the distributions form a one-parameter exponential family and, hence, the regularity conditions of the Information Inequality are satisfied. The Fisher information number is

$$I(\theta) = -E\left[\frac{\partial^2}{\partial\theta^2}\log[\theta(\theta + 1)x^{\theta-1}(1 - x)]\right]$$

$$= \frac{\theta^2 + (\theta + 1)^2}{\theta^2(\theta + 1)^2} > \frac{2(\theta + 3)}{\theta(\theta + 2)^2} = \frac{1}{\sigma_n^2}.$$

The inequality follows from cross-multiplying and then showing that the difference is $3\theta^3 + 6\theta^2 + 4\theta$, which is always positive for $\theta > 0$. Therefore T_n is inefficient because $\sigma_n^2 > 1/[I(\theta)]$.

A related discussion may be found in Johnson and Kotz (1970, Section 24.4, p. 46).

CHAPTER 10
HYPOTHESIS TESTING

Introduction

In Chapters 8 and 9, we dealt with the construction of and criteria for evaluating point estimators for some unknown parameter. Often it is desirable that a point estimate be accompanied by a measure of its precision. Instead of estimating the unknown parameter by some value, we would like to estimate an interval in which we believe the parameter lies. With this in mind, we define a two-sided confidence interval as follows. Suppose $X_1, \ldots, X_n$ is a sample from a distribution indexed by a real parameter θ in Ω. If $T_1(X)$ and $T_2(X)$ are two statistics that satisfy

$$P_\theta(T_1 \leq \theta \leq T_2) \geq 1 - \alpha \qquad \text{for all } \theta,$$

then the random interval $[T_1, T_2]$ is called a two-sided confidence interval for θ with confidence coefficient or level $(1 - \alpha)$. Appropriate modifications can easily be formulated for one-sided confidence intervals. More generally, the theory of confidence sets is discussed in Lehmann (1959).

Alternatively, we may be interested in making a yes or no inference based on the data about an unknown parameter. This process is known as "hypothesis testing." There is a close mathematical duality between the theory of confidence intervals and the theory of hypothesis testing; for this reason we will give more detail about only the theory of hypothesis testing. An excellent treatment of both subjects can be found in Lehmann (1959).

The classical formulation of hypothesis testing is as follows: We observe a random object X whose distribution P_θ is assumed to belong to a certain family of distributions $\mathbf{P} = \{P_\theta, \theta \text{ in } \Omega\}$. Usually X is a random vector, $X = (X_1, \ldots, X_n)$, whose components are independent and identically distributed random variables with a common distribution F_θ. Furthermore, we have a null hypothesis H—say, θ in Ω_0, some subset of Ω—and we would like to examine the consistency of the data with H. The null hypothesis H

is called "simple" if it completely specifies the distribution, such as $\theta = \theta_0$. Otherwise H is called "composite." The alternative K specifies θ in Ω_1, the complement of Ω_0.

The statistician has a two-decision problem: to accept or reject the null hypothesis H. A test of a statistical hypothesis H is any procedure for deciding whether to accept or reject H. If the sample space of possible values of X is S, a nonrandomized test can be described by the subset S_0 of S that consists of all sample points x for which H is rejected. This set S_0 is called the "critical region." In this case, the critical function can be defined by $\delta(x) = I_{S_0}(x)$, the indicator function of the critical region.

A richer theory leads us to include the notion of a randomized test. Based on the value x of the random object X, a (possibly) randomized test chooses among the two decisions of accepting or rejecting H by rejecting with probability $\delta(x)$. In this way, the function δ, called the "critical function," completely characterizes the test. When δ takes on only the values zero and one, the test is nonrandomized, in which case δ is simply the indicator function of the critical region.

To measure the performance of a test, we must consider the two types of errors that may be made. Rejection of the null hypothesis H when it is true is called a "type I error," and acceptance of H when it is false is called a "type II error." The power of a test against a simple alternative θ in Ω is the probability of rejecting H (and accepting θ) when θ is true. In terms of the critical function δ, we can define the power function as follows:

$$\beta(\theta, \delta) = E_\theta \delta(X) = \int \delta(x)\, dP_\theta(x)$$

which is the conditional probability of rejecting H given $X = x$, integrated with respect to the distribution of X when θ is true.

In the classical framework, a type I error is considered more serious, perhaps because the problem of finding a "best" test is often more tractable analytically. We begin by specifying α and then restricting attention to tests δ for which $\beta(\theta, \delta) \leq \alpha$ for all θ in Ω_0. Such a test is said to have α as its "level of significance." The smallest level of significance or, equivalently, the least upper bound probability of a type I error is called the "size" of the test. In choosing among competing tests δ of size less than or equal to some prespecified α, the goal is to find the test that maximizes the power $\beta(\theta, \delta)$ for all θ in Ω_1. We say that a level α test δ_0 is "inadmissible" if there exists a level α test δ_1 with $\beta(\theta, \delta_0) \leq \beta(\theta, \delta_1)$ for all θ in Ω_1 with strict inequality holding for some θ in Ω_1. A level α test δ^* is called "uniformly most powerful" (abbreviated UMP) for testing H: θ in Ω_0 against K: θ in Ω_1 if $\beta(\theta, \delta^*) \geq \beta(\theta, \delta)$ for all θ in Ω_1 for any other level α test δ. Of course, a uniformly most powerful test, if it exists, is always admissible.

When a uniformly most powerful test exists and also in more general situations, the test can often be specified by a test statistic. In this case, the critical region is of the form $\{x : T(x) \geq c\}$ for some statistic T, where c is chosen so that the size of the test is α. When the resulting test is uniformly most powerful, T is said to be "optimal." A general method for constructing test statistics is the likelihood ratio procedure. If X has a density function $f(x; \theta)$ with respect to some dominating σ-finite measure μ, then consider the likelihood ratio:

$$L(x) = \frac{\sup\{f(x; \theta) : \theta \text{ in } \Omega_1\}}{\sup\{f(x; \theta) : \theta \text{ in } \Omega_0\}}.$$

A likelihood ratio test rejects for large values of the statistic $L(x)$. A special case occurs when simple hypotheses are tested, so that $\Omega_0 = \{\theta_0\}$ and $\Omega_1 = \{\theta_1\}$ and $L(x) = f(x; \theta_1)/f(x; \theta_0)$. Note that the dominating measure μ can be taken to be $P_{\theta_0} + P_{\theta_1}$. A fundamental result, the Neyman-Pearson Lemma, says that the likelihood ratio test is uniformly most powerful (or just most powerful) in this special case of simple hypotheses. Precisely stated, in testing a simple hypothesis $H: f_0$ against a simple alternative $K: f_1$, a necessary and sufficient condition for a level α test δ to be most powerful at level α is that there exist a constant k such that

$$\delta(x) \overset{\text{a.e.}[\mu]}{=} \begin{cases} 1 & \text{when } f_1(x) > kf_0(x) \\ 0 & \text{when } f_1(x) < kf_0(x) \end{cases}$$

where "a.e.[μ]" indicates that this equality need only hold almost everywhere with respect to the dominating measure μ.

To construct uniformly most powerful tests when H is simple and K is composite, the basic procedure is to determine the most powerful test or tests for a fixed alternative (via the Neyman-Pearson Lemma) and then determine by inspection whether there exists a fixed test that is more powerful for all alternatives. Although such a uniformly most powerful test rarely exists when the family of distributions P_θ has real-valued densities $f_\theta(x)$ with respect to some dominating measure and θ is a real-valued parameter, and the problem is to test $H: \theta = \theta_0$ (or $\theta \leq \theta_0$) against $K: \theta > \theta_0$, a uniformly most powerful test will exist if the family of densities $p_\theta(x)$ has a property called "monotone likelihood ratio" (abbreviated MLR). A family of densities $p_\theta(x)$ that are distinct for different values of θ has monotone likelihood ratio if there exists a real-valued function $T(x)$ such that the likelihood ratio $p_{\theta_1}(x)/p_{\theta_0}(x)$ is a nondecreasing function of $T(x)$ for $\theta_1 > \theta_0$. For more details, see Section 3.3 of Lehmann (1959). For example, when the family of distributions is a one-parameter exponential family so that

$$\frac{dP_\theta(x)}{d\mu(x)} = p_\theta(x) = c(\theta)e^{Q(\theta)T(x)}h(x)$$

where Q is a strictly increasing function, then the densities $p_\theta(x)$ have monotone likelihood ratio in $T(x)$, and so uniformly most powerful tests will exist for one-sided alternatives.

When the hypothesis $H: f_\theta$, θ in Ω_0, is composite, then even if one first fixes an alternative K: g, the determination of a most powerful test is not always so straightforward. The idea is to reduce the problem to testing a simple hypothesis H_λ. Suppose λ is a probability distribution over Ω_0 and consider the simple hypothesis H_λ that the density is

$$h_\lambda(x) = \int_{\Omega_0} f_\theta(x)\, d\lambda(\theta).$$

Let β_λ be the power of the most powerful test δ_λ of H_λ against K: g. The probability distribution λ is called a "least favorable distribution" (at level α) if, for any other distribution λ' over Ω_0, we have $\beta_\lambda \leq \beta_{\lambda'}$.

For a distribution λ to have any relevance to our original hypothesis $H: f_\theta$, θ in Ω_0, the choice of λ should not provide any information about H. Indeed, if δ_λ is the (unique) most powerful level α test for H_λ and it is also level α for H, then δ_λ is also the (unique) most powerful level α test for H, and λ is least favorable. It is easily seen that this will be the case if λ assigns probability one to the set of θ in Ω_0 with $E_\theta \delta_\lambda(X) = \alpha$. For proofs and examples, see Sections 3.8 and 3.9 in Lehmann (1959).

Because uniformly most powerful tests rarely exist, it is natural to restrict attention to a certain class of tests and then hope there exists an optimal test within this class. In testing H: θ in Ω_0 against K: θ in Ω_1, it is undesirable that the probability of rejecting H when an alternative θ in Ω_1 is true be smaller than when H is true. This consideration leads us to restrict our search to level α tests δ with power functions $\beta_\delta(\theta)$ that satisfy:

$$\beta_\delta(\theta) \leq \alpha \qquad \text{if } \theta \text{ in } \Omega_0$$

$$\beta_\delta(\theta) \geq \alpha \qquad \text{if } \theta \text{ in } \Omega_1.$$

Such tests are called "unbiased." When a test with uniformly greatest power exists among the class of all unbiased tests, it is called "uniformly most powerful unbiased" (abbreviated UMPU). Fortunately, uniformly most powerful unbiased tests sometimes exist when uniformly most powerful tests do not.

To construct uniformly most powerful unbiased tests, it is necessary to characterize the totality of tests δ that satisfy $E_\theta \delta(X) = \alpha$ for all θ in ω, a subset of Ω. Such a test is called "similar" with respect to ω. Here we are particularly interested in ω being the common boundary between Ω_0 and Ω_1. A test δ that satisfies $E[\delta(X) \mid T(X) = t] = \alpha$ (a.e. with respect to $\mathbf{P}^T$, the family of distributions of T) is said to have "Neyman structure" with respect to T. If T is sufficient for $\mathbf{P} = \{P_\theta, \theta \text{ in } \omega\}$, then a necessary and sufficient condition for all similar tests to have Neyman structure with respect to T is that the family $\mathbf{P}^T$ be boundedly complete. Such considerations apply in constructing uniformly most powerful unbiased tests of certain one- and two-sided hypotheses that concern a real parameter where the family of distributions of X is a multiparameter exponential family. For the theory and applications of unbiasedness, see Chapters 4 and 5 in Lehmann (1959).

A second alternative when no uniformly most powerful test exists is to appeal to the principle of invariance. The situation may be described as follows: We observe a random object X with range $\mathbf{S}$ and distribution from $\mathbf{P} = \{P_\theta, \theta \text{ in } \Omega\}$. Consider testing H: θ in Ω_0 against K: θ in Ω_1. Suppose G is a group of transformations that act on the sample space $\mathbf{S}$. If X has distribution P_θ, let gX be a random object that takes on the value gx when $X = x$, so that gX has distribution $P_{\theta'}$, with θ' in Ω. The group G then induces a group of transformations $\bar{G}$ in Ω by setting $\bar{g}\theta = \theta'$. This leads to the following identity:

$$P_\theta\{gX \text{ in } A\} = P_{\bar{g}\theta}\{X \text{ in } A\}$$

where the subscript on each side of the equation refers to the distribution of X. We say that a parameter subset Ω' of Ω remains invariant or is preserved by $\bar{g}$ if $\bar{g}\Omega' = \Omega'$. Finally, we say that Ω' remains invariant under the group $\bar{G}$ if $\bar{g}\Omega' = \Omega'$ for all $\bar{g}$ in $\bar{G}$.

The problem of testing H: θ in Ω_0 against K: θ in Ω_1 is said to remain invariant under the group G if both Ω_0 and Ω_1 remain invariant under the induced group $\bar{G}$. When a problem remains invariant under a group of transformations G, it seems appealing to impose invariance restrictions on the tests used. For example, if $X = (X_1, \ldots, X_n)$ is a sample of n independent and identically distributed random variables, then the problem will remain invariant under the group of permutations of the coordinates of X. Unless the test is symmetric in the X_i, the acceptance or rejection of H will depend on the irrelevant recording or ordering of the observations.

With these considerations in mind, a test $\delta(X)$ is called "invariant" if

$$\delta(gx) = \delta(x) \qquad \text{for all } x \text{ in } \mathbf{X} \quad \text{and} \quad g \text{ in } G.$$

The problem is to find a uniformly most powerful invariant test—that is, a uniformly most powerful test within the class of invariant tests. In general, a function $T(x)$ is invariant if $T(gx) = T(x)$ for all x in $\mathbf{X}$ and g in G. $T(x)$ is said to be a "maximal invariant" if it is invariant and if $T(x_1) = T(x_2)$ implies $x_2 = gx_1$ for some g in G. In the language of group theory, $T(x)$ is a maximal invariant if it is constant on each orbit of G and takes on distinct values on distinct orbits. The problem of determining a uniformly most powerful invariant test is simplified because a test $\delta(X)$ is invariant if and only if it depends on the data X only through $T(X)$.

A test $\delta(x)$ is said to be "equivalent to an invariant test" if there exists an invariant test $\delta_0(x)$ such that

$$\delta(x) = \delta_0(x) \qquad \text{except on a set } N \text{ with } P_\theta(N) = 0 \text{ for all } \theta \text{ in } \Omega.$$

The test $\delta(x)$ is called "almost invariant" if

$$\delta(gx) = \delta(x) \qquad \text{except on a set } N_g \text{ (which may depend on } g\text{)}$$
$$\text{with } P_\theta(N_g) = 0 \text{ for all } \theta \text{ in } \Omega.$$

Such concepts are necessary to determine the relationship between uniformly most powerful invariant tests and uniformly most powerful unbiased tests. This must be done when a uniformly most powerful invariant test is also uniformly most powerful within the class of almost invariant tests. Ignoring measurability considerations, a function that is almost invariant under G is equivalent to an invariant function if there exists a σ-finite measure η over $(G, \mathbf{G})$, where $\mathbf{G}$ is a σ-field of subsets of G, such that $\eta(B) = 0$ implies $\eta(Bg) = 0$ for all g in G and B in $\mathbf{G}$. For more details, see Lehmann (1959, Section 6.5) and the references given there.

Sometimes uniformly most powerful unbiased tests exist when invariance considerations are not applicable, or a uniformly most powerful almost invariant test exists with respect to some group when a uniformly most powerful unbiased test does not exist. When both exist and the uniformly most powerful unbiased test is unique, then the uniformly most powerful almost invariant test is also unique and the two tests coincide (modulo null sets). Details may be found in Lehmann (1959, Section 6.6).

Our goal in constructing tests was to find a test with uniformly greatest power when this is possible. When uniformly most powerful tests do not exist, it may be desirable to find a test with a guaranteed minimum power. To this end, we see that any level α test δ with a continuous power function will satisfy $\inf_{\Omega_1} \beta(\theta, \delta) = \alpha$ if Ω_1 and Ω_0 are contiguous. Hence, in testing H: Ω_0 against K: Ω_1, we must assume that Ω is partitioned as $\Omega = \Omega_0 + \Omega_I + \Omega_1$, where Ω_I is the "indifference region." Any test δ that maximizes $\inf_{\Omega_1} \beta(\theta, \delta)$, the minimum power over Ω_1, is called a "maximin test." For theory and applications of this minimax property, see Lehmann (1959, Chapter 8).

10.1. An unreasonable level 0.05 test result.

Strict application of the principles of hypothesis testing can, at times, lead to unreasonable results. Consider, for example, the situation in which X has a Gaussian distribution $N(\theta, 1)$ with mean θ and variance one. Suppose we want to test the null hypothesis $H\colon \theta = 0$ against the alternative hypothesis $K\colon \theta = 1000$. The level $\alpha = 0.05$ likelihood ratio test will reject if $X > 1.645$, and by the Neyman-Pearson Lemma we know that this test is optimal.

Now consider the case in which $X = 2.1$ is observed. This optimal level 0.05 test would tell us to reject the null hypothesis and to accept the alternative hypothesis. However, accepting the alternative hypothesis (that the mean is 1000) is not very realistic or reasonable based on the data because if the mean truly had been 1000, it is incredibly unlikely that a value of 2.1 would have been observed. From a strictly commonsense approach, it would seem that the observed value 2.1 is much more consistent with the mean being zero than with the mean being 1000.

The paradox is partially resolved by recognizing that the specification of level 0.05 severely restricts the class of possible tests. Ordinarily, this specification limits the probability of a type I error (wrongly rejecting H) to at most 5%. In the present case, however, there is little ambiguity between the hypotheses, and the specification level 0.05 actually demands that we be wrong 5% of the time when H is actually true. Instead of being limited to a budgeted amount of error, we find ourselves forced to be wrong.

A better solution to this problem might therefore be to recognize that, because the hypotheses are so well separated, we have a right to be more demanding. Accordingly, a test at a much lower level ($\alpha = 0.01$, 0.001, or even smaller) would greatly decrease the probability of type I error, while maintaining very high power.

10.2. Two test statistics that differ almost surely, but nonetheless generate tests with identical power functions.

This example highlights the fact that test statistics are generally not unique, although the tests they generate may be unique. For many tests, rejection occurs when a test statistic exceeds a certain critical value. In these cases any strictly monotone function of the test statistic will result in a new test statistic that generates the same family of tests. The new critical values are simply the monotone function's values at the old critical values.

For concreteness, here is a definite example. Let X have the Gaussian distribution $N(\theta, 1)$ with mean θ and variance one. Consider testing the null hypothesis $H\colon \theta = 0$ against the alternative hypothesis $K\colon \theta = 1$. The random variable X is itself a test statistic, and the test that rejects when $X > 1.645$ is uniformly most powerful at level 0.05.

Another test statistic that is almost surely different from X is given, for example, by $Y = X + 3$. The level 0.05 test that rejects for large values of Y will reject when $Y > 4.645$ and does, in fact, reject when and only when the test based on X rejects. Because these two test statistics generate the same test, it follows that they share the same operating characteristics. In particular, both are uniformly most powerful and they have the same power function.

10.3. A hypothesis-testing problem where no nonrandomized tests at level less than 0.5 have any power.

Consider a single observation X of a Bernoulli trial with success probability p. In testing the null hypothesis H: $p = 0.5$ against the alternative hypothesis K: $p = 0.6$, any nonrandomized test must be one of the following four possibilities:

		Level of Test
Test 1	Always reject H	1.0
Test 2	Reject H when $X = 0$	0.5
Test 3	Reject H when $X = 1$	0.5
Test 4	Always accept H	0.0

The only one of these tests that has a level of significance less than 0.5 is the one that always accepts the null hypothesis H. This test has no power because the null hypothesis can never be rejected.

Similar examples may be constructed along these lines in which the extreme discreteness of the problem severely reduces the choice of possible nonrandomized tests. One advantage of randomization in testing is the ability to break up otherwise indivisible mass points, thus greatly extending the class of available testing procedures at a given level.

On the other hand, this phenomenon is usually not a problem in practice. This example is just an extreme case where we cannot test H with any power based on one observation.

10.4. A hypothesis-testing problem where the nonrandomized likelihood procedures have achievable levels of significance of only one and zero (Bickel and Doksum 1977, p. 196).

Let $X_1, \ldots, X_n$ be a sample (with replacement) from the discrete distribution that places mass $1/N$ on the integers $1, 2, \ldots, N$, where N is the population size. Consider testing H: $N = N_0$ against K: $N = N_1$, where $N_1 > N_0$. The likelihood ratio is as follows:

$$L(x_1, \ldots, x_n) = \begin{cases} \infty & \text{if } N_0 < \max(x_1, \ldots, x_n) \leq N_1 \\ \left(\dfrac{N_0}{N_1}\right)^n & \text{if } 1 \leq \max(x_1, \ldots, x_n) \leq N_0. \end{cases}$$

A likelihood ratio procedure will reject H if $L(x_1, \ldots, x_n)$ is large. By the Neyman-Pearson Lemma, such a test (with randomization permitted) is uniformly most powerful. The nonrandomized tests take a degenerate form, however, as indicated by the following complete list of possibilities:

Test A	Reject H if $N_0 \leq \max(x_1, \ldots, x_n) \leq N_1$.
Test B	Always reject H.

The size of test A is zero because under H it is impossible to observe a value larger than N_0. The size of test B is one. These are the only possible nonrandomized tests based on the likelihood ratio.

If we do not restrict ourselves to tests based on the likelihood ratio, then other nonrandomized tests are possible. Note that the likelihood ratio is constant with positive probability at $k = (N_0/N_1)^n$, and we can adjust j in the test defined by the critical region for which $j \leq \max(x_1, \ldots, x_n) \leq N_0$. With this method there are $N_0 + 1$ different sizes available for nonrandomized tests, and by the Neyman-Pearson Lemma these are most powerful at their respective levels. These nonrandomized tests based on the maximum order statistic have made use of randomness in the data that would have been "filtered out" by the likelihood ratio statistic. Thus these procedures are equivalent to randomized likelihood ratio procedures with the randomness coming from the sample itself rather than from an external source.

10.5. Nonexistence of a uniformly most powerful test for a simple hypothesis against only two alternatives.

Although it is unfortunate that uniformly most powerful tests rarely exist except in simple situations, we give an example of the nonexistence of a uniformly most powerful test for the sake of concreteness. Furthermore, this example clearly shows why the Neyman-Pearson Lemma cannot be extended in general to handle the case of testing a simple hypothesis against only two alternatives.

Let $X \sim N(\mu, 1)$ be an observation from a Gaussian distribution with unknown mean μ and variance one. First consider the simple case of testing the hypothesis H: $\mu = 0$ against K: $\mu = \mu_0$. The simple likelihood ratio test rejects for large values of

$$L(x) = \frac{\exp[-(x - \mu_0)^2/2]}{\exp(-x^2/2)} = \exp\left(\mu_0 x - \frac{\mu_0^2}{2}\right).$$

Equivalently, if $\mu_0 > 0$, then the simple likelihood ratio test rejects for large positive values of X. If $\mu_0 < 0$, however, then the simple likelihood ratio test rejects for large negative values of X. By the Neyman-Pearson Lemma, the test is uniformly most powerful. In detail, for example, if $\mu_0 > 0$, the level α test rejects if $X > z(1 - \alpha)$, where $z(1 - \alpha)$ is the upper 100α percentage point of the standard Gaussian distribution.

Now consider the compound (nonsimple) case of testing H: $\mu = 0$ against K: $\{\mu = \mu_0 \text{ or } \mu = \mu_1\}$ for any fixed $\mu_0 < 0$ and $\mu_1 > 0$. For a test to be uniformly most powerful in this situation, it must have power at least as great as the power function in the simple tests given earlier. But we know that the level α test that rejects when $X > z(1 - \alpha)$ has maximum power for detecting $\mu > 0$, whereas the level α test that rejects when $X < z(\alpha)$ has maximum power for detecting $\mu < 0$. Since these tests are different with different power functions, no single uniformly most powerful tests can exist for both possible alternatives μ_0 and μ_1.

By the same reasoning, no uniformly most powerful test exists in testing H against the much larger compound alternative $\mu \neq 0$. In fact, it is generally true that no uniformly most powerful test exists for testing H: $\theta = \theta_0$ against K: $\theta \neq \theta_0$ when $\{P_\theta\}$ is a one-parameter exponential family of probability distributions (as long as there are θ values on both sides of θ_0).

10.6. Nonexistence of a uniformly most powerful test for a simple hypothesis against a one-sided alternative.

In general, uniformly most powerful tests exist in testing a simple null hypothesis against a one-sided alternative when the sample is from a one-parameter exponential family of distributions. This holds more generally when the family of densities has monotone likelihood ratio (see, for example, Lehmann 1959, Theorem 2 of Chapter 3). A partial converse, from Pfanzagl (1968), says that if you have a one-parameter family (which obeys certain regularity conditions) for which there exist uniformly most powerful one-sided tests for all sample sizes for any particular level α in (0, 1), then the family must be a one-parameter exponential family of distributions.

Except in cases like these, however, a uniformly most powerful test need not exist. For example, suppose $X_1, \ldots, X_n$ is a sample from the Cauchy distribution with location parameter θ (which, incidentally, does not have monotone likelihood ratio because of its very long tails). The density is given by

$$f(x; \theta) = \frac{1}{\pi[1 + (x - \theta)^2]}, \qquad -\infty < x < \infty.$$

Consider testing $H\colon \theta = 0$ against $K\colon \theta > 0$. The likelihood ratio is given by

$$L(\mathbf{x}; 0, \theta) = \prod_{i=1}^{n} \frac{1 + x_i^2}{1 + (x_i - \theta)^2}.$$

Since the shapes of the critical regions generated by this likelihood ratio depend on θ, no single uniformly most powerful test can exist because of the uniqueness part of the Neyman-Pearson Lemma.

To see this for the special case of $n = 1$, we will determine all the possible shapes of the rejection region for various $\theta > 0$ and α. For testing $H\colon \theta = 0$ against $K_{\theta_1}\colon \theta = \theta_1$ with fixed $\theta_1 > 0$, the most powerful test rejects for large values of $L(x) = (1 + x^2)/[1 + (x - \theta_1)^2]$ and is unique up to sets of Lebesgue measure zero. First observe the somewhat surprising fact that $L(x) \to 1$ as $x \to \infty$ or as $x \to -\infty$. Also, $L(x)$ is less than, greater than, or equal to one according to whether x is less than, greater than, or equal to $\theta_1/2$, respectively. Furthermore, the derivative $L'(x)$ is less than, greater than, or equal to zero according to whether $-x^2 + \theta_1 x + 1$ is greater than, less than, or equal to zero. This quadratic has roots $r_1(\theta_1) = [\theta_1 + (\theta_1^2 + 4)^{1/2}]/2$ and $r_2(\theta_1) = [\theta_1 - (\theta_1^2 + 4)^{1/2}]/2$. A plot of the likelihood ratio is displayed in Figure 10.6.1. From this figure we see that $L(x)$ decreases from $-\infty$ to $r_2(\theta_1)$, then increases until $r_1(\theta_1)$, and then decreases to one from $r_1(\theta_1)$ to ∞. This is clearly not a monotonic likelihood ratio!

Now for fixed $\theta_1 > 0$ and α in $(0, 1)$, let $c(\theta_1, \alpha)$ satisfy $P_{\theta=0}\{L(X) > c(\theta_1, \alpha)\} = \alpha$. If α is small enough so that $c(\theta_1, \alpha) > 1$, the critical region will look as in Figure 10.6.2. Therefore, the possible shapes of rejection regions (defined uniquely up to sets of Lebesgue measure zero) are the following:

(i) An interval (c_1, c_2) where $\theta_1/2 < c_1 < c_2 < \infty$, corresponding to small values of α for fixed θ_1, as illustrated in Figure 10.6.2.

(ii) A union of two disjoint intervals $(-\infty, c_1)$ and (c_2, ∞) where $c_1 < r_2(\theta_1) < c_2 < \theta_1/2$, corresponding to α big enough so that $c(\theta_1, \alpha) < 1$. This case is illustrated in Figure 10.6.3.

(iii) An interval $(\theta_1/2, \infty)$, corresponding to the case in which $c(\theta_1, \alpha) = 1$.

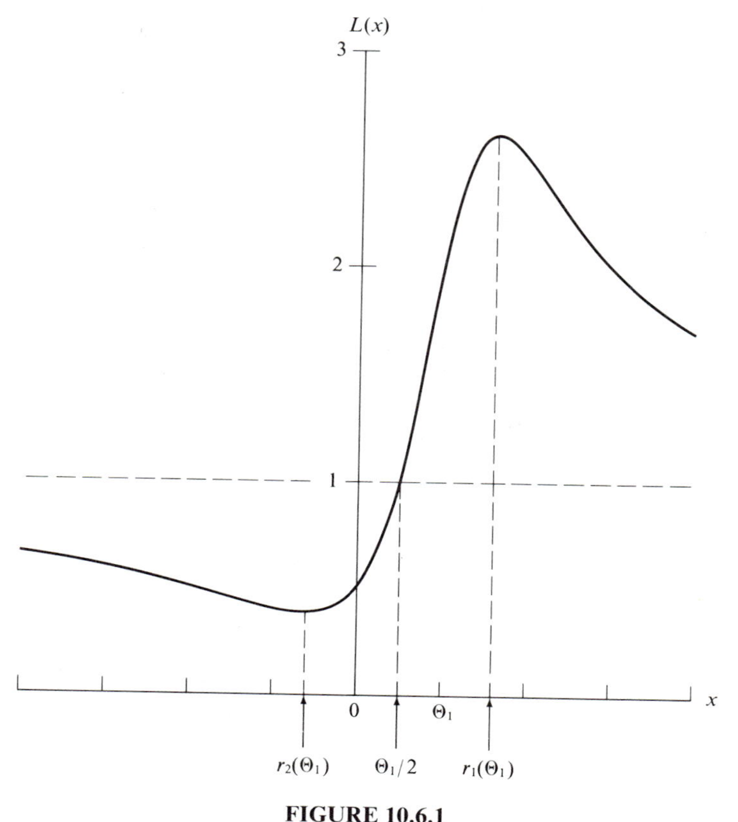

FIGURE 10.6.1

Because the function $c(\alpha, \theta_1)$ takes on values greater than and less than one, for a fixed value of α the critical regions of the most powerful tests of H against K_{θ_1} depend on the value of θ_1 (which, of course, is unknown). Therefore no uniformly most powerful test exists.

10.7. A most powerful one-sided test for a location parameter that does not reject for large values.

As in Example 10.6, consider the Cauchy family of densities defined as follows:

$$f_\theta(x) = \frac{1}{\pi[1 + (x - \theta)^2]}.$$

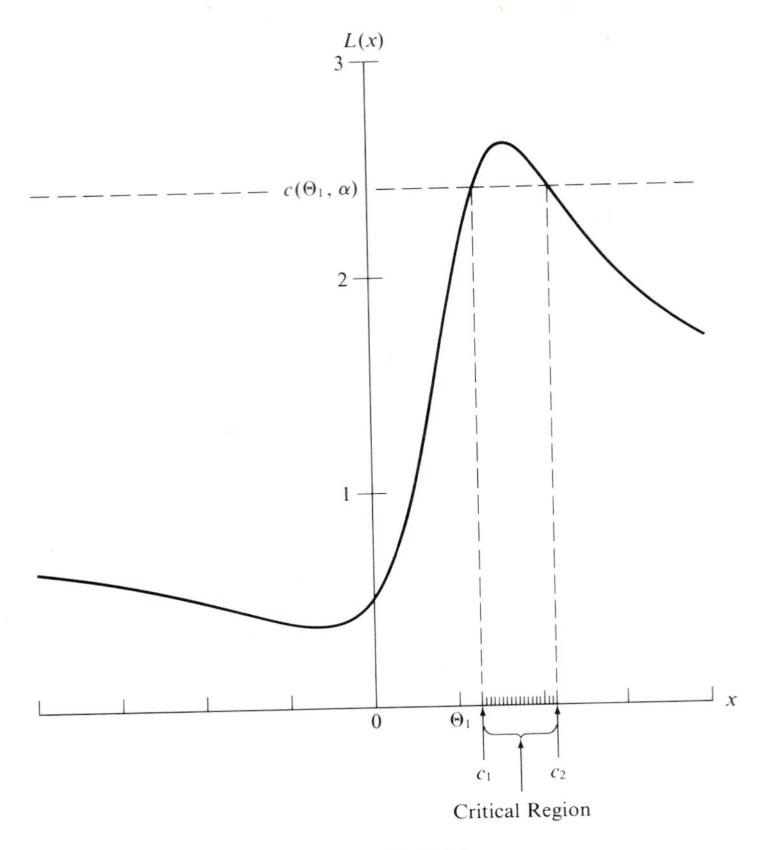

FIGURE 10.6.2

Based on an observed value of X from this family, we wish to test H: $\theta = \theta_0$ against K: $\theta = \theta_1$ where $\theta_1 > \theta_0$. By the Neyman-Pearson Lemma, the most powerful test rejects for large values of the likelihood ratio:

$$L(X) = \frac{1 + (X - \theta_0)^2}{1 + (X - \theta_1)^2}.$$

Rejecting for large values of $L(X)$ is not equivalent to rejecting for large values of X, however. To see this, it is enough to observe that $L(x) \to 1$ as $x \to \infty$ or as $x \to -\infty$. Figure 10.6.1 plotted the likelihood ratio statistic, showing how $L(x)$ increases but eventually decreases as x gets larger.

Although our intuition tells us that rejecting for large X is sensible when testing a location parameter, this may not be the best thing to do in all cases. Indeed, in the Cauchy case with sensible values of the level α, more information comes from the "shoulders" of the distribution than from the more extreme "tails." Intuitively, because any Cauchy distribution can occasionally generate a very extreme outlier, the behavior of the likelihood ratio

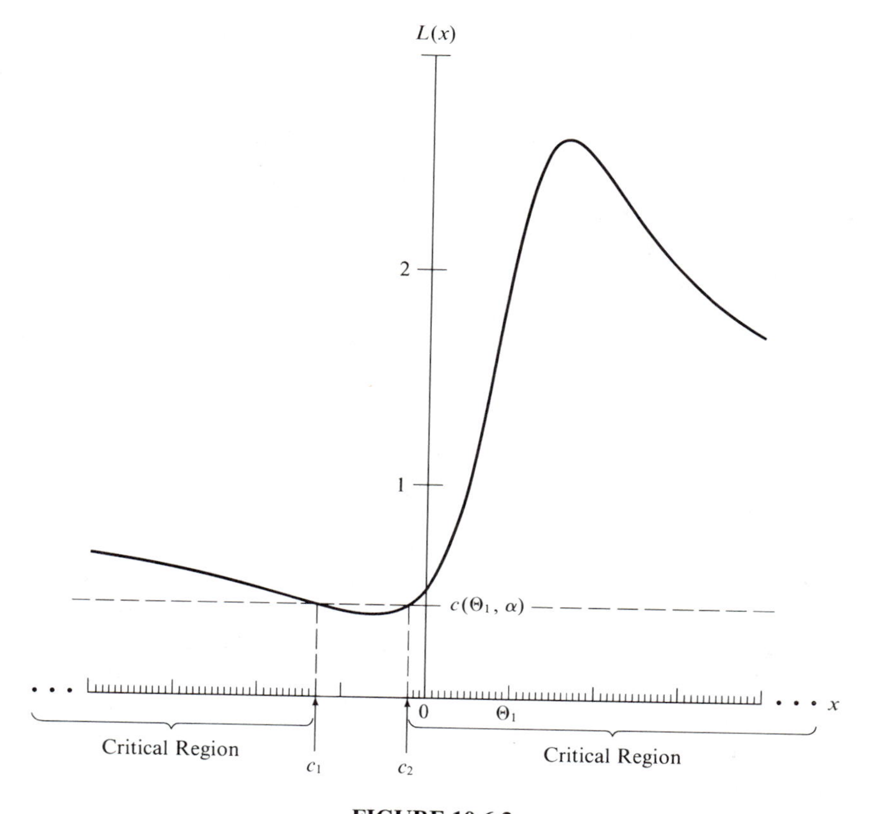

FIGURE 10.6.3

statistic reflects the difficulty in telling which of the two possibilities generated the very large value. This is because either could reasonably have produced it.

10.8. A family of distributions that is stochastically increasing but does not have monotone likelihood ratio.

Recall that a family of cumulative distribution functions $F_\theta(x)$ is stochastically increasing if the distributions are distinct and $\theta_1 > \theta_0$ implies that $F_{\theta_0}(x) \geq F_{\theta_1}(x)$ for all x. Intuitively, if X_i has distribution F_{θ_i} (for $i = 1, 2$) and $\theta_1 > \theta_0$, then X_1 tends to have larger values than X_0. In particular, any location family is a stochastically increasing family.

Even this special case of a location family, however, need not have monotone likelihood ratio. As noted in Example 10.6, the Cauchy density $1/\{\pi[1 + (x - \theta)^2]\}$ with location parameter θ does not have monotone likelihood ratio. Indeed, this is clear from an examination of Figure 10.6.1.

More general examples than the Cauchy distribution may be provided for Examples 10.6, 10.7, and 10.8 by using any location family with algebraic tails, which approach zero much more slowly than the Gaussian density does. Having algebraic tails means that for large x, the density satisfies the asymptotic relation $f_\theta(x) \sim (x - \theta)^{-k}$ for a fixed power k. In particular, the t family of distributions (as location families) with various degrees of freedom provides a general class of examples, with the Cauchy represented as a t with one degree of freedom. Interestingly, for any fixed number of degrees of freedom, no matter how large, the location family generated by the t-distribution does not have monotone likelihood ratio. Taking a limit as the number of degrees of freedom tends to infinity, however, we obtain the Gaussian family, which does have monotone likelihood ratio.

10.9. A family of distributions with monotone likelihood ratio that is not an exponential family.

We know that every one-parameter exponential family has monotone likelihood ratio; this example will show that the converse is false. Consider the family of uniform distributions on the n-dimensional cube $(0, \theta)^n$, with $\theta > 0$, generated by n independent and identically distributed $U(0, \theta)$ random variables $X_1, \ldots, X_n$. Such a family is not an exponential family of distributions because the support of the distribution depends on the parameter θ. For the case where $\theta_1 > \theta_0$, the likelihood ratio is easily computed as the ratio of the densities:

$$\frac{p_1(\mathbf{x})}{p_0(\mathbf{x})} = \begin{cases} \infty & \text{if } \theta_0 \le x_{(n)} < \theta_1 \\ \left[\dfrac{\theta_0}{\theta_1}\right]^n & \text{if } 0 < x_{(n)} < \theta_0 \end{cases}$$

where $x_{(n)} = \max(x_1, \ldots, x_n)$. Since this likelihood ratio is a nondecreasing function of $x_{(n)}$, it follows that the family of distributions has monotone likelihood ratio in $x_{(n)}$. Therefore, uniformly most powerful one-sided tests will exist even though the family is not a one-parameter exponential family. (See also Examples 10.13, 10.14, and 10.15.)

10.10. A location family that does not have monotone likelihood ratio, but a uniformly most powerful one-sided test exists.

Even a simple one-parameter location family, where everything is fixed and known except the location of the distribution, need not have monotone likelihood ratio. An example is provided by the Cauchy family of densities with location parameter θ and density $f(x; \theta) = 1/\{\pi[1 + (x - \theta)^2]\}$. As shown in the figures of Example 10.6, this family does not have monotone likelihood ratio in x. Furthermore, we have seen that no uniformly most powerful one-sided tests exist.

For an example of a location family that does not have monotone likelihood ratio but for which a uniformly most powerful one-sided test exists, consider the family of uniform distributions $U(\theta, \theta + 1)$ in the interval $(\theta, \theta + 1)$. In testing $H: \theta = \theta_0$ against $K: \theta > \theta_0$ based on a sample of size n, the level α test that rejects when $X_{(1)} > \theta_0 + c(\alpha)$ or $X_{(n)} > \theta_0 + 1$ is uniformly most powerful level α. The details are left to the reader.

More generally, location families with densities of the form $f(x - \theta)$ will have monotone likelihood ratio in x if and only if $-\log[f(x)]$ is a convex function. Such densities are called "strongly unimodal" and are discussed by Lehmann (1959, pp. 330–31).

10.11. An even density that generates a scale family with monotone likelihood ratio, but whose location family does not have monotone likelihood ratio.

When the probability density function $f(x)$ is an even function and the location family $f(x - \theta)$ has monotone likelihood ratio, it follows that the scale family $f(x/\sigma)/\sigma$ also has monotone likelihood ratio. This example shows that the converse is false (see Lehmann 1959, pp. 330–31).

Let $f(x) = 1/[\pi(1 + x^2)]$ be the standard Cauchy density. That the location family $f(x - \theta)$ does not have monotone likelihood ratio was seen in Examples 10.6, 10.7, 10.8, and 10.10. However, the scale family $f(x/\sigma)/\sigma$ actually does have monotone likelihood ratio in x^2. This follows because, if $\sigma_1 > \sigma_0$, then after some algebraic manipulation the likelihood ratio may be written in the following form:

$$\frac{f(x/\sigma_1)/\sigma_1}{f(x/\sigma_0)/\sigma_0} = \frac{\sigma_1}{\sigma_0}\left[1 - \frac{\sigma_1^2 - \sigma_0^2}{\sigma_1^2 + x^2}\right].$$

This form of the likelihood ratio shows that it is an increasing function of x^2. Thus uniformly most powerful tests for one-sided hypotheses exist for the Cauchy scale family $f(x/\sigma)/\sigma$, even though they do not exist for the Cauchy location family $f(x - \theta)$.

10.12. An example where an optimal nonrandomized solution in testing a simple hypothesis against a simple alternative does not reject at the largest likelihood ratio.

We will provide two examples here. The first will rely on the technicalities of sets of measure zero, whereas the second will not.

If the likelihood ratio takes on its largest value on a set of measure zero under both the hypothesis and the alternative (which can happen with continuous probability distributions), then the standard likelihood ratio test may be simply redefined so that it does not reject when this largest likelihood ratio is attained. The resulting test will be optimal because it differs from the likelihood ratio test only at a set of measure zero, and therefore it has the same operating characteristics.

The second example, which is more direct, considers the null hypothesis H: $p_0(x)$ and the alternative hypothesis K: $p_1(x)$ as defined by the following discrete distributions:

	X = 1	2	3
$p_0(x)$	$\frac{1}{10{,}000}$	$\frac{1}{20}$	$\frac{9499}{10{,}000}$
$p_1(x)$	$\frac{1}{100}$	$\frac{1}{10}$	$\frac{89}{100}$

If we denote the likelihood ratio by $r(x) = p_1(x)/p_0(x)$, then $r(1) = 100 > r(2) = 2 > r(3) = \frac{8900}{9499} \cong 0.937$. The three possible nonrandomized tests at level $\alpha = 0.05$ are "never reject," "reject if and only if $X = 1$," and "reject if and only if $X = 2$," which have powers equal to 0, 0.01, and 0.1, respectively. Hence the optimal nonrandomized test is to reject if and only if $X = 2$ even though $r(1) > r(2) > r(3)$. We would therefore not reject at the largest likelihood value, which occurs when $X = 1$.

By the Neyman-Pearson Lemma, however, the optimal most powerful test is given by the following randomized decision function:

$$\delta(x) = \begin{cases} 1 & \text{if } X = 1 \\ q & \text{if } X = 2 \\ 0 & \text{if } X = 3. \end{cases}$$

The randomization probability q is chosen so that the size of the test is $\alpha = 0.05$. The resulting test always rejects when $X = 1$ (which is the largest likelihood value), rejects with probability q if $X = 2$, and never rejects when $X = 3$. Solving $E_0\delta(X) = 0.05$, we find that $q = \frac{499}{500}$. This optimal randomized test has power equal to 0.1098, which, of course, is greater than the power of the optimal nonrandomized test.

Thus, although the Neyman-Pearson Lemma says to reject for large values of the likelihood ratio to find the most powerful test, this procedure is not optimal if we restrict our attention to nonrandomized tests. In practice, one usually needs to change only the level of the test slightly so that the optimal test within the class of (possibly) randomized tests is actually nonrandomized. (See also Example 10.4, however.)

10.13. Nonuniqueness of uniformly most powerful tests.

Let $\mathbf{X} = (X_1, X_2, \ldots, X_n)$ be a sample from the uniform distribution $U(0, \theta)$ with $\theta > 0$, and consider testing the null hypothesis H: $\theta = \theta_0$ against the alternative hypothesis K: $\theta > \theta_0$ at some level α. For any fixed $\theta_1 > \theta_0$, the likelihood ratio is given by the ratio of densities:

$$\frac{p_1(\mathbf{X})}{p_0(\mathbf{X})} = \left(\frac{\theta_0}{\theta_1}\right)^n \frac{I_{(0,\theta_1)}(X_{(n)})}{I_{(0,\theta_0)}(X_{(n)})}$$

where $X_{(n)} = \max(X_1, \ldots, X_n)$ is the largest order statistic. By the Neyman-Pearson Lemma, the most powerful test of H against K_{θ_1}: $\theta = \theta_1$ rejects H if $X_{(n)} \geq \theta_0$ and is defined arbitrarily if $0 < X_{(n)} < \theta_0$ so long as the size of the test is α. Any such test will be most powerful. Two particular examples are:

$$\delta_1(\mathbf{X}) = \begin{cases} 1 & \text{if } X_{(n)} \geq \theta_0 \\ \alpha & \text{otherwise} \end{cases}$$

$$\delta_2(\mathbf{X}) = \begin{cases} 1 & \text{if } X_{(n)} \geq \theta_0 \text{ or } X_{(n)} \leq \theta_0 \alpha^{1/n} \\ 0 & \text{otherwise.} \end{cases}$$

Note that δ_1 is randomized but δ_2 is not, and that the size of the test given by δ_2 is indeed α because

$$\begin{aligned} E_{\theta_0}\delta_2(\mathbf{X}) &= P_{\theta_0}\{X_{(n)} \geq \theta_0 \quad \text{or} \quad X_{(n)} \leq \theta_0 \alpha^{1/n}\} \\ &= P_{\theta_0}\{X_{(n)} \leq \theta_0 \alpha^{1/n}\} = P_{\theta_0}\{\text{all } X_i \leq \theta_0 \alpha^{1/n}\} \\ &= \left[\frac{\theta_0 \alpha^{1/n}}{\theta_0}\right]^n = \alpha. \end{aligned}$$

Infinitely many other uniformly most powerful tests exist, each having an acceptance region corresponding to a set in $(0, \theta_0)$ of Lebesgue measure $\alpha\theta_0$. Since all such tests are independent of θ_1, these tests are in fact uniformly most powerful.

This same example also shows that uniformly most powerful unbiased tests need not be unique.

10.14. A uniformly most powerful test that does *not* depend on the data only through the complete sufficient statistic.

Suppose $\mathbf{X} = (X_1, \ldots, X_n)$ (with $n > 1$) is a sample from the uniform distribution $U(0, \theta)$ with $\theta > 0$. From Example 10.13, $X_{(n)} = \max(X_1, \ldots, X_n)$ is complete and sufficient. However, in testing H: $\theta = \theta_0$ against K: $\theta >$

θ_0 at level α, any level α test that rejects when $X_{(n)} \geq \theta_0$ is uniformly most powerful. For example, consider the test given as follows:

$$\delta(\mathbf{X}) = \begin{cases} 1 & \text{if } X_{(n)} \geq \theta_0 \quad \text{or} \quad X_1 < \theta_0 \alpha \\ 0 & \text{otherwise.} \end{cases}$$

This test is uniformly most powerful at level α, but it does not depend on the data only through $X_{(n)}$, which is the minimal sufficient statistic. This example also shows that a uniformly most powerful unbiased test need not depend on the data only through a minimal sufficient statistic.

Of course, if a uniformly most powerful test $\delta(\mathbf{X})$ exists, then there exists a uniformly most powerful test that depends on any sufficient statistic S (which may possibly be vector-valued), which is given by

$$\delta^*[S(\mathbf{X})] = E[\delta(\mathbf{X}) \mid S(\mathbf{X})].$$

The corresponding existence result also holds for uniformly most powerful unbiased tests.

10.15. A uniformly most powerful two-sided test of a real parameter.

Although uniformly most powerful tests rarely exist, it is surprising when a two-sided uniformly most powerful test exists. For example, even in the case of exponential families, two-sided uniformly most powerful tests do not generally exist (see Lehmann 1959, Problem 3.26), whereas uniformly most powerful unbiased tests generally do exist in this case.

As in Examples 10.13 and 10.14, let $\mathbf{X} = (X_1, \ldots, X_n)$ be a sample from the uniform distribution $U(0, \theta)$ with $\theta > 0$, and consider testing the hypothesis H: $\theta = \theta_0$ against K: $\theta \neq \theta_0$ at level α. First consider a fixed alternative $\theta_1 < \theta_0$. The likelihood ratio in this case becomes

$$\frac{p_1(\mathbf{X})}{p_0(\mathbf{X})} = \begin{cases} 0 & \text{if } \theta_1 \leq X_{(n)} \leq \theta_0 \\ \left(\dfrac{\theta_0}{\theta_1}\right)^n & \text{if } 0 < X_{(n)} < \theta_1 \\ \infty & \text{if } X_{(n)} > \theta_0 \end{cases}$$

where $X_{(n)} = \max(X_1, \ldots, X_n)$ denotes the largest order statistic. The likelihood ratio has arbitrarily been defined to be ∞ when both of the densities vanish. Consider the following test:

$$\delta(\mathbf{X}) = \begin{cases} 1 & \text{if } X_{(n)} > \theta_0 \quad \text{or} \quad X_{(n)} \leq \theta_0 \alpha^{1/n} \\ 0 & \text{otherwise.} \end{cases}$$

Note that $E_{\theta_0}[\delta(\mathbf{X})] = \alpha$ and $\delta(\mathbf{X})$ clearly rejects for large values of the likelihood ratio. Moreover, since the test is independent of θ_1, it is uniformly most powerful for all alternatives $\theta < \theta_0$. From Example 10.13, however, the same test is uniformly most powerful for all alternatives $\theta > \theta_0$. Therefore this test is uniformly most powerful for two-sided alternatives.

10.16. Most powerful tests of varying sizes with critical regions that are not nested.

Usually the most powerful level α tests are nested in the sense that the corresponding rejection regions S_α satisfy the condition that if $\alpha_1 > \alpha_0$, then S_{α_0} is contained in S_{α_1}. If this is not the case, there will be certain values of the data for which we will reject the null hypothesis H at level α_0, but we will not reject H at a larger level α_1. We will give two examples. The first, a simple null hypothesis tested against a simple alternative, shows where nonnesting happens but an alternative optimal test does have properly nested critical regions. The second example shows that in a less restricted testing situation, nonnesting may be unavoidable.

First consider the following two discrete probability distributions for X under the null hypothesis H and the alternative hypothesis K:

	X		
	1	2	3
Under H	0.85	0.1	0.05
Under K	0.70	0.2	0.10

Note that the likelihood ratio takes on its largest value of two when X is either two or three. A most powerful test at level $\alpha = 0.05$ may therefore be given by rejecting when $X = 3$, and a most powerful test at level $\alpha = 0.1$ may be given by rejecting when $X = 2$. Note that these regions are neither nested nor overlapping!

There do exist most powerful tests in this situation that are nested. Instead of rejecting when $X = 2$, we might have rejected when $X = 3$ and rejected half of the time when $X = 2$. With this new choice of level 0.1 test, the regions are nested. Indeed, from applying the Neyman-Pearson Lemma with nested randomizations when necessary, it will always be possible to obtain nested rejection regions when testing a simple null against a simple alternative hypothesis.

The second example, from Lehmann (1959, Problem 3.29), shows how nonnesting may be unavoidable when the null hypothesis is not simple.

Suppose X takes on the values 1, 2, 3, and 4 with discrete probability densities f_0, f_1, and g defined by:

	X = 1	2	3	4
$f_0(x)$	$\frac{2}{13}$	$\frac{4}{13}$	$\frac{3}{13}$	$\frac{4}{13}$
$f_1(x)$	$\frac{4}{13}$	$\frac{2}{13}$	$\frac{1}{13}$	$\frac{6}{13}$
$g(x)$	$\frac{4}{13}$	$\frac{3}{13}$	$\frac{2}{13}$	$\frac{4}{13}$

Consider testing the null hypothesis H that the density of X is f_0 or f_1 against the alternative hypothesis K that the density of X is g. We will show that the most powerful test at level $\alpha = \frac{5}{13}$ rejects if and only if $X = 1$ or 3. Surprisingly, however, the most powerful test at level $\alpha = \frac{6}{13}$ rejects if and only if $X = 1$ or 2.

To find the most powerful tests for these values of H, we will apply Corollary 3.5 of Lehmann (1959), which involves finding a least favorable distribution for H. Define the density $\lambda_p = pf_0 + (1-p)f_1$, a mixture of f_0 and f_1. Its density and the ratio of the densities are given in the following table:

	X = 1	2	3	4
$\lambda_p(x)$	$\dfrac{4-2p}{13}$	$\dfrac{2+2p}{13}$	$\dfrac{1+2p}{13}$	$\dfrac{6-2p}{13}$
$\dfrac{g(x)}{\lambda_p(x)}$	$\dfrac{4}{4-2p}$	$\dfrac{3}{2+2p}$	$\dfrac{2}{1+2p}$	$\dfrac{4}{6-2p}$

In particular, when $p = \frac{1}{3}$ and when $p = \frac{9}{14}$, we have

	X = 1	2	3	4
$\lambda_{1/3}(x)$	$\frac{10}{39}$	$\frac{8}{39}$	$\frac{5}{39}$	$\frac{16}{39}$
$\dfrac{g(x)}{\lambda_{1/3}(x)}$	$\frac{6}{5}$	$\frac{9}{8}$	$\frac{6}{5}$	$\frac{3}{4}$
$\lambda_{9/14}(x)$	$\frac{19}{91}$	$\frac{23}{91}$	$\frac{16}{91}$	$\frac{33}{91}$
$\dfrac{g(x)}{\lambda_{9/14}(x)}$	$\frac{28}{19}$	$\frac{21}{23}$	$\frac{7}{8}$	$\frac{28}{33}$

In testing H_p: λ_p against K: g, a most powerful test rejects for large values of $g(x)/\lambda_p(x)$. Therefore, if $\alpha = \frac{5}{13}$, the most powerful test rejects when $X = 1$ or 3. However, if $\alpha = \frac{6}{13}$, then the most powerful test rejects when $X = 1$ or 2. The only thing that remains is to show that these tests are most powerful for the original problem of testing H: f_0 or f_1 against K: g. Since these tests have size equal to $\frac{5}{13}$ and $\frac{6}{13}$, respectively, for the original problem, the conditions of Corollary 3.5 of Lehmann (1959) are satisfied and the result follows.

10.17. Nonuniqueness of a least favorable distribution.

Consider the same situation as in Example 10.16. There was nothing particularly special in choosing $p = \frac{1}{3}$ and $p = \frac{9}{14}$ in deriving the most powerful tests of H_0: $\lambda_p = pf_0 + (1 - p)f_1$ against K: g that were also most powerful for the original problem of testing H: f_0 or f_1 against K: g. For example, consider $\alpha = \frac{5}{13}$. Since for any p, the test that rejects if and only if $X = 1$ or 3 is of level $\frac{5}{13}$ in testing H_p and is also of size $\frac{5}{13}$ for testing H, then by Theorem 3.7 of Lehmann (1959) we need only determine p so that $g(x)/\lambda_p(x)$ is larger at $X = 1$ and 3 than it is at $X = 2$ or 4. A direct calculation shows that this will be the case if and only if p is in the open interval from $\frac{2}{7}$ to $\frac{1}{2}$. Thus any λ_p with p in this interval will define a least favorable distribution in testing H: f_0 or f_1 against K: g at level $\frac{5}{13}$.

10.18. The least favorable distribution may depend on the size of the test.

In searching for a least favorable distribution to test a composite hypothesis H: $\{f_\theta$, with θ in $\Omega_0\}$ against some fixed alternative K: g, one is guided by intuition in searching for a distribution λ over Ω_0 that makes $h_\lambda(x) = \int f_\theta(x)\, d\lambda(\theta)$ as close as possible to g. The reason is that testing H_λ: h_λ against K should be no easier than testing H against K; otherwise, the most powerful test of H_λ against K would bear no relation to the original problem. Often geometrical considerations are helpful. Such considerations may not have any relation to the level of the proposed test, so it seems surprising that the least favorable distribution (even though it exists) may depend on α.

Again consider the same situation as in Examples 10.16 and 10.17. We found that λ_p with p in the interval $(\frac{2}{7}, \frac{1}{2})$ is least favorable at level $\alpha = \frac{5}{13}$. We also found that λ_p with $p = \frac{9}{14}$ defines a least favorable distribution in testing H against K at level $\frac{6}{13}$. In fact, by an argument analogous to the case $\alpha = \frac{5}{13}$, λ_p is least favorable in testing H against K at level $\alpha = \frac{6}{13}$ if and only if p is in the interval $(\frac{1}{2}, \frac{5}{7})$. Therefore, the least favorable distribution may indeed depend on α.

10.19. Nonexistence of a uniformly most powerful unbiased test in a two-parameter exponential family.

Suppose X and Y are independent Gaussian random variables with known variance one and means θ and η, respectively. Consider testing the simple null hypothesis H: $\theta = \eta = 0$ against the compound alternative hypothesis K: $\theta > 0$, $\eta > 0$. To find the most powerful test against a fixed alternative $\theta_1 > 0$, $\eta_1 > 0$, we must reject for large values of the likelihood ratio

$$\frac{\exp\{-(x - \theta_1)^2/2\}\exp\{-(y - \eta_1)^2/2\}}{\exp\{-x^2/2\}\exp\{-y^2/2\}}$$

which is equivalent to rejecting for large values of $\theta_1 X + \eta_1 Y$. Thus the test $\delta_{(\theta_1, \eta_1)}(X, Y)$ that rejects for large values of $\theta_1 X + \eta_1 Y$ is most powerful against the fixed alternative $(\theta, \eta) = (\theta_1, \eta_1)$. Furthermore, $\delta_{(\theta_1, \eta_1)}$ is an unbiased test against any alternative $\theta > 0$, $\eta > 0$. To see this, note that if $\delta_{(\theta_1, \eta_1)}(X, Y)$ rejects when $\theta_1 X + \eta_1 Y$ is greater than $c = c(\theta_1, \eta_1)$, then the power of the test against an alternative (θ, η) is

$$P_{\theta,\eta}\{\theta_1 X + \eta_1 Y > c\} = P_{\theta,\eta}\{\theta_1 X + \eta_1 Y - \theta_1\theta - \eta_1\eta > c - \theta_1\theta - \eta_1\eta\}$$

$$= P\left[Z > \frac{c - \theta_1\theta - \eta_1\eta}{(\theta_1^2 + \eta_1^2)^{1/2}}\right]$$

where Z has the standard Gaussian distribution with mean zero and variance one. Since θ, θ_1, η, and η_1 are nonnegative, this probability is at least $P\{Z > c/(\theta_1^2 + \eta_1^2)^{1/2}\}$, which by the definition of c is the level α of the test. Thus it follows that $\delta_{(\theta_1, \eta_1)}$ is an unbiased test of H against K.

If a uniformly most powerful unbiased test were to exist, it would have to have power at (θ_1, η_1) equal to the power of $\delta_{(\theta_1, \eta_1)}$ at (θ_1, η_1). But the tests $\delta_{(\theta_1, \eta_1)}$ are uniquely most powerful (almost surely with respect to $P_{\theta,\eta}$) and have different power functions, so it follows that no uniformly most powerful unbiased test exists.

Note that uniformly most powerful unbiased tests generally do exist for multiparameter exponential families (see, for example, Lehmann 1959, Section 4.4 and Chapter 5). For example, in testing H': $(\theta, \eta) = (0, 0)$ against K': $(\theta, \eta) \neq (0, 0)$ in the earlier Gaussian problem, a uniformly most powerful unbiased test does exist. This may seem paradoxical at first. However, the paradox can be resolved by observing that when the class of alternatives is enlarged, the number of competing tests gets smaller because they must be unbiased for all alternatives. Thus an optimal test from this smaller class can exist. Indeed, the most powerful tests derived earlier for a fixed alternative

(θ_1, η_1) are no longer unbiased against alternatives $(\theta, \eta) \neq (0, 0)$ and so are not competitors in the search for a uniformly most powerful unbiased test.

For an example of a two-parameter exponential family where the only unbiased level α test of H is the test that is identically equal to α, see Lehmann (1959, Problem 4.7, p. 151).

10.20. A hypothesis-testing problem where a uniformly most powerful test exists conditional on the value of an ancillary statistic, but not even a uniformly most powerful unbiased unconditional test exists.

Suppose we can measure an unknown parameter θ with either of two measuring instruments I_0 and I_1. Let the random variable C take the value zero or one according to whether I_0 or I_1 is used, and suppose these two possibilities happen with equal probability. Conditional on $C = c$, suppose X has a Gaussian distribution with mean θ and variance σ_c^2, where $\sigma_0^2 = 1$ and $\sigma_1^2 = 10$. Based on the data (C, X), we wish to test the null hypothesis H: $\theta = 0$ against the alternative K: $\theta > 0$.

Because C has a distribution independent of θ, it follows that C is ancillary. Furthermore, conditional on $C = c$, we are testing the mean of a Gaussian distribution with known variance against one-sided alternatives, and we know that the test that rejects for large values of X is uniformly most powerful. However, no uniformly most powerful test exists in the unconditional situation. To see this, fix an alternative $\theta_1 > 0$. The likelihood ratio becomes:

$$r(C, X) = \frac{\frac{1}{2}\exp\{-(X - \theta_1)^2/(2\sigma_C^2)\}/[(2\pi)^{1/2}\sigma_C]}{\frac{1}{2}\exp\{-X^2/(2\sigma_C^2)\}/[(2\pi)^{1/2}\sigma_C]}.$$

By the Neyman-Pearson Lemma, the most powerful level α test rejects for large values of $r(C, X)$ or, equivalently, when $(2X\theta_1 - \theta_1^2)/(2\sigma_C^2) > K$ for some constant $K = K(\alpha, \theta_1)$. Furthermore, this test is unbiased for all alternatives $\theta > 0$ because

$$P_\theta\left(\frac{2X\theta_1 - \theta_1^2}{2\sigma_C^2} > K\right) = P_\theta\left(X > \frac{2\sigma_C^2 K + \theta_1^2}{2\theta_1}\right)$$

$$\geq P_0\left(X > \frac{2\sigma_C^2 K + \theta_1^2}{2\theta_1}\right) = \alpha.$$

This test, which is unique almost surely with respect to P_θ, depends on θ_1, and so not even a uniformly most powerful unbiased test exists.

Often it is suggested that any inference about unknown parameters ought to be performed conditional on the values of ancillary statistics (for example, see Cox and Hinkley 1974, p. 38). Indeed, the ancillary statistic can often be viewed as an indicator of the experiment performed to produce the observations. Although the Conditionality Principle is intuitively appealing, this analysis shows that we can do better (in the sense of higher power) by not restricting attention to conditional tests in some situations.

In fact, the most powerful level α test constructed here is not the test defined by the conditional most powerful level α tests given C. Although this may not seem intuitive at first, the explanation lies in the fact that the most powerful (unconditional) level α test need not be a level α test conditional on C, and so we may budget our type I errors more freely to achieve greater power.

10.21. Existence of a uniformly most powerful unbiased test when a nuisance parameter is unknown even though no such optimal property holds when the nuisance parameter is known.

As in Example 10.20, we observe (X, C) in testing H: $\theta = 0$ against K: $\theta > 0$. Instead of supposing that C takes on the values zero and one with the known probabilities $\frac{1}{2}$ and $\frac{1}{2}$, suppose instead that $P\{C = 1\} = p$ and $P\{C = 0\} = 1 - p$, where p is unknown. In Example 10.20 it was shown that no uniformly most powerful unbiased test exists when it is known that $p = \frac{1}{2}$. When p is unknown, however, we will show that a uniformly most powerful unbiased test does indeed exist.

When p is unknown and θ has any fixed value, C is sufficient for p and is complete. To see this, if

$$pT(1) + (1 - p)T(0) = 0 \qquad \text{for all } p$$

then, since the coefficient of p on the left side must be zero, it would follow that $T(1) = T(0)$. But then $E[T(C)] = T(0) = 0$ would imply that T is identically zero, thus establishing completeness. By Theorem 4.2 in Lehmann (1959), all similar tests have Neyman structure. Therefore, the optimal uniformly most powerful unbiased test can be constructed conditional on the complete sufficient statistic. Compare this result with Example 10.20.

The explanation that an optimal uniformly most powerful unbiased test exists when p is unknown but not when p is known lies in the fact that we have further restricted ourselves to tests that are unbiased for all alternatives $\theta > 0$ and all values of p. Naturally it is easier to find an optimal test when the class of possibilities is smaller.

10.22. An inadmissible likelihood ratio test (Problem 6.18 from Lehmann 1959).

Suppose $P_1, \ldots, P_n$ with $n \geq 3$ are equally spaced points on the circle of radius 2 ($x^2 + y^2 = 4$), and that $Q_1, \ldots, Q_n$ are equally spaced points on the circle of radius 1 ($x^2 + y^2 = 1$), and let 0 denote the origin. For any fixed α in $[0, \frac{1}{2}]$, let (X, Y) be distributed over these $2n + 1$ points with densities defined for the hypotheses H and $K_{\mathbf{p}}$ as follows:

	P_i	Q_i	0
H	$\frac{\alpha}{n}$	$\frac{1-2\alpha}{n}$	α
$K_{\mathbf{p}}$	$\frac{p_i}{n}$	0	$\frac{n-1}{n}$

where $\Sigma p_i = 1$. Consider testing based on one observation (X, Y), the simple hypothesis H against the composite alternative K defined by all possible $\mathbf{p} = (p_1, \ldots, p_n)$, which are probabilities summing to one.

The problem remains invariant under the group of rotations of the plane by angles $2k\pi/n$ ($k = 0, 1, \ldots, n - 1$), where addition of angles results in $2k_1\pi/n + 2k_2\pi/n = 2\pi[(k_1 + k_2) \bmod n]/n$. Therefore the possible invariant tests are given in the table.

Test	*Size*	*Power*
1. Reject iff $(X, Y) = P_i$ for some i	α	$\frac{1}{n}$
2. Reject iff $(X, Y) = Q_i$ for some i	$1 - 2\alpha$	0
3. Reject iff $(X, Y) = 0$	α	$\frac{n-1}{n}$
4. Reject iff $(X, Y) \neq 0$	$1 - \alpha$	$\frac{1}{n}$
5. Always reject	1	1
6. Never reject	0	0
7. Reject iff $(X, Y) = P_i$ for some i or $(X, Y) = 0$	2α	1
8. Reject iff $(X, Y) = Q_i$ for some i or $(X, Y) = 0$	$1 - \alpha$	$\frac{n-1}{n}$

By inspection of the table, we see that the level α uniformly most powerful invariant test rejects if and only if $(X, Y) = 0$. This test has power $(n - 1)/n$.

On the other hand, the likelihood ratio test rejects for large values of the ratio of the maximized likelihoods:

$$r(X) = \frac{\sup_{[\theta \text{ in } K]} f_\theta(X)}{\sup_{[\theta \text{ in } H]} f_\theta(X)} = \frac{\sup_{\{(p_1, \ldots, p_n);\, p_i \geq 0,\, \Sigma p_i = 1\}} f_{K_{\mathbf{p}}}(X)}{f_H(X)}$$

If $X = P_i$, then $r(X) = 1/\alpha$; if $X = Q_i$, then $r(X) = 0$; and, if $X = 0$, then $r(X) = (n - 1)/(n\alpha)$. Hence the level α likelihood ratio test rejects if and only if $(X, Y) = P_i$ for some i and has power equal to $1/n$ for all alternatives. Since $1/n < (n - 1)/n$ when $n \geq 3$, the likelihood ratio procedure is inadmissible.

10.23. Nonuniqueness of a Bayes classification rule.

A Bayes rule for a classification problem need not be unique, as shown in this example. As in Bickel and Doksum (1977, Example 10.3.2), suppose $\Omega = \{\theta_0, \ldots, \theta_p\}$ and the action space is $\mathbf{A} = \{a_0, \ldots, a_q\}$. Let the loss of estimating θ_i by a_j be

$$L(a_j; \theta_i) = \begin{cases} 1 & \text{if } i \neq j \\ 0 & \text{if } i = j. \end{cases}$$

A Bayes rule for this classification problem maximizes the posterior risk

$$P(\boldsymbol{\theta} = \theta_i \mid X = x) = \frac{\pi_i p(x; \theta_i)}{\sum \pi_j p(x; \theta_j)}.$$

For example, let the distribution of X under each of two alternatives be given by the following table of $p(x; \theta)$:

	$x = 0$	$x = 1$	$x = 2$	
θ_0	$\frac{1}{2}$	$\frac{1}{4}$	$\frac{1}{4}$	(with probability $\pi_0 = \frac{1}{2}$)
θ_1	$\frac{1}{4}$	$\frac{1}{4}$	$\frac{1}{2}$	(with probability $\pi_1 = \frac{1}{2}$).

In this case, if we observe $X = 1$, a Bayes rule could choose either θ_0 or θ_1.

10.24. Nonuniqueness of a maximal invariant.

Let $G = \{g_c : c \text{ a real number}\}$ be the group of translations acting on Euclidean n space, $\mathbf{R}^n$, with $n \geq 3$, as defined by

$$g_c(\mathbf{X}) = (X_1 + c, \ldots, X_n + c) \qquad \text{for } \mathbf{X} = (X_1, \ldots, X_n) \text{ in } \mathbf{R}^n.$$

Clearly the set of differences $\mathbf{Y} = (X_1 - X_n, \ldots, X_{n-1} - X_n)$ is invariant under the group G. Furthermore, if $\mathbf{Y} = \mathbf{Y}'$, so that $X_i - X_n = X_i' - X_n'$, then $X_i' = X_i + c$ where $c = X_n' - X_n$. Thus $\mathbf{Y}$ is a maximal invariant.

By the same reasoning, it follows that $\mathbf{Z} = (X_2 - X_1, \ldots, X_n - X_1)$ is also a maximal invariant. Therefore, maximal invariants are not unique. However, all maximal invariants are equivalent in the sense that their sets of constancy—namely, the orbits of G—coincide.

10.25. A test whose power function is invariant, but the test is not.

In general, a test $\delta(X)$ that is almost invariant under a group G of transformations of X has a power function that is invariant under the group $\bar{G}$ induced by G in the parameter space. A partial converse is true as well, under the assumption that the family of distributions of X is boundedly complete (see Lehmann 1959, Lemma 6.2). This counterexample from Lehmann (1959) shows that the converse is false in general.

Suppose X_1, X_2, and X_3 are independent Gaussian random variables with mean θ and variance σ^2. The test $\delta(\mathbf{X})$ that rejects when

$$|X_2 - X_1| > 1 \quad \text{and} \quad \frac{X_1 + X_2 + X_3}{3} < 0$$

or

$$|X_3 - X_2| > 1 \quad \text{and} \quad \frac{X_1 + X_2 + X_3}{3} \geq 0$$

is not (almost) invariant under the additive group G of transformations defined by $X_i' = X_i + c$. For example, the set $A = \{(x_1, x_2, x_3) : |x_1 - x_2| < 1, |x_3 - x_2| > 1, -10 < x_i < 0\}$ has positive probability for all (θ, σ^2), and $\delta(A) = 0$. However, $\delta(A + 10) = 1$ where $A + 10 = \{(x_1 + 10, x_2 + 10, x_3 + 10) : (x_1, x_2, x_3) \text{ is in } A\}$.

Since the distribution of $X_i - X_j$ (with $i \neq j$) is independent of θ, and X_1, X_2, and X_3 are independent, the power function of the preceding test is independent of θ and so the power function is invariant under the induced group $\bar{G}$ whose elements act as $\bar{g}[(\theta, \sigma^2)] = (\theta + c, \sigma^2)$.

10.26. A uniformly most powerful invariant test that is not a likelihood ratio test.

Although a uniformly most powerful invariant test (when it exists) often coincides with the likelihood ratio test, this need not always be the case. This is readily seen from Example 10.22, where, in fact, the uniformly most powerful invariant test renders the likelihood ratio test inadmissible.

10.27. An inadmissible uniformly most powerful invariant test (Cox and Hinkley 1974, p. 170).

When a problem remains invariant under a group of transformations, attention should be restricted to invariant tests; otherwise, the test might depend on an arbitrary choice of the coordinate system. Even when invariance is applicable, however, this example from C. Stein shows that it can lead to inadmissible tests.

Suppose $X = (X_1, X_2)$ and $Y = (Y_1, Y_2)$ have independent bivariate Gaussian distributions with mean zero and nonsingular covariance matrices Σ and $c\,\Sigma$, respectively, where both c (which is positive) and Σ are unknown. We wish to test the null hypothesis H: $c = 1$ against the alternative K: $c > 1$. The problem remains invariant under the group of nonsingular linear transformations of X and Y, for if

$$\begin{bmatrix} X_1' & Y_1' \\ X_2' & Y_2' \end{bmatrix} = A \begin{bmatrix} X_1 & Y_1 \\ X_2 & Y_2 \end{bmatrix}$$

where A is a nonsingular matrix, it follows that (X_1', X_2') and (Y_1', Y_2') are independent bivariate Gaussian distributions with mean zero and covariance matrices $\Sigma' = A\,\Sigma\,A'$ and $cA\,\Sigma\,A' = c\,\Sigma'$, respectively.

Note that the matrix

$$Z = \begin{bmatrix} X_1 & Y_1 \\ X_2 & Y_2 \end{bmatrix}$$

is nonsingular with probability one. Thus, for any two values z_1 and z_2, there exists (with probability one) a nonsingular matrix A with $z_1 = Az_2$, where we may simply define $A = z_1 z_2^{-1}$. Mathematically speaking, all possible sample points are in the same orbit. That is, if we consider the action on $\mathbf{R}^2$ of the group of transformations G that consist of all nonsingular linear transformations, the orbit of points traced out in the sample space is the entire sample space (provided we, without loss of generality, restrict attention to the sample space of points Z that are nonsingular).

Hence we see that no useful invariant test statistic exists under nonsingular linear transformations. In other words, the only possible invariant level α test is the trivial one that rejects H randomly (regardless of the data) with probability α, and so this test is uniformly most powerful invariant with power function identically α.

It should be pointed out that such a result is not due entirely to the difficult nature of the problem because we now show this test to be inadmissible. For example, if we make use of only (X_1, Y_1), then X_1 and Y_1 are independently distributed as $N(0, \sigma^2)$ and $N(0, c\sigma^2)$, respectively. Since the ratio $Y_1^2/(cX_1^2)$ has an $F_{1,1}$ distribution, the test that rejects when Y_1^2/X_1^2 exceeds the upper 100α percentage point of the $F_{1,1}$ distribution has an increasing power function as c increases from one to ∞. Thus this test has power everywhere larger than α when $c > 1$. In fact, this test is uniformly most powerful for tests based on (X_1, Y_1) (see Lehmann 1959, Problem 3.33). Clearly, this makes use of the sample, showing that an appeal to invariance may be too strong a criterion in some cases.

Uniformly most powerful invariant procedures are often admissible. In particular, if the problem remains invariant under a finite group of transformations and a uniformly most powerful invariant test exists, then it is admissible. This counterexample shows that the result does not extend to infinite groups.

10.28. A problem invariant under two groups such that the uniformly most powerful invariant tests are different under each group.

We give an example where two groups, G_1 and G_2, of transformations leave a problem invariant, uniformly most powerful invariant tests exist, and yet these two tests differ with positive probability. As in Example 10.27 with scaled unknown covariance matrices, consider testing the null hypothesis $H: c = 1$ against the alternative $K: c > 1$. Let G_1 be the group of transformations g defined by

$$g\begin{bmatrix} X_1 & Y_1 \\ X_2 & Y_2 \end{bmatrix} = \begin{bmatrix} X_1' & Y_1' \\ X_2' & Y_2' \end{bmatrix} = \begin{bmatrix} b & 0 \\ a_1 & a_2 \end{bmatrix}\begin{bmatrix} X_1 & Y_1 \\ X_2 & Y_2 \end{bmatrix}$$

where $b \neq 0$ and $a_2 \neq 0$. These transformations (defined by nonsingular lower triangular matrices) form a subgroup of the group considered in Example 10.27 and also leave the problem invariant.

We now find the maximal invariant in steps. Setting $b = 1$ gives a subgroup of G_1, and (X_1, Y_1) is maximal invariant for this subgroup. Multiplying X_1 and Y_1 by an arbitrary nonzero constant defines a second subgroup, which, together with the first, generates the whole group. This then gives Y_1^2/X_1^2 as a maximal invariant for G_1. Since Y_1^2/X_1^2 has c times an $F_{1,1}$ distribution, which is easily verified to have monotone likelihood ratio, the test that rejects for large values of Y_1^2/X_1^2 is uniformly most powerful invariant.

On the other hand, if we interchange the first and second coordinates and look at the group G_2 of upper triangular matrices with nonzero diagonal elements, by the same reasoning it would follow that the uniformly most powerful invariant test with respect to G_2 would reject for large values of Y_2^2/X_2^2. This test clearly differs from the one derived earlier with respect to G_1.

Of course, if we look at the group generated by G_1 and G_2 together, we find ourselves back in the situation of Example 10.27, where the only invariant tests are constant and the uniformly most powerful invariant test is inadmissible.

10.29. An almost invariant test that is not equivalent to an invariant test.*

Suppose the sample space consists of all pairs (y, z) of real numbers. Given any finite set of, say, m real numbers, $F = \{a_1, \ldots, a_m\}$ in $(0, 1)$ and a permutation $\{j_1, \ldots, j_m\}$ of the integers $\{1, \ldots, m\}$, consider the transformation g defined by

$$g(y, z) = \begin{cases} (y, z) & \text{if } z \text{ is not in } F \\ (z, a_{j_i}) & \text{if } z = a_i. \end{cases}$$

Let G be the group of transformations g of this form for all possible m, all possible sets $\{a_1, \ldots, a_m\}$, and all possible permutations j_i. Furthermore, suppose Y and Z are independent and identically distributed Gaussian random variables with distribution $N(\theta, 1)$. Then the identity function $\mathbf{I}(y, z) = (y, z)$ is almost invariant. To see this, for any g in G we will have $\mathbf{I}(y, z) = g(y, z)$ except for at most a finite set of values—say, $\{a_1, \ldots, a_m\}$. For all θ, $\int_{\{a_1, \ldots, a_m\}} dN(\theta, 1) = 0$, so $\mathbf{I}$ is almost invariant.

The function $T\colon (y, z) \to y$ is a maximal invariant for the following

*This example was communicated by E. Lehmann who attributes the idea to Berk (1970).

reasons. T is invariant because permuting any number of z values does not change any y values. To show that T is a maximal invariant, we must show that $T(y_1, z_1) = T(y_2, z_2)$ implies that there exists some g in G that takes (y_1, z_1) to (y_2, z_2). But $T(y_1, z_1) = T(y_2, z_2)$ implies $y_1 = y_2$. We may define such a g by taking $m = 2$, $F = \{z_1, z_2\}$, and the permutation of (1, 2) to be (2, 1). This shows that T is maximal invariant.

Therefore, although the identity map $\mathbf{I}(y, z) = (y, z)$ is almost invariant, it is not equivalent to an invariant function because any invariant function must be of the form $h(T(y, z)) = h(y)$. Thus we have shown that the identity is almost invariant but is not equivalent to an invariant function. This can be readily translated into a test that is almost invariant but not equivalent to an invariant test. For example, consider the test given by

$$\delta(y, z) = \begin{cases} 0 & \text{if } z \leq 0 \\ z & \text{if } 0 < z < 1 \\ 1 & \text{if } z \geq 1. \end{cases}$$

This is an almost invariant test that is not equivalent to a test that depends on y alone. Also see Example 10.30.

10.30. A uniformly most powerful invariant test that is not uniformly most powerful almost invariant.

As in Example 10.29, let (Y, Z) be independent Gaussian random variables with distribution $N(\theta, 1)$, and consider testing the null hypothesis H: $\theta = 0$ against the alternative hypothesis K: $\theta > 0$. The group G defined in that example using all possible permutations of finitely many Z values leaves the problem invariant with Y as a maximal invariant. Hence the level α uniformly most powerful invariant test is given by δ_1, which rejects when $Y > c_\alpha$, where c_α is the upper 100α percentage point of the standard Gaussian distribution.

This test has power function

$$\beta_1(\theta) = P_\theta\{Y > c_\alpha\} = P\{Y - \theta > c_\alpha - \theta\} = 1 - H(c_\alpha - \theta)$$

where H is the cumulative distribution function of the standard Gaussian distribution. However, consider δ_2, the level α test that rejects when $Y + Z > c_\alpha 2^{1/2}$. This test is almost invariant and has power function

$$\beta_2(\theta) = P_\theta\{Y + Z > c_\alpha 2^{1/2}\} = P_\theta\left\{\frac{Y + Z - 2\theta}{2^{1/2}} > \frac{c_\alpha 2^{1/2} - 2\theta}{2^{1/2}}\right\}$$

$$= 1 - H(c_\alpha - 2^{1/2}\theta).$$

Since $H(c_\alpha - 2^{1/2}\theta) < H(c_\alpha - \theta)$ for $\theta > 0$, it follows that $\beta_2(\theta) > \beta_1(\theta)$ for all $\theta > 0$. Because this second test has larger power everywhere, it follows that δ_1 is not uniformly most powerful almost invariant.

10.31. A hypothesis-testing problem where both uniformly most powerful unbiased and uniformly most powerful invariant tests exist, but they do not coincide.

Although a uniformly most powerful almost invariant test will coincide with a unique uniformly most powerful unbiased test, this is false if we restrict attention to strictly invariant tests. As in Examples 10.29 and 10.30, suppose Y and Z are independent Gaussian random variables with distribution $N(\theta, 1)$. In testing the null hypothesis $H: \theta = 0$ against $K: \theta > 0$, it was seen in Example 10.30 that the test that rejects for large values of Y is uniformly most powerful invariant. However, the test that rejects for large values of $Y + Z$ is uniformly most powerful and is uniformly most powerful unbiased as well. Notice that this test is unique almost everywhere.

Because this test is almost invariant, it is uniformly most powerful almost invariant as well. Thus the uniformly most powerful unbiased and the uniformly most powerful almost invariant tests coincide, in accordance with Theorem 6.6 in Lehmann (1959).

10.32. Nonexistence of a uniformly most powerful invariant test, even though a uniformly most powerful unbiased test exists.

Suppose $X_1, \ldots, X_n$ is a sample from the Gaussian distribution $N(\theta, \sigma^2)$, where both parameters are unknown. In testing the null hypothesis $H: \sigma = \sigma_0$ (with $\sigma_0 > 0$) against the alternative hypothesis $K: \sigma \neq \sigma_0$, the problem remains invariant under location changes defined by the group G whose elements g_c act as follows:

$$g_c(x_1, \ldots, x_n) = (x_1 + c, \ldots, x_n + c)$$

where c is any real number. If B is a Borel set in the real numbers and η denotes Lebesgue measure, then by making the correspondence $B' = \{g_c : c \text{ in } B\}$, a subset of G, and setting $\eta'(B') = \eta(B)$, we have:

$$\eta'(B') = 0 \quad \text{implies} \quad \eta'(B'g_c) = 0 \qquad \text{for any } c.$$

Thus the conditions of Theorem 6.4 of Lehmann (1959) are satisfied, implying that any almost invariant test is equivalent to an invariant one.

We know that the statistic $S = (\bar{X}, \Sigma (X_i - \bar{X})^2)$ is sufficient and complete for (θ, σ^2), and the group G induces transformations of S by the following rule:

$$g_c S = (\bar{X} + c, \sum (X_i - \bar{X})^2).$$

This leaves $\Sigma (X_i - \bar{X})^2$ as a maximal invariant. By Theorem 6.5 of Lehmann, we can therefore restrict attention to tests based on $\sum (X_i - \bar{X})^2$, for if a uniformly most powerful almost invariant test exists, then a uniformly most powerful invariant test will exist that depends on the X_i only through $\Sigma (X_i - \bar{X})^2$.

Note that $\Sigma (X_i - \bar{X})^2/\sigma^2$ has a chi-squared distribution with $(n - 1)$ degrees of freedom. It is then easily verified that the family of distributions of $\Sigma (X_i - \bar{X})^2$ depends on σ^2 alone and has monotone likelihood ratio in $\Sigma (X_i - \bar{X})^2$. Therefore for any fixed alternative σ_1, the most powerful test of the null hypothesis $H: \sigma = \sigma_0$ against $K: \sigma = \sigma_1$ based on this statistic will reject for large values of $\Sigma (X_i - \bar{X})^2$ if $\sigma_1 > \sigma_0$ and will reject for small values if $\sigma_1 < \sigma_0$. Therefore no uniformly most powerful test based on the maximal invariant $\Sigma (X_i - \bar{X})^2$ exists, and so no uniformly most powerful invariant test exists.

The reader is referred to Section 5.2 of Lehmann (1959) to see that a uniformly most powerful unbiased level α test does exist. The uniformly most powerful unbiased test has an acceptance region given by

$$c_1 < \frac{\sum (X_i - \bar{X})^2}{\sigma_0^2} < c_2$$

for constants c_1 and c_2 that satisfy the relationship

$$\int_{c_1}^{c_2} \phi_{n-1}(y)\, dy = \frac{1}{n-1} \int_{c_1}^{c_2} y\phi_{n-1}(y)\, dy = 1 - \alpha$$

where ϕ_{n-1} denotes the density of the chi-squared distribution with $(n - 1)$ degrees of freedom.

Note that a uniformly most powerful invariant test did not exist for this example, even though invariance considerations did apply to the situation. There are many problems—for example, the comparison of two binomial or Poisson populations—where uniformly most powerful unbiased tests do exist but invariance considerations are not even applicable. More details may be found in Lehmann (1959, Section 4.5).

10.33. A uniformly most powerful invariant test that is not maximin.

Often a uniformly most powerful invariant test has the maximin property, but this is not always true. See Lehmann (1959, Sections 8.3 and 8.4) for the appropriate conditions to ensure that this will be the case.

As in Example 10.27, suppose (X_1, X_2) and (Y_1, Y_2) have independent bivariate Gaussian distributions with mean θ and covariance matrices Σ and $c\,\Sigma$, respectively. In testing the null hypothesis H: $c = 1$ against the alternative hypothesis K: $c > 1$, we found that a level α uniformly most powerful invariant test exists and that it rejects H (regardless of the data) with probability α. However, the test based on X_1 and Y_1 alone, which rejects when $Y_1^2/X_1^2 > K$, has a power function that is a strictly increasing function of c because $Y_1^2/(cX_1^2)$ has an $F_{1,1}$ distribution. Its minimum power over the set of alternatives $c > c_1$, for any $c_1 > 1$, is greater than α. Therefore this uniformly most powerful invariant test is not maximin.

APPENDIX
ADDITIONAL COUNTEREXAMPLES

New counterexamples are being discovered all the time. Some are used to disprove conjectures (and occasionally a published theorem), others are used to establish a result as "best possible," while still others are neater and cleaner replacements of older counterexamples.

We present here a smorgasbord of selected counterexamples from the relatively recent statistical literature, grouped according to primary topic, to supplement the more detailed examples found in Chapters 1–10.

A.1. Probability and measure.

Ghosh and Meeden (1977) provide a basic exposition of the nonattainability of Chebyshev bounds. Schwanger (1984) gives an example in which the Bonferroni bounds are not helpful and successive bounds do not converge. Weirich (1983) provides counterexamples to the usual interpretation of conditional probability. Leblanc and van Fraassen (1979) use a counterexample to show that not every Popper probability function is a Carnap probability function, although the converse has been proven. Hamilton and Anderson (1978) give an example of a consistent set of conditional probabilities that is not realized by any random field. Losert (1982) provides counterexamples to some conjectures about doubly stochastic measures. Maharam (1982) gives an example of an orthogonal family of measures on the Borel sets of the unit square such that no uncountable subfamily is uniformly orthogonal (provided the continuum hypothesis is assumed).

Billingsley (1968, p. 234, Remark 2) shows that a single probability measure on a metric space equipped with the Borel σ-field need not be tight. In connection with "tightness," see also Billingsley's Problems 1 and 8 (pp. 40–41). Darst and Zink (1967) give an example of a perfect measurable space that is not a Lusin space. Janssen (1979) uses a counterexample to show that the Lebesgue decomposition of a stochastic kernel into a continuous and a noncontinuous part need not produce kernels. Mauldin, Preiss, and von Weizsacker (1983) provide some counterexamples concerning the separation by a measurable set of two orthogonal measure convex sets of probability measures. Wakker (1981) includes a counterexample to show that certain results about sigma algebras do not extend to algebras.

A.2. Random variables and distributions.

Melnick and Tenebien (1982) use counterexamples to show (1) that uncorrelated normal variables are not necessarily independent, (2) that if each of a finite set of linear combinations of a bivariate distribution is normal, then the bivariate distribution is not necessarily normal, and (3) that linear combinations of normal variables need not be normal. Pierce and Dykstra (1969) provide an example of n random variables such that any proper subset is jointly normal and independent, yet taken together they are neither normal nor independent. Hamedani (1984) gives an example of a nonnormal multivariate distribution for which each proper subset of the variables is multivariate normal and an infinite number of linear combinations of them is normal. Davis (1967) shows that trouble can arise with multivariate generalizations of the t- and F-distributions; in particular, these may depend on the population covariance matrix. Ord (1968) shows that the discrete Student's t-distribution has all finite odd-order moments zero, although the distribution need not be symmetric.

Brockett and Kemperman (1982) use counterexamples to show that convolutions of high order need not be unimodal. Wells (1978) uses a counterexample to distinguish between two concepts of multivariate unimodality. Kaplansky (1945) gives examples to show that positive kurtosis need not imply peakedness about the mean. Kanter (1976) argues that the current proof of the unimodality of all stable densities is invalid.

Lachenbruch and Brogan (1971) discuss distributions that have no moments but whose percentiles are easily obtained. Stoops and Barr (1971) indicate which moments of which order statistics exist for samples from the Cauchy distribution; for example, all order statistics except the extremes have first moments even though the average does not. Dudewicz and Dann (1972) show that there do not exist a pair of n-sided dice such that each sum is equally probable. Arnold (1979) gives an example of independent identically distributed nonuniform random variables whose sum modulo m is uniformly distributed.

Wolfe (1975) gives an example of a characteristic function such that an odd-order derivative exists at zero but does not exist at a sequence of numbers tending to zero. Perlman and Wichura (1975) show that a seemingly true proposition concerning conditional distributions is in fact false in general. Bondesson (1983) shows that certain classes of distributions that are important in reliability theory are not preserved under operations such as mixing and convolution. Bradley (1977) shows how "bizarre" distribution shapes can easily occur in practice.

A.3. Independence.

Krewski and Bickis (1984) give a nontrivial example of an infinite collection of events that is exhaustive; they also show that a finite collection of independent events cannot be exhaustive unless one of the events occurs with probability one. Geisser and Mantel (1962) give practical examples of pairwise independent, but not independent random variables. Wong (1972) gives two examples of n events ($n > 2$) for which any $n - 1$ of them are mutually independent, yet taken together they are not independent. Wang (1979) gives examples of dependent random variables such that every proper subset is independent. Gibbons (1968) provides examples to distinguish independence from mutual exclusivity and to distinguish independence from zero correlation. Burdick (1972) shows that there exist two independent random variables where one is symmetric and the other is not, but their sum is symmetric. Dawid (1979) gives some counterexamples involving the conditional independence of random variables. Gupta (1979) provides some counterexamples to statements about independence in competing risk analysis.

A.4. Sequences of random variables.

Plane and Gordon (1982) show that the Central Limit Theorem does not apply to a finite population in the case of sampling without replacement. Mirham (1969) provides an example of an infinite sum of independent random variables with finite moments that does not tend to a normal distribution. Herrndorf (1984) gives a counterexample to a conjecture of Newman and Wright regarding a Central Limit Theorem for dependent random variables. Jain (1976) gives an example of a sequence of independent and identically distributed Banach-space valued random variables that satisfies the Central Limit Theorem but not the law of the iterated logarithm. Kuelbs (1976) finds a sequence of independent and identically distributed Banach-space valued random variables that satisfies the law of the iterated logarithm but fails to obey the Central Limit Theorem. Landers and Rogge (1977) give a counterexample in the field of Central Limit Theorems with random summation indices.

Beck and Warren (1976) give counterexamples to the strong law of large numbers for random variables that take values in a Banach space. Reich (1982) gives some counterexamples to the distribution theory of weighted sums of independent random variables. Freedman and Diaconis (1982a) include two counterexamples: The first shows that there is an exchangeable process that is location symmetric but not conditional location symmetric; the second shows that a weighted sum of independent and identically distributed nonsymmetric random variables may nonetheless be symmetric. Kiefer (1976) gives a counterexample to a conjecture concerning the minimum number of fixed-length sequences with fixed total probability. Roussas (1972, pp. 7–10) gives counterexamples related to measures of closeness of sequences of probability measures: contiguity, $L1$ convergence, tightness, and mutual absolute continuity.

A.5. Stochastic processes and martingales.

Hahn (1977) constructs counterexamples to verify a conjecture concerning sample continuity and the Central Limit Theorem for stochastic processes. Steele (1980) gives an example to show that left limits are necessary to establish the continuity of sample functions of a stochastic process. Shanbhag and Westcott (1977) use a counterexample to show that a Cox process may be infinitely divisible even though the measure it is directed by is not infinitely divisible. Mineka (1968) gives an example of a mixture of recurrent random walks that is not recurrent. Gaenssler (1983, pp. 13–15, 36–38) includes counterexamples on empirical processes. Bramson and Neidhardt (1981) show that there exist subsets of the circle that are small in the Lebesgue sense but tend to be visited often by certain stochastic processes. Berman (1974) shows that a stochastic process that is badly discontinuous and unbounded everywhere can nonetheless be very well behaved in the sense that the complete graph is determined from an arbitrary window view.

Berry and Wang (1982) exhibit an optimal stopping problem that has an unusual and unexpected solution with structures similar to islands and peninsulas. Dubins (1973) notes that functions of stopping times are not necessarily themselves stopping times and determines functions of n-tuples of stopping times that are always stopping times. Dubins and Sudderth (1975) show that stationary gambling strategies are not adequate by using an example in which, for every initial state, there is a strategy with a high probability of winning, but for some initial state every stationary strategy does poorly. Fisher and Ross (1968) give an example of a denumerable Markovian decision process with an optimal nonstationary rule (but no optimal stationary rule) to answer a question of Derman. Dudley (1972, 1973) assumes the continuum hypothesis to construct a counterexample to settle an open question on measurable stochastic processes. Ross (1971) gives a counterexample to show that a good stationary policy need not always exist in a Markov decision process. Haigh (1983) gives some counterexamples on the maintainability of Markov manpower structures. Hendricks (1972) uses a counterexample to establish the truth of an outstanding conjecture in the Hausdorff dimension of images of a process with stable components. Blackwell and Dubins (1975) show that the Borel-measurable Axiom of Choice is false in an investigation of conditional distributions for discrete and continuous time processes. Kurtz and Wainger (1973) show that the Yaglom limit need not exist for an age-dependent subcritical branching process.

Monroe (1976) gives an example of a right continuous martingale that is "about as bad as a martingale can be" and whose quadratic variation does not converge. Gilat (1972) provides an example of a Markovian martingale that converges in distribution but not in probability, and an example that converges in probability but not almost surely. Berman (1976) gives an example of a weak martingale that is not a martingale but is bounded and has countably many paths.

Dubins and Freedman (1979) give an example of an exchangeable stochastic process that cannot be represented as a mixture of sequences of independent and identically distributed random variables; two counterexamples relating to the Kolmogorov Consistency Theorem are also given.

A.6. Estimation.

Buehler (1982) and the related discussion include examples of ancillary statistics and indicate which properties they satisfy. Godambe (1982) relates a paradox involving the ancillarity principle. Canner (1969) provides examples in which minimum variance unbiased mean and regression estimates can fall either above or below all of the data points. Stein (1950) gives an example for which there is no locally minimum variance unbiased estimate. Torgersen (1971) gives a counterexample involving translation invariant estimators to show that a theorem is incorrect.

Freedman and Diaconis (1982b) show that M-estimators, a class of robust estimators based on a generalization of maximum likelihood, may be inconsistent in some cases. Lambert (1982) shows that the Student's t-test is not qualitatively robust. Rukhin (1982) finds conditions that show that adaptive procedures (which are simultaneously optimal against several families of distributions) need not exist.

Stigler (1980) discusses an example from Edgeworth for which a single observation is a better estimate of the mean than the average of two observations. Basu (1956) shows that an estimator that is asymptotically inefficient may have a greater probability concentration than a Fisher efficient estimator. Farrell (1967) shows that there does not always exist a uniformly consistent sequence of estimators of a density function. Ghosh, Parr, Singh, and Babu (1984) show that the bootstrap variance estimator for the sample median is not consistent in general. Additional counterexamples on the inconsistency of the bootstrap are given by Beran (1984) and by Bickel and Freedman (1981).

Koenker and Bassett (1984) offer four examples of unusual statistical behavior in least absolute deviation estimators: (1) weak but not strong consistency, (2) consistency without Cesaro identifiability, (3) a slow rate of convergence to a "bizarre" limiting distribution, and (4) consistency despite a violation of a usual design condition. Dempster, Schatzoff, and Wermuth (1977) demonstrate the superiority of shrinkage estimators over least squares in certain regression problems.

DeGroot and Goel (1980) show that it is, in fact, possible to estimate the correlation coefficient from a bivariate normal sample without using any information about the pairing of the two coordinates. Chanda (1962) shows that a serial correlation can be smaller than -1 or larger than 1. Armstrong (1967) gives an entertaining example to show that factor analysis can give misleading results. Burdick and Herr (1980) provide examples of the problems encountered in unbalanced two-way analysis of variance. Green and Doll (1974) give an example to show how the use of dummy variables can lead to misleading inferences about seasonality. Cleveland and McGill (1983) indicate, with an example, how optical illusions in the perception of color can affect the interpretation of a statistical illustration. Mosteller, Siegel, Trapido, and Youtz (1981) show that when people are asked to indicate a regression line by eye, they tend to choose a line that is closer to the first principal axis than to the regression line.

A.7. Sufficiency.

Sampson and Spencer (1976) discuss practical techniques for showing that a given statistic is not sufficient or that a given sufficient statistic is not minimal, and they include examples. Landers (1974) provides examples to show that assumptions about the component families cannot be left out in a theorem that concerns the existence of a minimal sufficient statistic in a product family. Pitcher (1957) gives examples for which no necessary and sufficient statistic (or subfield) exists. Landers and Rogge (1972) provide two counterexamples to show that "the existence of a minimal sufficient sigma field is neither necessary nor sufficient for the existence of a minimal sufficient statistic." Hasegawa and Perlman (1974) give a counterexample to a particular construction of a minimal sufficient subfield, but then they provide a nonconstructive proof of its existence. Bhattacharyya, Johnson, and Mehrotra (1977) show that with censored sampling, "most of the common parametric models fail to possess completeness of minimal sufficient statistics." Stigler (1972) gives an example of a "just barely complete" complete family of probability measures such that if any one of its infinite number of members is removed, the resulting class is not complete. Burkholder (1961) gives an example of a nonsufficient subfield that contains a sufficient subfield, thus solving a problem posed by Bahadur.

A.8. Likelihood procedures.

Bahadur (1958) gives an example of an inconsistent maximum likelihood estimate, but when noise is added to the data, the maximum likelihood estimate is consistent. Rao (1962) discusses various criticisms of maximum likelihood estimation. Plante (1970) discusses counterexamples and likelihood theory. Solari (1969) gives an example for which there are two solutions to the likelihood equation, but no maximum likelihood estimator exists because of saddle-point-type behavior. Tarone and Gruenhage (1975) give two examples from the literature in which univariate likelihood results are incorrectly applied in the multivariate context. Hoyt (1969) gives two examples of maximum likelihood estimates of the mean of a distribution that are not the sample mean.

Ghosh and Sinha (1982) give an example of a curved exponential family for which the maximum likelihood estimate is not third-order efficient. Reiss (1973) includes counterexamples to show that sufficiency conditions cannot be removed in theorems regarding the consistency of maximum likelihood estimators for families of unimodal densities. Oliver (1972) shows an example in which the maximum likelihood estimator is one of the observations in the sample but does not correspond to a particular order statistic. Ferguson (1978) gives an example using the Cauchy distribution for which the maximum likelihood location estimator decreases as one of the sample data values increases.

A.9. Bayes methods.

Dempster (1963b) gives an example of a paradox in Bayesian or fiducial inference of a covariance matrix. Shafer (1982a) argues that the fifth of Bayes's six propositions (one of two that deal with the "rule of conditioning") does not stand up under a careful analysis. Shafer (1982b) and the associated discussion explore Lindley's Paradox that "a sharp null hypothesis may be strongly rejected by a sampling-theory test of significance and yet be awarded high odds by a Bayesian analysis based on a small prior probability for the null hypothesis and a diffuse distribution of one's remaining probability over the alternative hypothesis." Stigler (1983) suggests that Bayes may not have been the one to discover Bayes's Theorem and explores the evidence. Stone and Springer (1965) present a Bayesian paradox in which the more carefully chosen of two prior distributions leads to less precise inferences. Guttman and Tan (1974) show that a method for determining a prior distribution need not lead to a unique result.

Stone (1976) analyzes two examples of inconsistency when improper Bayesian priors are used. Freedman and Diaconis (1983) show that an inconsistent Bayes estimate results when a mixture of a tail-free prior and a point mass is used in sampling from an unknown probability distribution on the integers. Flacke and Therstappen (1979) show that Bayesian sufficiency need not imply sufficiency in general. Blackwell and Ramamoorthi (1982) show that even though every classically sufficient statistic is Bayes sufficient in a Borel setting, the converse need not apply. Their example involves hypothesis testing, and they claim that "Bayesians, but not classicists, can achieve zero error probabilities." Bickel and Blackwell (1967) use counterexamples to establish the limits of an incompatibility result in unbiased Bayesian estimation.

Brown (1979) shows that a generalized Bayes rule for a prior with "light" tails can be inadmissible. Skibinsky and Cote (1963) give examples to show that some standard estimates can be inadmissible in the presence of prior information. Berger (1976) considers the inadmissibility of generalized Bayes estimators and shows that under general conditions, the best invariant estimator of the first coordinate of a location vector with at least four components is inadmissible.

Brillinger (1962) provides a survey of examples relating to fiducial inference. Dempster (1963a) discusses inconsistencies in the fiducial arguments of Fisher. Dawid and Stone (1982) report "an inconsistency which may arise when partial information becomes available" within the context of functional models and fiducial inference.

A.10. Admissibility.

Gutmann (1982a) shows that Stein's paradox (that using admissible procedures in each of several problems may be inadmissible) cannot happen when the sample spaces are finite. Berger (1982) shows that, when estimating a multivariate normal mean with quadratic loss, significant improvements over the maximum likelihood estimator with minimax estimators are possible only in small regions of the parameter space, thus shedding further light on Stein's paradox. Baranchik (1973) shows that the Stein paradox also holds in multiple regression estimation. Saxena and Alam (1982) show that the positive part of $X - p$, where X has a noncentral chi-squared distribution with p degrees of freedom, dominates the maximum likelihood estimator under the squared error loss function. Gutmann (1982b) shows that Stein's paradox is impossible when parameters must be estimated in sequence and future data cannot be used to help estimate the present parameter. Egerton and Laycock (1981) provide some counterexamples to show that stochastic shrinkage and ridge regression do not always produce improvements. (Could this be a counterexample to a trend based on counterexamples?)

Fox (1981) gives an example of an inadmissible best invariant point estimator of a location parameter. Marden (1983) shows that many popular tests based on the likelihood ratio matrix are inadmissible but that this is not necessarily serious. Perng (1970) provides examples of inadmissibility of some "good" translation invariant procedures. Moors (1981) shows that linearly invariant estimators are generally inadmissible when the parameter space is truncated. Sackrowitz and Strawderman (1974) show that the maximum likelihood estimator is often inadmissible when the parameters of m independent binomial distributions with known ordering are estimated. Zidek (1969) shows that the best invariant estimator of an extreme normal quantile is inadmissible.

Makani (1977) gives examples that show that, given two independent estimation problems, an estimator based on data from only one problem can be admissible as an estimate of the parameter of the other problem. Brown (1968) shows that some usual estimators of scale are inadmissible. Blackwell (1951) gives two examples for which the Girschick-Savage estimate is not admissible in translation parameter estimation. Biyani (1980) shows that the Yates-Grundy estimator is inadmissible in unequal probability sampling. The Yates-Grundy procedure estimates the variance of the Horvitz and Thompson estimate of the total of a finite population based on a weighted sampling procedure.

A.11. Hypothesis testing.

Hill (1975) gives an example for which the likelihood ratio in favor of a true hypothesis actually tends to zero in the limit of large sample size. Jogdeo and Bohrer (1973) give examples and counterexamples to show that ordering restrictions (such as monotone likelihood ratio) cannot be omitted in the theory of optimum tests. Solomon (1975) shows that Neyman-Pearson tests and generalized likelihood ratio tests are not necessarily equivalent when two simple hypotheses are tested. Shaffer (1980) gives a counterexample to an apparently reasonable interpretation of a stagewise multiple-test procedure. Sclove, Morris, and Radhakrishnan (1972) show that estimation preceded by testing is not an optimal procedure within the context of any linear hypothesis concerning a mean vector, including regression models.

Suissa and Shuster (1984) show that a biased test beats Fisher's exact test for the equality of two binomial proportions in a large region of the parameter space. Hilgers (1982) gives a counterexample to show that the Wilcoxon-Mann-Whitney test is not necessarily a test of equality of distributions. Basu (1980) and the related discussion provide some problems associated with the Fisher randomization test. Cover (1973) discusses the possibility of constructing a procedure to decide whether or not the mean of a sequence of random variables is a rational number or not, and concludes (somewhat surprisingly) that it can be done to some extent.

A.12. Confidence intervals.

Joshi (1967) shows that the usual confidence sets for the mean of a multivariate normal distribution are inadmissible. Pratt (1963) discusses the construction of shorter confidence intervals for the mean of a (univariate) normal distribution with known variance that are more efficient than the standard calculations. Hwang and Casella (1982) find explicit confidence procedures that dominate the usual one in the case of estimating a multivariate normal mean by centering about the positive-part James-Stein estimator. Robinson (1975, 1977) gives some counterexamples to the theory of confidence intervals.

Meeks and D'Agostino (1983) show that if confidence limits are computed only after rejecting a null hypothesis, then the ordinary confidence limits are not valid. Kiefer (1977) gives an example of a confidence interval with confidence coefficient less than one, but for which we sometimes know with certainty that the unknown parameter is inside the interval. Pierce (1973) gives some examples of problems with conditional coverage of confidence regions. Miller (1964) gives three counterexamples to show that the jackknife method of constructing confidence intervals or tests does not always produce satisfactory results.

A.13. Categorical data analysis.

Conover (1974) gives examples against the use of the Yates continuity correction in 2×2 contingency table analysis. Haber (1980) shows that, for testing independence in a 2×2 contingency table, both Pearson's chi-squared test and the Yates continuity correction "may give misleading outcomes even for moderate sample sizes and expected values." Tate and

Hyer (1973) find that the chi-squared test may not be as accurate as is generally accepted. Holland (1973) shows that there is no covariance stabilizing transformation for the trinomial distribution. Whittemore (1978) gives counterexamples regarding the collapsibility of multidimensional contingency tables. Joag-Dev and Proshan (1983) show an example in which negative and positive associations "exist side by side in models of categorical data analysis."

A.14. Miscellaneous.

Bryson (1976) discusses misunderstandings surrounding an incorrect prediction of the 1936 U.S. presidential election. Francis (1973) compares several widely used computer programs for the analysis of variance, and found in some cases that the results were "either misleading or sometimes completely wrong." Siegel (1983) gives examples of two-way tables that require a long time to be analyzed by median-polish methods. Hoffman (1960) gives an example to show that a uniqueness property of triangular partially balanced incomplete block designs fails only when $n = 8$, even though it holds true for all smaller and larger values. Wiorkowski (1972) shows that in knockout tournaments the second-best player is not so likely to come in second as one might expect. Israel (1981) presents a counterexample to a conjecture showing that stronger players need not win more knockout tournaments.

Berry and Fristedt (1980) show that myopic strategies are not optimal in general for the two-armed bandit problem. Blackwell (1968) gives an example of a Borel set that does not contain a graph from its projection, and relates it to dynamic programming problems for which there is a good strategy but not a good Borel-measurable strategy. Rizvi and Shorrock (1979) provide counterexamples to statements about the convexity of matrix functions. Van den Berg (1983) disproves a conjecture in percolation theory using a counterexample. Root (1971) gives a counterexample to a statement in renewal theory. Samuel-Cahn (1975) uses a counterexample to show that a formula of Fisher is false in its full generality. Koehn and Thomas (1975) provide some counterexamples to the original formulation of Basu's Lemma. Kiefer (1973) provides a counterexample to a result that would (if true) have been a generalization of the Shannon-McMillan Theorem of information theory. Pfanzagl (1971) gives a counterexample to show that a lemma involving continuity is false. Krafft and Schaefer (1984) use a counterexample to disprove a conjecture regarding an equivalence of two conditions in random replacement sampling plans.

REFERENCES

Abell, G. 1980. *Realm of the universe*. 2d ed. Philadelphia: Saunders.

Anderson, T. W. 1958. *An introduction to multivariate statistical analysis*. New York: Wiley.

Armstrong, J. S. 1967. Derivation of theory by means of factor analysis or Tom Swift and his electric factor analysis machine. *The American Statistician* 21, no. 5:17–21.

Arnold, B. C. 1979. Nonuniform decompositions of uniform random variables under summation modulo *m*. *Bollettino della Unione Matematica Italiana* 16-A:100–102.

Ash, R. 1972. *Real analysis and probability*. New York: Academic Press.

Bahadur, R. R. 1958. Examples of inconsistency of maximum likelihood estimates. *Sankhya* 20:207–10.

Baranchik, A. J. 1973. Inadmissibility of maximum likelihood estimators in some multiple regression problems with three or more independent variables. *Annals of Statistics* 1:312–21.

Barnett, V. D. 1966. Evaluation of the maximum likelihood estimator where the likelihood equation has multiple roots. *Biometrika* 53:151–65.

Basu, D. 1956. The concept of asymptotic efficiency. *Sankhya* 17:193–96.

———. 1964. Recovery of ancillary information. *Sankhya* (*A*) 26:3–16.

———. 1980. Randomization analysis of experimental data: the Fisher randomization test (with discussion). *Journal of the American Statistical Association* 75:575–82.

Beck, A., and P. Warren. 1976. Counterexamples to strong laws of large numbers for Banach space-valued random variables. In *Theorie ergodique*, edited by A. Dold and B. Eckmann, 1–14. Berlin: Springer-Verlag.

Beran, R. 1984. Bootstrap methods in statistics. *Jahresbericht der Deutschen Mathematiker-Vereinigung* 86:14–30.

Berger, J. O. 1976. Inadmissibility results for generalized Bayes estimators of coordinates of a location vector. *Annals of Statistics* 4:302–33.

———. 1982. Selecting a minimax estimator of a multivariate normal mean. *Annals of Statistics* 10:81–92.

Berk, R. H. 1970. A remark on almost invariance. *Annals of Mathematical Statistics* 41:733–35.

Berkson, J. 1980. Minimum chi-square, not maximum likelihood! *Annals of Statistics* 8:457–87.

Berman, J. 1976. An example of a weak martingale. *Annals of Probability* 4:107–108.

Berman, S. M. 1974. A Gaussian paradox: determinism and discontinuity of sample functions. *Annals of Probability* 2:950–53.

Berry, D. A., and B. Fristedt. 1980. Two-armed bandits with a goal II dependent arms. *Advances in Applied Probability* 12:958–71.

Berry, D. A., and P. Wang. 1982. Optimal stopping regions with islands and peninsulas. *Annals of Statistics* 10:634–36.

Bhattacharyya, G. K., R. A. Johnson, and K. G. Mehrotra. 1977. On the completeness of minimal sufficient statistics with censored observations. *Annals of Statistics* 5:547–53.

Bickel, P. J., and D. Blackwell. 1967. A note on Bayes estimates. *Annals of Mathematical Statistics* 38:1907–11.

Bickel, P. J., and K. A. Doksum. 1977. *Mathematical statistics: Basic ideas and selected topics*. San Francisco: Holden-Day.

Bickel, P. J., and D. A. Freedman. 1981. Some asymptotic theory for the bootstrap. *Annals of Statistics* 9:1196–1217.

Billingsley, P. 1968. *Convergence of probability measures*. New York: Wiley.

———. 1979. *Probability and measure*. New York: Wiley.

Biyani, S. H. 1980. On inadmissibility of the Yates-Grundy variance estimator in unequal probability sampling. *Journal of the American Statistical Association* 75:709–12.

Blackwell, D. 1951. On the translation parameter problem for discrete variables. *Annals of Mathematical Statistics* 22:393–99.

———. 1968. A Borel set not containing a graph. *Annals of Mathematical Statistics* 39:1345–47.

Blackwell, D., and L. E. Dubins. 1975. On existence and non-existence of proper, regular, conditional distributions. *Annals of Probability* 3:741–52.

Blackwell, D., and M. A. Girshick. 1954. *Theory of games and statistical decisions*. New York: Wiley.

Blackwell, D., and R. V. Ramamoorthi. 1982. A Bayes but not classically sufficient statistic. *Annals of Statistics* 10:1025–26.

Blackwell, D., and C. Ryll-Nardzewski. 1963. Non-existence of everywhere proper conditional distributions. *Annals of Mathematical Statistics* 34:223–25.

Bondesson, L. 1983. On preservation of classes of life distributions under reliability operations: some complementary results. *Naval Research Logistics Quarterly* 30:443–47.

Bradley, J. V. 1977. A common situation conducive to bizarre distribution shapes. *The American Statistician* 31:147–50.

Bramson, M., and A. Neidhardt. 1981. Thin but unavoidable sets. *Annals of Probability* 9:154–56.

Breiman, L. 1968. *Probability*. Reading, Mass.: Addison-Wesley.

Brillinger, D. R. 1962. Examples bearing on the definition of fiducial probability with a bibliography. *Annals of Mathematical Statistics* 33:1349–55.

Brockett, P. L., and J. H. B. Kemperman. 1982. On the unimodality of high convolutions. *Annals of Probability* 10:270–77.

Brown, L. 1968. Inadmissibility of the usual estimators of scale parameters in problems with unknown location and scale parameters. *Annals of Mathematical Statistics* 39:29–48.

Brown, L. D. 1979. Counterexample—an inadmissible estimator which is generalized Bayes for a prior with "light" tails. *Journal of Multivariate Analysis* 9:332–36.

Bryson, M. C. 1976. The Literary Digest poll: making of a statistical myth. *The American Statistician* 30:184–85.

Buehler, R. J. 1982. Some ancillary statistics and their properties (with discussion). *Journal of the American Statistical Association* 77:581–89.

Burdick, D. L. 1972. A note on symmetric random variables. *Annals of Mathematical Statistics* 43:2039–40.

Burdick, D. S., and D. G. Herr. 1980. Counterexamples in unbalanced two-way analysis of variance. *Communications in Statistics: Theory and Methods* 9:231–41.

Burkholder, D. L. 1961. Sufficiency in the undominated case. *Annals of Mathematical Statistics* 32:1191–1200.

Canner, P. L. 1969. Some curious results using minimum variance linear unbiased estimators. *The American Statistician* 23, no. 5:39–40.

Chanda, K. C. 1962. On bounds of serial correlations. *Annals of Mathematical Statistics* 33:1457–60.

Chen, R., and L. Shepp. 1983. On the sum of symmetric random variables. *The American Statistician* 37:237.

Chow, Y., and H. Teicher. 1978. *Probability theory*. New York: Springer-Verlag.

Chung, K. L. 1974. *A course in probability theory*. 2d ed. New York: Academic Press.

Churchill, E. 1946. Information given by odd moments. *Annals of Mathematical Statistics* 17:244–46.

Cleveland, W. S., and R. McGill. 1983. A color-caused optical illusion on a statistical graph. *The American Statistician* 37:101–105.

Conover, W. J. 1974. Some reasons for not using the Yates continuity correction on 2 × 2 contingency tables (with discussion). *Journal of the American Statistical Association* 69:374–76.

Conway, J. 1978. *Functions of one complex variable*. 2d ed. New York: Springer-Verlag.

Cover, T. M. 1973. On determining the irrationality of the mean of a random variable. *Annals of*

Statistics 1:862–71.
Cox, D. R. 1958. Some problems connected with statistical inference. *Annals of Mathematical Statistics* 29:357–72.
Cox, D. R., and D. V. Hinkley. 1974. *Theoretical statistics*. London: Chapman and Hall.
———. 1978. *Problems and solutions in theoretical statistics*. London: Chapman and Hall.
Darst, R. B., and R. E. Zink. 1967. A perfect measurable space that is not a Lusin space. *Annals of Mathematical Statistics* 38:1918.
David, H. A. 1970. *Order statistics*. New York: Wiley.
Davis, A. W. 1967. A counter-example relating to certain multivariate generalizations of t and F. *Annals of Mathematical Statistics* 38:613–15.
Dawid, A. P. 1979. Some misleading arguments involving conditional independence. *Journal of the Royal Statistical Society, Series B* 41:249–52.
Dawid, A. P., and M. Stone. 1982. The functional-model basis of fiducial inference. *Annals of Statistics* 10:1054–67.
DeGroot, M. H., and P. Goel. 1980. Estimation of the correlation coefficient from a broken random sample. *Annals of Statistics* 8:264–78.
Dempster, A. P. 1963a. Further examples of inconsistencies in the fiducial argument. *Annals of Mathematical Statistics* 34:884–91.
———. 1963b. On a paradox concerning inference about a covariance matrix. *Annals of Mathematical Statistics* 34:1414–18.
Dempster, A. P., M. Schatzoff, and N. Wermuth. 1977. A simulation study of alternatives to ordinary least squares (with discussion). *Journal of the American Statistical Association* 72:77–91.
deVeaux, D. 1976. Tight upper and lower bounds for correlation of bivariate distributions arising in air pollution modeling. Technical Report 5. Stanford University, Department of Statistics.
Dubins, L. E. 1973. Which functions of stopping times are stopping times? *Annals of Probability* 2:313–16.
Dubins, L. E., and D. A. Freedman. 1979. Exchangeable processes need not be mixtures of independent identically distributed random variables. *Zeitschrift fuer Wahrscheinlichkeitstheorie und Verwandte Gebiete* 48:115–32.
Dubins, L. E., and W. D. Sudderth. 1975. An example in which stationary strategies are not adequate. *Annals of Probability* 3:722–25.
Dudewicz, E. J., and R. E. Dann. 1972. Equally likely dice sums do not exist. *The American Statistician* 26, no. 4:41–42.
Dudley, R. M. 1972. A counterexample on measurable processes. *Proceedings of the Sixth Berkeley Symposium on Mathematical Statistics and Probability* 2:57–66.
———. 1973. Correction to "A counterexample on measurable processes." *Annals of Probability* 1:191–92.
Efron, B., and C. Morris. 1973a. Stein's estimation rule and its competitors—an empirical Bayes approach. *Journal of the American Statistical Association* 68:117–30.
———. 1973b. Combining possibly related estimation problems. *Journal of the Royal Statistical Society, Series B*, 35:379–421.
Egerton, M. F., and P. J. Laycock. 1981. Some criticisms of stochastic shrinkage and ridge regression, with counterexamples. *Technometrics* 23:155–59.
Farrell, R. H. 1967. On the lack of a uniformly consistent sequence of estimators of a density function in certain cases. *Annals of Mathematical Statistics* 38:471–74.
Feller, W. 1959. Non-Markovian processes with the semigroup property. *Annals of Mathematical Statistics* 30:1252–53.
———. 1968. *An introduction to probability theory and its applications*. Vol. 1. 3d ed. New York: Wiley.
———. 1971. *An introduction to probability theory and its applications*. Vol. 2. 2d ed. New York: Wiley.
Ferguson, T. S. 1962, 1963. Location and scale parameters in exponential families of distributions. *Annals of Mathematical Statistics* 33:986–1000. Correction 34:1603.
———. 1967. *Mathematical statistics*. New York: Academic Press.
———. 1978. Maximum likelihood estimates of the parameters of the Cauchy distribution for samples of size 3 and 4. *Journal of the American Statistical Association* 73:211–13.

———. 1982. An inconsistent maximum likelihood estimate. *Journal of the American Statistical Association* 77:831–34.
Fisher, L., and S. M. Ross. 1968. An example in denumerable decision processes. *Annals of Mathematical Statistics* 39:674–75.
Flacke, W., and N. Therstappen. 1979. Bayesian sufficient statistics and variance. *Annales de l'Institut Henri Poincare, Section B* 15:303–14.
Fox, M. 1981. An inadmissible best invariant estimator: the I.I.D. case. *Annals of Statistics* 9:1127–29.
Francis, I. 1973. A comparison of several analysis of variance programs. *Journal of the American Statistical Association* 68:860–65.
Freedman, D. A. 1971. *Markov chains*. San Francisco: Holden-Day.
———. 1974. The Poisson approximation for dependent events. *Annals of Probability* 2:256–69.
Freedman, D. A., and P. Diaconis. 1982a. DeFinetti's theorem for symmetric location families. *Annals of Statistics* 10:184–89.
———. 1982b. On inconsistent *M*-estimators. *Annals of Statistics* 10:454–61.
———. 1983. On inconsistent Bayes estimates in the discrete case. *Annals of Statistics* 11:1109–18.
Fu, J. C. 1984. The moments do not determine a distribution. *The American Statistician* 38:294–95.
Gaenssler, P. 1983. Empirical processes. Institute of Mathematical Statistics Lecture Notes—Monograph Series, 3.
Geisser, S., and N. Mantel. 1962. Pairwise independence of jointly dependent variables. *Annals of Mathematical Statistics* 33:290–91.
Gelbaum, B. R., and J. Olmsted. 1964. *Counterexamples in analysis*. San Francisco: Holden-Day.
Ghosh, J. K., and B. K. Sinha. 1982. Third order efficiency of the MLE—a counterexample. *Calcutta Statistical Association Bulletin* 31:151–58.
Ghosh, M., and G. Meeden. 1977. On the non-attainability of Chebyshev bounds. *The American Statistician* 31:35–36.
Ghosh, M., W. C. Parr, K. Singh, and G. J. Babu. 1984. A note on bootstrapping the sample median. *Annals of Statistics* 12:1130–35.
Gibbons, J. D. 1968. Mutually exclusive events, independence and zero correlation. *The American Statistician* 22, no. 5:31–32.
Gilat, D. 1972. Convergence in distribution, convergence in probability and almost sure convergence of discrete martingales. *Annals of Mathematical Statistics* 43:1374–79.
Girshick, M. A., and L. J. Savage. 1951. Bayes and minimax estimators for quadratic loss functions. In *Proceedings of the Second Berkeley Symposium*, 53–73. Berkeley: University of California Press.
Godambe, V. P. 1982. Ancillarity principle and a statistical paradox. *Journal of the American Statistical Association* 77:931–33.
Gradshteyn, I. S., and I. M. Ryzhik. 1965. *Table of integrals, series, and products*. New York: Academic Press.
Green, R. D., and J. P. Doll. 1974. Dummy variables and seasonality—a curio. *The American Statistician* 28, no. 2:60–62.
Gupta, R. C. 1979. Some counterexamples in the competing risk analysis. *Communications in Statistics: Theory and Methods* 8:1535–40.
Gutmann, S. 1982a. Stein's paradox is impossible in problems with finite sample space. *Annals of Statistics* 10:1017–20.
———. 1982b. Stein's paradox is impossible in the nonanticipative context. *Journal of the American Statistical Association* 77:934–35.
Guttman, I., and W. Y. Tan. 1974. Inconsistencies in the Villegas method of determining a prior distribution. *Annals of Statistics* 2:383–86.
Haber, M. 1980. A comparison of some continuity corrections for the chi-squared test on 2×2 tables. *Journal of the American Statistical Association* 75:510–15.
Hahn, M. G. 1977. Conditions for sample-continuity and the central limit theorem. *Annals of Probability* 5:351–60.
Haigh, J. 1983. Maintainability of manpower structures—counterexamples, results and conjectures. *Journal of Applied Probability* 20:700–705.
Halmos, P. R. 1950. *Measure theory*. New York: Van Nostrand.
———. 1960. *Naive set theory*. New York: Van Nostrand.
Hamedani, G. G. 1984. Nonnormality of linear combinations of normal random variables. *The Amer-*

ican Statistician 38:295–96.

Hamilton, M., and W. J. Anderson. 1978. A consistent system of conditional probabilities which is not compatible with any random field. *Canadian Journal of Statistics* 6:95–101.

Hasegawa, M., and M. D. Perlman. 1974. On the existence of a minimal sufficient subfield. *Annals of Statistics* 2:1049–55.

Hendricks, W. J. 1972. Hausdorff dimension in a process with stable components—an interesting counterexample. *Annals of Mathematical Statistics* 43:690–94.

Herrndorf, N. 1984. An example on the central limit theorem for associated sequences. *Annals of Statistics* 12:912–17.

Heyde, C. C. 1963. On a property of the lognormal distribution. *Journal of the Royal Statistical Society, Series B* 25:392–93.

Hilgers, R. 1982. On the Wilcoxon-Mann-Whitney-Test as nonparametric analogue and extension of *t*-test. *Biometrical Journal* 24:3–15.

Hill, B. M. 1975. Aberrant behavior of the likelihood function in discrete cases. *Journal of the American Statistical Association* 70:717–19.

Hodges, J. L., Jr., and E. L. Lehmann. 1950. Some problems in minimax point estimation. *Annals of Mathematical Statistics* 21:182–97.

Hoffman, A. J. 1960. On the uniqueness of the triangular association scheme. *Annals of Mathematical Statistics* 31:492–97.

Holland, P. W. 1973. Covariance stabilizing transformations. *Annals of Statistics* 1:84–92.

Hoyt, J. P. 1969. Two instructive examples of maximum likelihood estimates. *The American Statistician* 23, no. 2:14.

Hwang, J. T., and G. Casella. 1982. Minimax confidence sets for the mean of a multivariate normal distribution. *Annals of Statistics* 10:868–81.

Ibragimov, I. A., and R. Z. Has'minskii. 1981. *Statistical estimation: asymptotic theory*. Translated by S. Kotz. New York: Springer-Verlag.

Israel, R. B. 1981. Stronger players need not win more knockout tournaments. *Journal of the American Statistical Association* 76:950–51.

Jain, N. C. 1976. An example concerning CLT and LIL in Banach space. *Annals of Probability* 4:690–94.

James, W., and C. Stein. 1961. Estimation with quadratic loss. *Proceedings of the Fourth Berkeley Symposium* 1:361–79.

Janssen, A. 1979. Some remarks on the decomposition of kernels. *Proceedings of the American Mathematical Society* 73:328–30.

Joag-Dev, K., and F. Proshan. 1983. Negative association of random variables, with applications. *Annals of Statistics* 11:286–95.

Jogdeo, K., and R. Bohrer. 1973. Some simple examples and counterexamples about the existence of optimum tests. *Journal of the American Statistical Association* 68:679–82.

Johnson, N. L., and S. Kotz. 1970. *Continuous univariate distributions*—2. New York: Wiley.

Joshi, V. M. 1967. Inadmissibility of the usual confidence sets for the mean of a multivariate normal population. *Annals of Mathematical Statistics* 38:1868–75.

Kanter, M. 1976. On the unimodality of stable densities. *Annals of Probability* 4:1006–1008.

Kaplansky, I. 1945. A common error concerning kurtosis. *Journal of the American Statistical Association* 40:259.

Karlin, S. 1958. Admissibility for estimation with quadratic loss. *Annals of Mathematical Statistics* 29:406–36.

Kempthorne, O. 1966. Some aspects of experimental inference. *Journal of the American Statistical Association* 61:11–34.

Kendall, M. G., and A. Stuart. 1979. *The advanced theory of statistics*. 4th ed. Vol. 1. New York: Macmillan (Hafner Press).

———. 1977. *The advanced theory of statistics*. 3d ed. Vol. 2. New York: Macmillan (Hafner Press).

Khaleelulla, S. M. 1982. *Counterexamples in topological vector spaces*. Berlin: Springer-Verlag.

Kiefer, J. C. 1957. Invariance, minimax sequential estimation, and continuous time processes. *Annals of Mathematical Statistics* 28:573–601.

———. 1973. A counterexample to Perez's generalization of the Shannon-McMillan theorem. *Annals of Probability* 1:362–64.

———. 1976. On the minimum number of fixed length sequences with fixed total probability. *Annals of Probability* 4:335–37.

———. 1977. Conditional confidence statements and confidence estimators. *Journal of the American Statistical Association* 72:789–808.

Koehn, U., and D. L. Thomas. 1975. On statistics independent of a sufficient statistic: Basu's lemma. *The American Statistician* 29:40–42.

Koenker, R. W., and G. W. Bassett. 1984. Four (pathological) examples in asymptotic statistics. *The American Statistician* 38:209–12.

Koopman, B. O. 1950. Necessary and sufficient conditions for Poisson's distribution. *American Mathematical Society Proceedings* 1:813–23.

Kowalski, C. J. 1973. Non-normal bivariate distributions with normal marginals. *The American Statistician* 27:103–106.

Krafft, O., and M. Schaefer. 1984. On Karlin's conjecture for random replacement sampling plans. *Annals of Statistics* 12:1528–35.

Kraft, C. H., and L. M. LeCam. 1956. A remark on the roots of the maximum likelihood equation. *Annals of Mathematical Statistics* 27:1174–77.

Krewski, D., and M. Bickis. 1984. A note on independent and exhaustive events. *The American Statistician* 38:290–91.

Kuelbs, J. 1976. A counterexample for Banach space valued random variables. *Annals of Probability* 4:684–89.

Kurtz, T. G., and S. Wainger. 1973. The nonexistence of the Yaglom limit for an age dependent subcritical branching process. *Annals of Probability* 1:857–61.

Lachenbruch, P. A., and D. R. Brogan. 1971. Some distributions on the positive real line which have no moments. *The American Statistician* 25, no. 1:46–47.

Lambert, D. 1982. Qualitative robustness of tests. *Journal of the American Statistical Association* 77:352–57.

Landers, D. 1974. Minimal sufficient statistics for families of product measures. *Annals of Statistics* 2:1335–39.

Landers, D., and L. Rogge. 1972. Minimal sufficient sigma fields and minimal sufficient statistics. Two counterexamples. *Annals of Mathematical Statistics* 43:2045–49.

———. 1977. A counterexample in the approximation theory of random summation. *Annals of Probability* 5:1018–23.

Leblanc, H., and B. C. van Fraassen. 1979. On Carnap and Popper probability functions. *Journal of Symbolic Logic* 44:369–73.

LeCam, L. 1953. On some asymptotic properties of maximum likelihood estimates and related Bayes estimates. *University of California Publications in Statistics* 1:277–329.

Lehmann, E. L. 1959. *Testing statistical hypotheses.* New York: Wiley.

———. 1983. *Theory of point estimation.* New York: Wiley.

Lehmann, E. L., and H. Scheffe. 1950. Completeness, similar regions, and unbiased estimation. *Sankhya* 10:305–40.

Lindley, D. V., and M. R. Novick. 1981. The role of exchangeability in inference. *Annals of Statistics* 9:45–58.

Losert, V. 1982. Counterexamples to some conjectures about double stochastic measures. *Pacific Journal of Mathematics* 99:387–97.

Lukacs, E. 1970. *Characteristic functions.* New York: Macmillan (Hafner Press).

Maar, J. R. 1973. Counterexamples in probability and mathematical statistics. Ph.D. thesis, George Washington University.

Maharam, D. 1982. Orthogonal measures: an example. *Annals of Probability* 10:879–80.

Makani, S. M. 1977. A paradox in admissibility. *Annals of Statistics* 5:544–46.

Marden, J. I. 1983. Admissibility of invariant tests in the general multivariate analysis of variance problem. *Annals of Statistics* 11:1086–99.

Mauldin, R. D., D. Preiss, and H. von Weizsacker. 1983. Orthogonal transition kernels. *Annals of Probability* 11:970–88.

Meeks, S. L., and R. B. D'Agostino. 1983. A note on the use of confidence limits following rejection of a null hypothesis. *The American Statistician* 37:134–36.

Melnick, E. L., and A. Tenebien. 1982. Misspecifications of the normal distribution. *The American Statistician* 36:372–73.

Miller, R. G. 1964. A trustworthy jackknife. *Annals of Mathematical Statistics* 35:1594–1605.

Mineka, J. 1968. A mixture of recurrent random walks need not be recurrent. *Annals of Mathematical Statistics* 39:1753–54.

Mirham, G. A. 1969. A cautionary note regarding invocation of the central limit theorem. *The American Statistician* 23, no. 5:38.

Monroe, I. 1976. Almost sure convergence of the quadratic variation of martingales: a counterexample. *Annals of Probability* 4:133–38.

Mood, A., F. Graybill, and D. Boes. 1974. *Introduction to the theory of statistics*. New York: McGraw-Hill.

Moore, D. S. 1971. Maximum likelihood and sufficient statistics. *American Mathematical Monthly* 78:50–52.

Moors, J. J. A. 1981. Inadmissibility of linearly invariant estimators in truncated parameter spaces. *Journal of the American Statistical Association* 76:910–15.

Mosteller, F., A. F. Siegel, E. Trapido, and C. Youtz. 1981. Eye fitting straight lines. *The American Statistician* 35:150–52.

Neyman, J., and E. Scott. 1948. Consistent estimators based on partially consistent observations. *Econometrika* 16:1–32.

Oliver, E. H. 1972. A maximum likelihood oddity. *The American Statistician* 26:43–44.

Ord, J. K. 1968. The discrete Student's *t* distribution. *Annals of Mathematical Statistics* 39:1513–16.

Padmanabhan, A. R. 1977. Ancillary statistics which are not invariant. *The American Statistician* 31:124.

Patel, J. K., C. H. Kapadia, and D. B. Owen. 1976. *Handbook of statistical distributions*. New York: Dekker.

Perlman, M. D., and M. J. Wichura. 1975. A note on substitution in conditional distribution. *Annals of Statistics* 3:1175–79.

Perng, S. K. 1970. Inadmissibility of various "good" statistical procedures which are translation invariant. *Annals of Mathematical Statistics* 41:1311–33.

Pfanzagl, J. 1968. A characterization of the one parameter exponential family by existence of uniformly most powerful tests. *Sankhya A* 30:147–56.

———. 1971. A counterexample to a lemma of L. D. Brown. *Annals of Mathematical Statistics* 42:373–75.

Pierce, D. A. 1973. On some difficulties in a frequency theory of inference. *Annals of Statistics* 1:241–50.

Pierce, D. A., and R. L. Dykstra. 1969. Independence and the normal distribution. *The American Statistician* 23, no. 4:39.

Pitcher, T. S. 1957. Sets of measures not admitting necessity and sufficient statistics or subfields. *Annals of Mathematical Statistics* 28:267–68.

Plane, D. R., and K. R. Gordon. 1982. A simple proof of the nonapplicability of the central limit theorem to finite populations. *The American Statistician* 36:175–76.

Plante, A. 1970. Counter-examples and likelihood. In *Foundations of statistical inference*, edited by V. P. Godambe and D. A. Sprott, 357–71. Toronto: Holt, Rinehart and Winston of Canada.

Pollard, D. 1984. *Convergence of stochastic processes*. Berlin: Springer-Verlag.

Portnoy, S. 1977. Asymptotic efficiency of minimum variance unbiased estimators. *Annals of Statistics* 5:522–29.

Pratt, J. W. 1963. Shorter confidence intervals for the mean of a normal distribution with known variance. *Annals of Mathematical Statistics* 34:574–86.

Rao, C. R. 1962. Apparent anomalies and irregularities in maximum likelihood estimation. *Sankhya* 24:73–95.

———. 1973. *Linear statistical inference and its applications*. New York: Wiley.

Reich, J. I. 1982. When do weighted sums of independent random variables have a density—some results and examples. *Annals of Probability* 10:787–98.

Reiss, R. 1973. On the measurability and consistency of maximum likelihood estimates for unimodal densities. *Annals of Statistics* 1:888–901.

Rizvi, M. H., and R. W. Shorrock. 1979. A note on matrix-convexity. *Canadian Journal of Statistics* 7:39–41.

Robinson, G. K. 1975, 1977. Some counterexamples to the theory of confidence intervals. *Biometrika* 62:155–61. Correction 64:655.

Root, D. 1971. A counterexample in renewal theory. *Annals of Mathematical Statistics* 42: 1763–66.

Ross, S. M. 1971. On the nonexistence of epsilon-optimal randomized stationary policies in average cost Markov decision models. *Annals of Mathematical Statistics* 42: 1767–68.

Roussas, G. G. 1972. *Contiguity of probability measures: Some applications in statistics.* Cambridge: Cambridge University Press.

———. 1973. *A first course in mathematical statistics.* Reading, Mass.: Addison-Wesley.

Royden, H. L. 1968. *Real analysis.* 2d ed. New York: Macmillan.

Rukhin, A. 1982. Adaptive procedures in multiple decision problems and hypothesis testing. *Annals of Statistics* 10: 1148–62.

Sackrowitz, H., and W. Strawderman. 1974. On the admissibility of the M.L.E. for ordered binomial parameters. *Annals of Statistics* 2: 822–28.

Sampson, A., and B. Spencer. 1976. Sufficiency, minimal sufficiency, and the lack thereof. *The American Statistician* 30: 34–35.

Samuel-Cahn, E. 1975. Remark on a formula by Fisher. *Journal of the American Statistical Association* 70: 720.

Saxena, K. M. L., and K. Alam. 1982. Estimation of the non centrality parameter of a chi squared distribution. *Annals of Statistics* 10: 1012–16.

Schwanger, S. J. 1984. Bonferroni sometimes loses. *The American Statistician* 38: 192–97.

Sclove, S. L., C. Morris, and R. Radhakrishnan. 1972. Non-optimality of preliminary-test estimators for the mean of a multivariate normal distribution. *Annals of Mathematical Statistics* 43: 1481–90.

Serfling, R. J. 1980. *Approximation theorems of mathematical statistics.* New York: Wiley.

Shafer, G. 1982a. Bayes's two arguments for the rule of conditioning. *Annals of Statistics* 10: 1075–89.

———. 1982b. Lindley's paradox (with discussion). *Journal of the American Statistical Association* 77: 325–34.

Shaffer, J. P. 1980. Control of directional errors with stagewise multiple test procedures. *Annals of Statistics* 8: 1342–47.

Shanbhag, D. N., and M. Westcott. 1977. A note on infinitely divisible point processes. *Journal of the Royal Statistical Society, Series B* 39: 331–32.

Siegel, A. F. 1979. The noncentral chi-squared distribution with zero degrees of freedom and testing for uniformity. *Biometrika* 66: 381–86.

———. 1983. Low median and least absolute residual analysis of two-way tables. *Journal of the American Statistical Association* 78: 371–74.

Simmons, G. 1977. An unexpected expectation. *Annals of Probability* 5: 1157–58.

Skibinsky, M., and L. Cote. 1963. On the inadmissibility of some standard estimates in the presence of prior information. *Annals of Mathematical Statistics* 34: 539–48.

Solari, M. E. 1969. The "maximum likelihood solution" of the problem of estimating a linear functional relationship. *Journal of the Royal Statistical Society, Series B* 31: 372–75.

Solomon, D. L. 1975. A note on the non-equivalence of the Neyman-Pearson and generalized likelihood ratio tests for testing a simple null versus a simple alternative hypothesis. *The American Statistician* 29: 101–102.

Solovay, R. M. 1970. A model of set-theory in which every set of reals is Lebesgue measurable. *Annals of Mathematics* 92: 1–56.

Steele, J. M. 1980. A counterexample related to a criterion for a function to be continuous. *Proceedings of the American Mathematical Society* 79: 107–109.

Steen, L., and J. Seebach. 1970. *Counterexamples in topology.* New York: Holt, Rinehart and Winston.

Stein, C. 1950. Unbiased estimates with minimum variance. *Annals of Mathematical Statistics* 21: 406–15.

———. 1956. Inadmissibility of the usual estimator for the mean of a multivariate normal distribution. *Proceedings of the Third Berkeley Symposium on Mathematical Statistics and Probability* 1: 197–206.

Stigler, S. M. 1972. Completeness and unbiased estimation. *The American Statistician* 26, no. 2: 28–29.

———. 1980. An Edgeworth curiosum. *Annals of Statistics* 8: 931–34.

———. 1983. Who discovered Bayes's theorem? *The American Statistician* 37: 290–96.

Stone, M. 1976. Strong inconsistency from uniform priors. *Journal of the American Statistical Association* 71: 114–16.

Stone, M., and B. G. F. Springer. 1965. A paradox involving quasi prior distributions. *Biometrika*

52:623–27.

Stoops, G., and D. Barr. 1971. Moments of certain Cauchy order statistics. *The American Statistician* 25, no. 5:51.

Suissa, S., and J. J. Shuster. 1984. Are uniformly most powerful unbiased tests really best? *The American Statistician* 38:204–207.

Tarone, R. E., and G. Gruenhage. 1975. A note on the uniqueness of roots of the likelihood equations for vector-valued parameters. *Journal of the American Statistical Association* 70:903–904.

Tate, M. W., and L. A. Hyer. 1973. Inaccuracy of the chi-squared test of goodness of fit when expected frequencies are small. *Journal of the American Statistical Association* 68:836–41.

Torgersen, E. N. 1971. A counterexample on translation invariant estimators. *Annals of Mathematical Statistics* 42:1450–51.

van den Berg, J. 1983. A counterexample to a conjecture of J. M. Hammersley and D. J. A. Welsch concerning first-passage percolation. *Advances in Applied Probability* 15:465–67.

Wagner, C. H. 1982. Simpson's paradox in real life. *The American Statistician* 36:46–48.

Wakker, P. 1981. Agreeing probability measures for comparative probability structures. *Annals of Statistics* 9:658–62.

Wang, Y. H. 1979. Dependent random variables with independent subsets. *American Mathematical Monthly* 86:290–92.

Weirich, P. 1983. Conditional probabilities and probabilities given knowledge of a condition. *Philosophy of Science* 50:82–95.

Wells, D. R. 1978. A monotone unimodal distribution which is not central convex unimodal. *Annals of Statistics* 6:926–31.

Whittemore, A. S. 1978. Collapsibility of multidimensional contingency tables. *Journal of the Royal Statistical Society, Series B* 40:328–40.

Williamson, R., R. Crowell, and H. Trotter. 1972. *Calculus of vector functions*. 3d ed. Englewood Cliffs, N.J.: Prentice-Hall.

Wiorkowski, J. J. 1972. A curious aspect of knockout tournaments of size 2 to the *m*. *The American Statistician* 26, no. 3:28–30.

Wolfe, S. J. 1975. On derivatives of characteristic functions. *Annals of Probability* 3:737–38.

Wong, C. K. 1972. A note on mutually independent events. *The American Statistician* 26, no. 2:27–28.

Zacks, S. 1971. *The theory of statistical inference*. New York: Wiley.

Zidek, J. V. 1969. Inadmissibility of the best invariant estimator of extreme quantile of the normal law under squared error loss. *Annals of Mathematical Statistics* 40:1801–1808.

Zygmund, A. 1947. A remark on characteristic functions. *Annals of Mathematical Statistics* 18:272–76.

INDEX